园冶

Yuan Ye Illustrated
— Classical Chinese Gardens Explained
I

吴肇钊　陈艳　吴迪　著

By Wu Zhaozhao, Chen Yan and Wu Di

马劲武　译（英文）

English Translation by Jinwu Ma

图释

【上卷】

中国建筑工业出版社

图书在版编目(CIP)数据

园冶图释 / 吴肇钊，陈艳，吴迪著；马劲武译. —北京：中国建筑工业出版社，2012.10
 ISBN 978-7-112-14704-5

Ⅰ.①园… Ⅱ.①吴… ②陈… ③吴… ④马… Ⅲ.①造园林-研究-中国-古代②《园冶》-注释 Ⅳ.①TU986

中国版本图书馆CIP数据核字（2012）第223086号

责任编辑：郑淮兵 杜一鸣 马 彦
责任校对：刘梦然 陈晶晶

园冶图释

吴肇钊 陈艳 吴迪 著
马劲武 译（英文）

*

中国建筑工业出版社出版、发行（北京西郊百万庄）
各地新华书店、建筑书店经销
北京方舟正佳图文设计有限公司制版
北京方嘉彩色印刷有限责任公司印刷

*

开本：787×1092毫米 1/8 印张：51½ 字数：1066千字
2012年11月第一版 2012年11月第一次印刷
定价：528.00元（共三卷）
ISBN 978-7-112-14704-5
 （22735）

版权所有 翻印必究
如有印装质量问题，可寄本社退换
（邮政编码 100037）

目 录

序一——孟兆祯 6
序二——陈晓丽 9
序三——王绍增 11
序四——刘秀晨 13
序五——甘伟林 王泽民 15
序六——王金满 18
自序一 20
自序二 22
《园冶图释》缘起 24
《园冶》冶叙——阮大铖 ... 26
《园冶》题词——郑元勋 ... 26
《园冶》自序——计成 28
兴造论 001
园说 013

一 相地 025
（一）山林地 045
（二）城市地 055
（三）村庄地 081
（四）郊野地 084
（五）傍宅地 095
（六）江湖地 109

二 立基 121

三 屋宇 141
（一）门楼 152
（二）堂 153
（三）斋 154
（四）室 157
（五）房 158
（六）馆 159
（七）楼 161
（八）台 169
（九）阁 174

（十）亭 176
（十一）榭 190
（十二）轩 191
（十三）卷 194
（十四）广 195
（十五）廊 199
（十六）五架梁 201
（十七）七架梁 202
（十八）九架梁 203
（十九）草架 204
（二十）重椽 204
（二十一）磨角 205
（二十二）地图 207

四 装折 209
（一）屏门 210
（二）仰尘 211
（三）户槅 212
（四）风窗 222

五 栏杆 225
栏杆图式 226

六 门窗 245
门窗图式 246

七 墙垣 253
（一）白粉墙 254
（二）磨砖墙 254
（三）漏砖墙 254
（四）乱石墙 258

八 铺地 259
（一）乱石路 260
（二）鹅子地 260

（三）冰裂地 260
（四）诸砖地 261
砖铺地图式 261

九 掇山 265
（一）园山 272
（二）厅山 276
（三）楼山 279
（四）阁山 281
（五）书房山 282
（六）池山 283
（七）内室山 284
（八）峭壁山 285
（九）山石池 286
（十）金鱼缸 286
（十一）峰 289
（十二）峦 292
（十三）岩 293
（十四）洞 294
（十五）涧 295
（十六）曲水 296
（十七）瀑布 297

十 选石 301

十一 借景 311
《园冶》自识 329
海外交流 331
主要参考文献 365
跋一 366
跋二 369
译后记 370
鸣谢 374

CONTENTS

Preface 1, by Meng Zhaozhen 7
Preface 2, by Chen Xiaoli 10
Preface 3, by Wang Shaozeng 12
Preface 4, by Liu Xiuchen 14
Preface 5, by Gan Weilin,
Wang Zemin 16
Preface 6, by Wang Jinman 19
Preface by Author 1, by Wu Zhaozhao
.... 21
Preface by Author 2, by Chen Yan,
Wu Di 23
The Origin of *Yuan Ye Illustrated* 25
Preface of *Yuan Ye*, by Ruan Dacheng ... 26
Inscription of *Yuan Ye*,
by Zheng Yuanxun 27
Preface of *Yuan Ye*, by Ji Cheng 28
On Design and Construction 001
On Garden Design 013

1. Siting 025
(1) Hilly Forest Land 045
(2) Urban Land 055
(3) Village Land 081
(4) Suburban Land 084
(5) Residence Side Land 095
(6) River and Lake Land 109

2. Base 121

3. Buildings 141
(1) Men Lou 152
(2) Tang 153
(3) Zhai 154
(4) Shi 157
(5) Fang 158
(6) Guan 159
(7) Lou 161
(8) Tai 169
(9) Ge 174
(10) Ting 176
(11) Xie 190
(12) Xuan 191
(13) Juan 194
(14) Guang 195
(15) Lang 199
(16) Five-Frame Girder 201
(17) Seven-Frame Girder 202
(18) Nine-Frame Girder 203
(19) Cao Jia 204
(20) Chong Chuan (Double Rafter) 204
(21) Mo Jiao 205
(22) Floor Plan 207

4. Decorations 209
(1) Screen Doors 210
(2) Yang Chen 211
(3) Hu Ge 212
(4) Feng Chuang 222

5. Railings 225
Railing Patterns 226

6. Doors & Windows 245
Door & Window Illustrations 246

7. Walls 253
(1) White Plaster Wall 254
(2) Ground Brick Wall 254
(3) Lattice Brick Wall 254
(4) Random Masonry Wall 258

8. Pavings 259
(1) Random Masonry Path 260
(2) Cobblestone Path 260
(3) Ice Crack Path 260
(4) Brick Path 261
Brick Paving Patterns 261

9. Rockery 265
(1) Yuan Rockery 272
(2) Ting Rockery 276
(3) Lou Rockery 279
(4) Ge Rockery 281
(5) Study Rockery 282
(6) Chi Rockery 283
(7) Indoor Rockery 284
(8) Cliff Rockery 285
(9) Rockery Pond 286
(10) Gold Fish Pond 286
(11) Peaks 289
(12) Luan 292
(13) Yan 293
(14) Caves 294
(15) Jian 295
(16) Meandering Streams 296
(17) Waterfalls 297

10. Rock Selection 301

11. Scene Leveraging 311
Postscript of *Yuan Ye* 329
Overseas Exchange Activities 331
Major Reference 365
Postscript 1 368
Postscript 2 369
**Postscript by
the English Translator** 372
Acknowledgement 374

目 次（日本語）

序一　孟兆禎
序二　陳曉麗
序三　王紹増
序四　劉秀晨
序五　甘偉林　王沢民
序六　王金満
自序一
自序二
「園冶図釈」縁起
冶叙──阮大鋮
題詞──鄭元勲
自序──計成
興造論
園説

一　相地
（一）山林地
（二）城市地
（三）村荘地
（四）郊野地
（五）傍宅地
（六）江湖地

二　立基

三　屋宇
（一）門楼
（二）堂
（三）斎
（四）室
（五）房
（六）館
（七）楼
（八）台
（九）閣
（十）亭
（十一）榭
（十二）軒
（十三）巻
（十四）広
（十五）廊
（十六）五架梁
（十七）七架梁
（十八）九架梁
（十九）草架
（二十）重椽
（二十一）磨角
（二十二）地図

四　装折
（一）屏門
（二）仰塵
（三）戸楄
（四）風窓

五　欄杆
欄杆図式

六　門窓
門窓図式

七　墻垣
（一）白粉墻
（二）磨磚墻
（三）漏磚墻
（四）乱石墻

八　舗地
（一）乱石路
（二）鵝子路
（三）氷裂路
（四）諸磚地

磚舗地図式

九　掇山
（一）園山
（二）廳山
（三）楼山
（四）閣山
（五）書房山
（六）池山
（七）内室山
（八）峭壁山
（九）山石池
（十）金魚缸
（十一）峰
（十二）巒
（十三）岩
（十四）洞
（十五）澗
（十六）曲水
（十七）瀑布

十　選石

十一　借景

自識
海外交流
参考文献
跋一
跋二
後書き
謝辞

序一

  首先要祝贺吴肇钊君又有新作问世。中国园林中的置石与掇山是充分反映中华民族"人杰地灵"的专门技艺，就石灰岩储量而言，世界第一的国家并未创造掇山工艺，唯中华民族有这项国粹。刘敦桢先生曾说"假山是中国园林最灵活和最具体的手法"，古代园林甚至有"无园不石"之说。但是，掇山置石的设计和技艺是最精湛的，也是最"难"的。世界造园名著《园冶》虽全面、精到地诠释了园林艺术的理论，而相对而言，"掇山"是这本书的精髓，阅读起来也很难懂，因为掇山、置石都是具体的艺术形象，真的"有法无式"那就很难学，实际上是"有成法，无定式"。园林设计中的绘画表现掇山也是很难的。肇钊君此作所缘是由于自己数十年设计实践，深知其重要与艰难并存，所以才大胆地站出来尝试"先难而后得"的成果。这也是我要感谢三位作者和所有致力于此书的人们所付出的辛劳。

  中国园林虽然来自中国文学与绘画，但完全用传统的山水画还不一定能准确地表达。肇钊在水墨山水画的基础上探讨了运用现代的材料和文房用具来表达自然山水的新途径，我认为基本上是成功的，因为他追求尽可能的美，他在这方面非一日之功。20世纪60年代末他在无锡设计了一处假山，我给他评了个"及格"。日后伴随他在扬州设计片石山房、卷石洞天，进步明显，达到了优秀的水平，这当然与他合作的假山师傅的技艺水平是不可分的。在如此深厚积累的基础上当然可以"承前启后"，与时俱进地摸索一下。

  这也是"摸石头过河"，摸索前进。书中绝大多数的地方我都去过，看了图释后我内心默默点头，肇钊君也是"读万卷书，行万里路"的学习方法，主要是外师造化，同时也学习前辈巨匠山水画的传统，然后才自内心产生他自己的画法。我想研究此道的前辈和专家们还很多，希望广大读者多提宝贵意见。

孟兆桢
2011.6.10.

（中国工程院院士）

Preface 1

First I would like to congratulate Mr. Wu Zhaozhao for his new book publication. The rockery and mound building art in Chinese gardens fully reflects the talent of the people, as this quintessential national art was developed in a country that is not blessed with the most reserves of limestone, of which a rockery is made. Mr. Liu Dunzhen once said, "The rockery art is the most flexible and concrete technique in Chinese gardens." For some ancient gardens, there is even a saying that goes, "A garden cannot live without a rockery." However, the art of rockery is exquisite and is the most challenging one. Although the internationally renowned book *Yuan Ye* explains and interprets garden art theory comprehensively and with precision and subtlety, relatively speaking, as "rockery" is the essence of this book, it is still difficult to understand. This is because mound building and rockery design are both concrete art forms and it is hard to learn something that "has rules but no forms". Actually "there is well established rules but no fixed forms." It is also difficult to render a rockery in landscape paintings. The work done here by Mr. Wu Zhaozhao is based on his decades of design practices. He deeply understands its importance and its challenge and boldly attempts this "gain after pain" approach and accomplishment. That is why I'd like to express my appreciation to the three authors and all of those who dedicated their work towards the completion of this book.

Although Chinese gardens derive from Chinese literature and painting, it is hard to accurately express or present them completely with the traditional landscape paintings. Based on the fundamentals of Chinese ink landscape paintings, Zhaozhao explores using modern materials and stationary to express and render natural landscape. Personally I think it is successful, as he pursues beauty at length and he has been honing his skills for a long time. In the end of 1960s he designed a rockery in Wuxi and I gave him a "pass" score. Since then, with his design work Pianshi Shanfang and Juanshi Dongtian in Yangzhou, he made significant progress and reached an outstanding level. Of course his accomplishments could not be achieved without the skilled rockery masters, whom he cooperated with. Based on such deep accumulations, he is well qualified "to serve as a connection between the past and the future", and to keep abreast of the times.

This is also an example of "trial and error", or proceeding while exploring. I have been to most of the places cited in the book, and after reading it, I gave a nod in my heart. Zhaozhao "reads hundreds of volumes and travels thousands of miles". His method is to learn from the nature and the landscape painting traditions of great masters of the past, and then come up with his own way. I think there must be a lot of respected predecessors and experts on these studies and I solicit great ideas or feedback on this work.

Meng Zhaozhen (Academician, China Academy of Engineering)
June 10, 2011

前書き

　はじめには呉剣肇君の新作のご誕生について祝います。

　中国の造園における置石や掇山は中華民族のいわゆる「人傑地霊」という主旨を充分に反映する専門的な庭園づくり作法になります。それはたどえ石灰岩の埋蔵量が世界一に誇る国でも見かけられなかったことで、中華民族特有の国粋だといえましょう。「築山というのは中国造園における最も融通が利き、尚且つ最も具体的な手法でもある。」と劉敦楨先生がかつてこう指摘されたことがあり、古典造園にもいわば「石がなければ園にならず」という説まであります。とはいえ、掇山と石置のデサイン及び工事を実施する際の技術も最も素晴らしく、「難しい」といえましょう。

　世界的な造園名作である「園冶」が全面的、精確的に造園のセオリーについて解説してくれましたが、「掇山」はその精髓の核になるため、ながなが心得難い。それは、置石や掇山共に具体的な芸術具象であり、正に「作法があっても格式無しき」であるため、実に習べ難い。事実のことをいえば「既製の作法があるが、定番の格式はない」。それに造園デサインにおいて、いかに絵を通じて掇山の効果をパフォーマンスすることも至難の業だというまでもありません。剣肇君がこの大作を自ら数十年のデサイン経験をキッタケにして、その重要性と難しさを認識したうえでこそ、大胆に「難から取得へ」とチャレンジし、この成果を修められました、と私は考えております。この場を借りて、三人の作者やこの作に力を注いだすべての人々に感謝の意を表す次第であります。

　中国の造園は中国古来の文学及び絵から生み出されたのだとよく耳にしますが、実際には完全に伝統的な山水絵の技を用いても十分に造園の姿を描くことが無理なことに過ぎないでしょう。そこで、剣肇君が水墨山水絵の作法に基づいて近代的な原材料及び文房具を生かして、新たに自然山水絵の道を試しながら歩んできて、基本的に成功を挙げられたと私は認識しています。彼氏はできる限りの美を追求しつづきたものでもあり、この立派なできあがりは絶対に「一日の功」ではありえません。二十世紀六十年代の末に剣肇君が無錫でデサインした築山を見せてくれた際に私は「合格」と点をつけてやりました。その後、氏があいつづいて揚州の「片石山房」や「巻石洞天」などなどの実積を積んできて、徐々に「優秀」のレベルに辿ってきました。それは勿論彼氏と協力にしあっていた築山職人の腕のすごさにもかかわっていました。このような深い現場経験のおかげで、氏が今日のように前人の成果を受け継ぎ新たな発展を日増しに進むことができたのに違いありません。

　このような試しは「石をたたいて橋を渡る」ことにたどえられ、試しながらとにかく前へ進んでゆくことが得られます。

　この本に関わっている場所は私がほとんど行ったことがあります。そしてこれらの図解を読みながら、剣肇君が「万巻の本を読み、千里の旅を歩む」という「外師造化」な勉強方法をよく感心します。自然を師匠にすると同時に前代山水絵の名家にも習い、そして内心で独自の画法を集大成してつくりだすことだと理解しています。

　この分野において大勢な先輩や専門家がおられると思いますが、何より読者たちのご意見をうかがいたいこともありましょう。

孟兆禎（中国工程院院士（アカデミメンバー））

2011 年 6 月 10 日

（日文翻译：孟凡）

序二

计成的《园冶》是我国最早关于园林营造的专著，它系统地记录和总结了我国古代园林的主流——江南文人园的造园经验和理论。它完整地阐述了崇尚"天人合一"的中国古人从思维方式，策划方针，规划原理，设计路线，建筑分类与结构，装修图案与技艺，建筑材料与施工技术，植物的使用与意境的塑造等一整套营造室外人居环境的实践和学问，随着世界对人与自然关系关注度的日益增强，《园冶》在人类历史上的价值必将日益受到全人类日隆的推崇。

然而《园冶》是用骈俪体书写的古文，其难读和歧解之多也是举世公认的，这极大限制了中国园林理论的应用推广，正确解析这部奇书成为中国园林理论界一大研究难点，也是长期以来中国学界以及若干外国学者关注的重点。

我认为理解《园冶》离不开造园实践。中国古典园林创作是"入境"式的，设计师靠不断在场地或工地巡视，在头脑中浮现丰富的构图或场景来作出抉择，所以其效果常常具有"随手拈来"之感，其设计是多角度，动态，因地制宜的，是通过"人境合一"达到"天人合一"的。这和西方主要靠平立剖图纸来推敲的设计路线是有根本差别的。

吴肇钊先生具有多年园林营造的实践，以及对中国古代文化的执着热爱和相当造诣，他经过20多年的积累和数年的专心著述，终于在《园冶》完稿380周年时完成了巨著《园冶图释》。通过图释，让《园冶》从很难读懂变成几乎中学以上都可以看懂，将对中国古代园林思想艺术的继承发扬产生非常好的作用。此外，本书以中英文对照的方式出版，除了有利于将对中国园林推向世界，还可以通过形象思维帮助世界理解中国，将中国的"人与自然和谐相处"的思想用形象语言远播到全人类。

整理古代创作理论，"借古开今"，是推动民族形式发展的重要途径。诚如本书作者所感悟的那样："撰写《园冶图释》的过程，不仅仅是几十年研究成果与实践经验的汇总，还确有继续学习、受益匪浅的新感受。在此期间完成的设计作品，又有新的探索与追求，均得到极佳的效果。"《园冶图释》不是简单的注释，其间洋溢着自古及今造园者们的营造智慧，其科技成就的提出是为了更好地展望造园的未来。

陈晓丽

（中国风景园林学会理事长）

Preface 2

Ji Cheng's *Yuan Ye* is the earliest monograph on gardening construction, which systematically recorded and summarized the gardening experience and theories of the mainstream of China's ancient garden – Jiangnan literati garden. It elaborated the "human-universe harmony" advocating Chinese ancient people's complete series of outdoor living environment shaping practices and knowledge from the ways of thinking, planning policies, planning principles, design routes, architecture categorization and structures, decoration patterns with skills, building materials and construction techniques, plants use, and artistic conception building. With the world's increasing concern about the relationship between human and nature, the value of Yuan Ye in human history will definitely be praised highly by the whole world.

However, *Yuan Ye* was written in classical parallelism style, and the difficulties in reading it and the ambiguous interpretations of it are also universally acknowledged, which greatly limits the application and popularization of Chinese gardening theories. Proper understanding this masterpiece has become the difficult point of research in the Chinese landscape architecture theory circle and also the focus of concerns of Chinese scholars and some foreign scholars for a long time.

I think the understanding of *Yuan Ye* is inseparable from gardening practice. Chinese classical garden creation is "immersive": designers make choices from the abundant compositions or scenes emerged in the brains through visiting the sites constantly, so its effect is often felt as scenes picked up randomly, and its design is multi-angle, dynamic, and adaptive to local conditions, and the "human-universe harmony" is reached through the "human-environment harmony". This is the fundamental difference from the Western design route that mainly relies on flat and vertical sectional drawings to design.

Mr. Wu Zhaozhao enjoys many years of gardening construction practices, as well as his persistent love and considerable attainments of ancient Chinese culture. After over 20 years of accumulation and years of concentration on writing, the masterpiece *Yuan Ye Illustrated* is finally completed upon the 380th anniversary of the completion of *Yuan Ye*. Drawings make the secondary and higher level people can understand the originally hard-to-read *Yuan Ye*, which will function favorably to the inheritance and development of ancient Chinese landscape architecture ideas and arts. In addition, the book is a Chinese and English bilingual publication, which will not only benefit to promote Chinese garden to the world, but also help the world to understand China through thinking in images and spread the Chinese thinking of "human-nature harmonious co-existence" to all humans through the figurative language.

It is an important way to promote the development of national form through sorting out ancient creation theories to "make use of the past to benefit the present". As the author's inspiration showed: "The process of writing *Yuan Ye Illustrated* is not just a summary of a few decades of research and practical experience, but also the new feelings of continuous learning and many benefits. Design works completed during this period had new exploration and pursuit and all achieved good effects." *Yuan Ye Illustrated* is not simple annotation, but filled with ancient to present gardeners' gardening wisdom, and the presentation of its scientific achievements is to better prospect the future of gardening.

Chen Xiaoli
Director of Chinese Society of Landscape Architecture

序三

整整380年前（1631年）的秋末，计成为汪士衡（明朝官员，事迹已不可考）营造寤园之暇，在园中的扈冶堂完成了《园冶》一书。三年后在即将刻板付印之际，他又在书后添加了一百多字的"自识"，表明"暇著斯《冶》，欲示二儿，但觅梨栗而已。故梓行，合为世便"的心迹。但他怎样都不会想到，此举将在数百年后对解决人类的一个基本问题——人与自然的关系——作出重要贡献。

我把风景园林看作经营人类室外生活空间的活动，亦即从生活角度处理人与自然的关系。中国园林的创造过程始终离不开意象与体验的互动，这与当下流行从西方舶来的主要依靠平面图的创作路线大相径庭。

源于苏格拉底、柏拉图的希腊逻辑哲学和希伯来一神论的西方思想，有一个暗藏的理念：人有权代行"上帝"的工作。于是，掌握自然以快速地发展，改造自然求无限地开拓，成为应有之义。而中国人的想法是：人不过是自然中的一分子，虽然他很特殊，拥有更大的自由权，但也要顺应自然，越是万年大业越要"顺天"而行，而不是一味使用自然，对付自然。

因此，西方人可以通过一张平面图来决定一块土地的命运，而中国人的造园始终不离对现场的观察和对环境的体验，从中找到自己可以有所作为的地方。此中人类的作为无非是为了创造生境、画境和意境，这一切都应在造园者心中化作具体的意象，并通过与现实环境不断"互校"的过程来逐步实现。

其实，《园冶》所阐述的处理人与自然关系的方法应该是根本性的方法。假设人类还有十万年的寿命，只拥有几千年文字历史的人类何以自命已经接近真理？如果是一百万年呢？一千万年呢？难道人类不该永远小心翼翼地对待大自然吗？我想，不需要很长时间人类就会明白这个道理，那时的人类会自动转向东方文明寻求出路，待到《园冶》成书400年之际，其所蕴含的思想和方法很可能就已经在世界上大放异彩。

从1921年陈植先生于本多静六教授处见到《园冶》至今已近100年，其间经朱启钤、陶兰泉、阚铎和城建出版社的校勘出版，特别是陈植的注释，陈从周、孟兆祯等的提倡和讲授，《园冶》在中国的复兴颇有成效。然而毕竟《园冶》是用骈体古文写成，对于从我那一代就彻底遵循西化体系培养教育出来的大部分中国人来说，很难读懂，又有一些似懂非懂的释本扰乱其中，更显出正本清源的普及之重要。

《园冶》其实就是一部忠实记录计成创作过程的书，过程中头脑中闪现的总是诗画，而创作思想都糅入这些诗画中。顺着这个思路去读《园冶》，就不会感到太难，更不会"失之毫厘，差之千里"。吴君肇钊正是循诸长辈的引导，抓住了这个要点，结合自己的创作实践去理解《园冶》，充分发挥其绘画的优长来阐述《园冶》，积二十多年之功完成了《园冶图释》大作，并听从中国建筑工业出版社的建议以双语出版（聘请马劲武君翻译成英文），以有利于将《园冶》推广普及，进而推向世界，实功耀千秋之举。想当初，《园冶》刻印者刘炤曾誉之为"夺天工"，而《园冶图释》的出版，实在是光大《园冶》之"定乾坤"的一步。

应肇钊君之请，我对本书作了校核，个别地方对《园冶》的断句和异体字做出了调整，不当之处，还望关心《园冶》诸君不吝赐教。

王绍增 2011年秋杪于广州暨南花园
（《中国园林》学刊主编，原华南农业大学园林系系主任）

Preface 3

At the end of autumn exactly 380 years ago (1631 AD), while building Wu Garden for Wang Shiheng, a Ming dynasty official whose biography cannot be found, Ji Cheng completed "Yuan Ye" at Huye Hall in the garden. When it was about to print after three years, he added at the end of the book about a hundred words "notes", indicating that "during the writing of the book, I would like to share it with my two sons; however, since they were only interested in child plays and too young to understand, I decided to publish it to share with others." What he did not realize is that, after hundreds of years, his writings would help significantly to address a fundamental issue of human being – the relationship between man and nature.

Landscape architecture is, in my view, an activity for man to manage outdoor lives, that is, a relationship of man and nature from life's perspective. The creation of a Chinese landscape garden cannot be realized without the interaction between conception and experience, and this is in sharp contrast against the imported Western approach of design relying mainly on floor plans.

The Western Greek philosophy originated from Socrates and Plato and that from Hebrew monotheism imply that man has the right to play God. Therefore, it is obligatory for man to learn about nature so that he could develop rapidly and transform our environment for unlimited and unconstrained development. Contrast to the Western philosophy, the Chinese thinks that man is only a part of nature, and although he is special and has a greater degree of freedom, he should conform to nature. The greater an undertaking is, the more conforming the man should be, and he may not exploit or confront nature.

Westerners can use a plan drawing to determine the destiny of a piece of land; however, the Chinese landscape garden designers keep observing sites and experiencing environment, seeking entry points for possible accomplishment. Here man's non-accomplishment, if any, is nothing more than pursuing ideal state in life, art, and mind. All these, in a garden designer's mind, should be turned into concrete imagery, and through constant "mutual verification" with the real world, realize progressively.

In fact, the relationship between man and nature elaborated in *Yuan Ye* should be a fundamental one. Assuming that human kind has a hundred thousand years of life span, how can man with only a few thousands of years of history claim to have approached the truth? What if with a million years of life span? How about ten million years? Shouldn't we tend nature with extreme care? I think it would not take long before man could understand this reasoning and by that time, man would turn to Eastern civilization for answers. When it is 400 years anniversary for *Yuan Ye*'s publication, its philosophy and methodology would shine and prevail around the world.

It has been exactly 100 years since Mr. Chen Zhi saw *Yuan Ye* at Professor Honda Seiroku's place in 1921. During which time through Zhu Qiqian, Tao Lanquan, Kan Duo, and Urban Construction Press' collating and publication, especially with Chen Zhi's notes and Chen Congzhou and Meng Zhaozhen and others' advocating and teaching, we see *Yuan Ye* come back to life in China again. However, since it was written in the old parallel prose style, for most of us Chinese readers who completely went through western style education system, it is really hard to comprehend. To add insult to injury, there are several other versions of interpretation around. Therefore, it is particularly important to thoroughly overhaul and popularize the work.

Yuan Ye is a faithful record of its author Ji Cheng's landscape garden design process. During his design, there were always poetries and landscape paintings flashed in his mind, and his design ideas always mingled with the poetries and paintings. To understand *Yuan Ye* along this line of thought, one does not feel too difficult to comprehend, and would not commit "a miss is as good as a mile" error. It is based on these senior scholars' guidance with the afore-mentioned thought in mind that Wu Zhaozhao authored the book. To better comprehend *Yuan Ye*, he combined with his own design practice and fully took advantage of his mastery in landscape painting. He has been working for over twenty years on the work and accepted the recommendations from the China Architecture & Building Press to have it published bilingually via translation work by Mr. Ma Jinwu. This would promote and popularize *Yuan Ye* beyond China and would benefit generations to come. Remember that when the engraver and printer Liu Zeng hailed the work as "wonderful workmanship excelling nature"? Well, the publication of *Yuan Ye Illustrated* would carry it further forward to really have some impact in the world.

I accepted Mr. Wu Zhaozhao's invitation to write this preface, and proofread the work, and I had at some point adjusted on punctuations and variant Chinese characters. If I had made any mistakes in the process, I would like those who care about *Yuan Ye* correct me.

Wang Shaozeng
Editor-in-Chief, Chinese Landscape Architecture
Ex-Chair, Department of Landscape Architecture, South China Agriculture University

序四

一个在园林现实与虚幻中耕耘并游走一生的人

书案前这部三卷成册的《园冶图释》样本，是吴肇钊先生等三位作者付出巨大心血精心绘制撰写完成的。肇钊把这部书从深圳寄来，我不仅先睹为快，也深感作为同学好友、园林同行、从孩提到年迈50年友情的一份厚重。《园冶》是我们园林学人一辈子攻读的首本，是我国风景园林学科理论与实践最经典的传统总结，也是对规划、建筑、园林三位一体的人居环境体系的世界性贡献。过去曾有一些读本，如《园冶注释》《园冶图说》《园冶文化论》《园冶园林美学研究》《中国园林美学》等，这本《图释》则是更多地从实践运用的角度完成的。它图文并茂、深入浅出，是肇钊和他的同事几十年工作历程的心得，可喜可贺。他索序于我，于是欣然命笔。

我和肇钊是50年前北京林业大学（原北京林学院）园林系的校友。上学时虽然我高他一届，但是志趣和对专业的执着，却让我们几个年轻学子成为挚友。至今，我眼前浮现的还是那个眼睛明亮、皮肤黝黑、开朗好学的小伙。那时我们在学校美工社一起学画创作，在校园一起滑冰游泳，我还和他所在的六二班同学一起到京郊怀柔参加"四清"，向村民教唱红歌……他幽默、聪明、俏皮，还记得他自编自演的三句半惟妙惟肖，令人捧腹。"文革"期间他练就一手好画功，是同学中艺术悟性最好的，巨幅油画作品入选1966年全国美展。我则由于更多偏爱音乐和他们班的同学一起创作并集全校之力演出了"忆苦思甜"大合唱。那时学生生活很艰苦，纯真的心却把我们这些热爱生活、钟情艺术的年轻人连在一起。

毕业之后大家天各一方，彼此的信息却是清楚的。他从扬州辗转深圳乃至全国各地，几十年下来不知设计了多少园林项目，你走到北京、天津、广州、江苏、湖北、深圳、海南、香港，甚至到了美国华盛顿、德国斯图加特，到处都可以听到他的名字，看到他的作品。虽然我们各干各的，却都成为一辈子坚守园林"死不改悔"的一帮人。我们总是在各地的方案评审、专业会议中相遇，光阴荏苒经历各异，事业却成为我们共同的追求和联系的纽带。我们彼此尊重，常常谈到专业中前沿的、有争议的问题。改革开放和城市化大潮把城市园林一次次推到浪尖，我们共同享受和体验着繁忙与困惑。直至有一天听说他身体不好做了大手术，为他捏了一把汗，送去真诚的祝福。这个一辈子不得闲暇的执着者，今天却又一次拿起笔，干出那么大部头的著作，令人钦佩、感动、受到鼓舞。肇钊，50年了，你怎么就没改一改脾气和那颗要强的心，你是好样的。

热爱生活、追逐艺术、思维敏锐、钟情园林，活跃在园林设计领域的那个黑小伙，如今也成了白发人。值得一提的是肇钊还有一颗充满激情、时尚前卫、活得有滋有味的年轻人的心，他有时着装近乎古怪，发型在变幻中引领潮流，在创作思维中又不停地拿捏着传统与时尚、文脉与创新的关系并吐故纳新，在园林的现实与虚幻中耕耘并游走。活得如此之精彩。这些并没有改变他善良的为人和对朋友的真诚，他是一个与众不同的人。我这样勾勒一个内心充实、心态不老的挚友是出于真诚。谢谢你，肇钊和你的同事所做的有益工作。

刘秀晨

（国务院参事、全国政协委员、中国风景园林学会副理事长兼秘书长）

Preface 4

A Man Who Cultivates and Wanders in the Real and Unreal Garden Art for Life

The three volume draft of *Yuan Ye Illustrated* in front of me is the result of painstakingly hard work over three years by Wu Zhaozhao and other two authors. Zhaozhao mailed the book from Shenzhen so that I could take the pleasure of reading it first. At the same time, I could also feel the weight of fifty years of friendship fostered and developed from young to old as good friends, classmates, and colleagues in landscape architecture. *Yuan Ye* is the first and most important book that we should read as scholars in landscape architecture. It is the most classic summarization of traditional Chinese landscape architecture theory and practice, and it also contributes globally to human habitation system from the perspective of planning, architecture, and landscape architecture trinity. There were some similar books, such as *Yuan Ye Annotated*, *Yuan Ye Explained by Graphics*, *Yuan Ye – On Culture*, *The Architectural Aesthetics Study*, and *Aesthetics of Chinese Gardens* etc.; however, this *Yuan Ye Illustrated* stresses more on practical implementations. The authors did a wonderful job in using rich graphics with texts to explain profound theories in simple languages. This is what Mr. Wu Zhaozhao and his colleagues have been studying and accumulating over decades of work. I congratulate them. Mr. Wu asked me to write a preface and I gladly accepted.

Zhaozhao and I were students in the Department of Landscape Architecture in Beijing Forestry University (then Beijing Forestry College) fifty years ago. Although I was one year earlier than him, our common interests and passions for the major made us a few young students close friends. Up to this day, I still remember vividly the bright-eyed, dark skinned, cheerful and studious young guy. We were learning drawing and painting at the fine art interest group, and swam and skated together on campus. I and his 1962 class students went to Huairou, a suburban area of Beijing, to have participated "Si Qing" (Note: a socialist education movement), and there we taught local villagers to sing Hong Ge (Note: "red songs" that eulogize China and communist-led Chinese revolutions) among many other activities. He's humorous, smart and playful, and I still remember that his self-written and self-performed San Ju Ban (Note: Three and a Half Sentences, a form of Chinese stand-up comedy) made us all laugh out loud. During the "Great Cultural Revolution", he perfected his painting skill. He is a top talent among us with the best wit for artistic comprehension. His super-sized canvas painting was selected as one of the entries in 1966's national art show. I, on the other hand, favored music and with students from his class and collected all the talent on campus to have performed a "Yiku Sitian" (Note: contrasting past misery with present happiness, a Chinese political movement at that time) choir. Our college life was difficult; however, our innocent hearts bound our life-and-art-loving youths together.

After graduation, we went different ways, but kept track of each other's whereabouts. He left for Yangzhou, then Shenzhen, and then everywhere in the country. After several decades of work he had completed countless garden design projects. Wherever you go, Beijing, Tianjin, Guangzhou, Jiangsu, Hubei, Shenzhen, Hainan, Hong Kong, or even Washington D.C. in the U.S. and Stuttgart in Germany, you can hear his name and see his work. Despite our different jobs, we all become life-long "diehard" garden aficionados. We often meet at various project review and professional meetings at different places. Throughout all these years with different experiences, the cause has become our common pursuit and base for strong bonding. We respect each other and often discuss cutting edge or controversial issues. The reform and opening of China and the mega trend of urbanization have pushed urban landscape architecture to the fore front, and we share the enjoyment of busy professional works as well as the experiences of confusions and frustrations. Then all of a sudden I heard him undergoing a major surgery, and I was nervous and sent him my sincere wishes. That this persistent busy fellow picked up his pen, pencil and brush again and produced this major work is really admirable, moving and inspiring. Zhaozhao, after all these fifty years, you still have the same disposition and drive to excel. You are the man!

That life-loving, art-pursuing, sharp-minded, garden aficionado and hyper active dark skinned young guy has now become a white-haired man. It is worth mentioning that Zhaozhao still has a young heart that is passionate, fashionable and avant-garde and he enjoys life with great relish. His attire sometimes is almost eccentric and his hair style leads the trend among all the variations. His design idea walks on a fine line between the tradition and the vogue, and lingers about innovation and cultural context. Getting rid of the stale and taking in the fresh, he wanders in the real and the unreal world of garden art. What a wonderful life! All these haven't changed a bit in his kind-heartedness towards people and empressement towards friends. He is a unique person. Here with sincerity I'm portraying a close friend who has inner wealth and a youthful mentality. Thank you, Zhaozhao and your colleagues for your helpful work!

Liu Xiuchen

(State Department counselor, member of the National Political Consultative Conference, Vice Director and Secretary-General of Chinese Society of Landscape Architecture)

序五

中国古典园林在历史上有两次对外冲击，第一次主要是盛唐时期，中国园林向朝鲜半岛和日本的传播。自西汉开始，中国文化通过朝鲜半岛传入日本，盛唐时期，双方的使节、学者、僧人频繁往来，园林、建筑风格和技术随着文化传播一起传入朝鲜和日本，17世纪《园冶》一书传入日本，对日本近代园林产生深刻影响。第二次是18世纪中国园林对欧洲的影响。从17世纪至19世纪初，欧洲各国都出现过"中国热"，到了18世纪初，海外贸易发展迅速，欧洲许多商人和传教士来到中国，他们对前所未知而又水准很高的中国文化，觉得十分新奇，于是欧洲出现了中国热，中国造园艺术首先影响英国，在英国兴起一种新的园林称自然风致园或图画式园林，这种自然风致园或图画式园林又以庄园府邸园林为代表。18世纪中叶，法国掀起更加深刻的"中国热"，通过英国的介绍，自然风致园或图画式园林在法国也流行起来，称作英中式花园。不久之后，中国式花园传遍整个欧洲，在英国、法国、比利时、瑞典等国也出现了仿中式的亭、塔、厅堂、楼阁等园林建筑。

当代中国园林的对外传播是中国古典园林第三次对外冲击。当代中国园林的对外传播应该从"明轩"说起，1978年春，美国纽约大都会艺术博物馆为陈列中国家具、字画，拟在该馆二楼平台建造一座中式庭院，派遣该馆远东部顾问方闻先生来华考察，陈从周先生推荐将苏州网师园一隅"殿春簃"移植到该博物馆，通过考察和论证，各方同意了这一方案。"明轩"是当代在海外建造的第一座中国园林，自"明轩"起掀起了世界的中国园林热；"明轩"所创造的在国内进行构件制作，在海外进行安装的施工方法，也为中国园林大量出口到海外提供了方便。从1980年至今32年的时间中，在海外建造的中国园林已超过100座，遍布美国、加拿大、德国、墨西哥、英国、法国、荷兰、瑞典、瑞士、奥地利、西班牙、比利时、马耳他、埃及、刚果、几内亚、澳大利亚、新加坡、泰国和中国香港等20多个国家和地区。中国园林被誉为"华夏魂"、"中华民族之光"、"常驻文化大使"。

在当代中国园林的对外传播中，中外园林建设有限公司（即原中建园林公司和中外园林建设总公司）是名副其实的主力军，在100座海外中国园林中，约有80座是中外园林建设的。而这其中，吴肇钊先生又是名副其实的主将之一。1992年中外园林承担建造德国斯图加特中国园任务，委托吴肇钊先生为设计师并全程督导施工，后此园命名"清音园"，在1993年德国斯图加特国际园艺展获德国园艺家协会金奖和联邦政府奖。在此之后，吴肇钊先生正式调入中外园林，海外造园一发不可收拾，连续在美国、加拿大、日本、中国香港等地主持建造多个中国园，获得诸多荣誉。正如吴肇钊先生自序中所言，海外造园实践，更促使他深入研究古典园林艺术，提高手绘古典园林效果图技法。

和吴肇钊先生一起工作多年，他和别的园林设计师所不同的一是他一直没有脱离造园施工实践，他对所设计的项目，常常亲临指导施工，特别是假山施工，造诣颇深。二是他在项目设计和施工的同时，一直没有放弃学术研究，二十多年来，不断有论文发表和论著出版。他的论著《夺天工》、《吴肇钊景园建筑画集》、《中国园林立意·创作·表现》等就是他园林设计和施工实践的总结，又分别在艺术、技术、材料等方面有新的突破。这种求实和创新的治学精神在当前浮躁学风颇为盛行的环境下显得更加难能可贵。他50年来在学习和实践中，不断加深对《园冶》的理解，加上他深厚的绘画功底，《园冶图释》这部巨著的完成和出版就是顺理成章的了。

《园冶图释》的出版，对于理解《园冶》的精髓提供了新的形式，特别是中英文双语出版，对于外国同行和对中国古典园林感兴趣的外国朋友会有极大帮助。我们相信《园冶图释》在中国古典园林第三次对外冲击中将起到不可低估的作用。

（中国风景园林学会原常务副理事长、原中外园林建设总公司总经理甘伟林）

（中国风景园林学会园林工程分会理事长原中外园林建设总公司总经理王泽民）

2012年暮春

Preface 5

In the history Chinese classical gardens gave two shocks to the external world. The first one was during the prosperous period of Tang Dynasty and the Chinese Arts of Landscape Architecture was spread to the Korean Peninsula and Japan during that time. The Chinese Culture was introduced into Japan since Xi Han Dynasty through the Korean Peninsula, so in Tang Dynasty, the ambassadors, scholars and monks of the two Countries contacted and exchanged frequently in between about the styles of landscape architectures, as well as technology. So the Chinese classical garden style was spread into Korea and Japan along with the spreading of Chinese Culture. In 17th Century the book of *Yuan Ye* was firstly introduced into Japan and impacted deeply on the Japanese modern gardens. The second Shock to the world was in 18th century, that Chinese classical garden art strongly influenced Europe landscape. From the 17th century to the early 19th century, there were many European countries lifting "Chinese Boom". In the early 18th century, the overseas trade had developed rapidly. Many European traders and missionaries came to China and they felt strange and interested in the previously unknown and profound standards of the Chinese Culture. So the whole Europe rose with "Chinese Boom". The Chinese Garden Art firstly affected to the UK, especially in Britain emerged a new garden style known as the Natural Landscape Gardens or the Pictorial Gardens, among which the Manor House Garden and Cottage Garden were most outstanding as the representatives. In Mid-18th century, introduced by the British, France set off the more profound "Chinese Boom" by Natural Landscape gardens or the Pictorial Garden styles were popular in France, and were called the English-Chinese Garden. Soon after, the Chinese-style gardens spread throughout Europe. You could find the imitation of Chinese kiosks, towers, halls, pavilions and other garden architectures in the United Kingdom, France, Belgium, Sweden and other countries.

The external communication of the contemporary Chinese Landscape Architecture Art did the third shocks by the Chinese classical gardens to the world which started in the spring of 1978 for "Ming Court" in the Metropolitan Museum of Art, New York. The Metropolitan Museum of Art, for the purposes of displaying a set of Chinese furniture and Chinese paintings, they want a Chinese garden to be built in the museum on the second floor platform. They sent their special adviser of Far East Art Department, Mr. Wen Fang to visit China. Mr. Chen Congzhou recommended to modeling the Late Spring Studio (Dianchunyi), a courtyard inside the Master of Fishing Nets Garden in Suzhou into their Museum. Throughout inspections and verification, both sides agreed to this plan and named as "Ming Court" which is the first Chinese garden built overseas in contemporary. After "Ming Court", the world had a "Chinese Garden Boom". The method of producing all the garden components domestically and installing overseas provide a convenient opportunity for building the Chinese gardens abroad. From 1980 to the present, it is about 32 years, there are more than 100 Chinese gardens had been built spreading all over the world such as the United States, Canada, Germany, Mexico, the United Kingdom, France, the Netherlands, Sweden, Switzerland, Austria, Spain, Belgium, Malta, Egypt, Congo, Guinea, Australia, Singapore, Thailand and Hong Kong, China more than 20 countries and regions. The Chinese gardens now are known as "the Soul of China", "Lightness of the Chinese Nation" as well as "Permanent Mission of Cultural Ambassador".

LAC, Landscape Architecture Co. Ltd., (the original name is Landscape Architecture Company of China) is truly the main force in the external communication in the contemporary Chinese Landscape with the world. Among the 100 overseas Chinese gardens, about 80 of which were constructed by LAC. And Mr. Wu Zhaozhao is worthy of the name of the leading person of LAC. In 1992 a Chinese Garden "Qingyin Garden" was built by LAC in Stuttgart, Germany and LAC was appointed Mr. Wu Zhaozhao as the chief designer and full superviser of this project and the Garden was awarded the Gold Prize of Horticulturist Society of German in the International Garden Exhibition Stuttgart '93 Germany. After this, Mr. Wu Zhaozhao officially worked for LAC and continuously in charge of many Chinese garden construction works in the United States, Canada, Japan, Hong Kong, and received numerous honors. As stated in the preface of Mr. Wu Zhaozhao written by him, the overseas gardening practice has prompted his study of Chinese classic garden art while improving his hand-painted techniques of Chinese classical garden renderings paintings.

Working together with Mr. Wu Zhaozhao for many years, he is different from other landscape architects that he has not been out of the garden construction practice, who often in person to guide and supervise the construction of the project designed by himself, especially with attainments for rockery construction. At the same time he did not give up academic research for over twenty years while he is keeping up with the designing and construction works. He published continuously many papers and books. His books of *Win the works of the Heaven*、 *The Art Of Wu Zhao Zhao* and *Conception, creation, performance of Chinese Landscape* are both his summary of the landscape design and construction practices, as well as having a new breakthrough in the art, technology and materials respectively. The spirit of his persuading the realistic and innovative in the study is more than valuable in the present impulsive environment. His study and practice of 50 years, deepening the understanding about the book of *Yuan Ye*, coupled with his strong drawing skills, *Yuan Ye Illustrated* completion and publication of this masterpiece is without any doubt.

The publication of *Yuan Ye Illustrated* provides new form to understand for the essence of *Yuan Ye*, especially both in

Chinese and English bilingual version. It will be the great help for their foreign counterparts and foreign friends interested in the Chinese classical gardens. We believe that *Yuan Ye Illustrated* will play a very important role in the third shocks of Chinese Landscape Classical gardens to the world.

Gan Weilin
Former executive vice president of the Chinese Society of Landscape Architecture
Former General Manager of Landscape Architecture Co. of China

Wang Zemin
Former executive president of Landscape Engineering Branch of the Chinese Society
of Landscape Architecture
Former General Manager of Landscape Architecture Co. of China

Late spring of 2012

序六

园林艺术是人类文明的璀璨结晶，中国历来被称为"世界园林之母"，不论是精巧秀美的江南古典园林，还是宏伟壮丽的京城皇家园林，无不引发中外人士的由衷赞叹，并成为广为效法或仿建的经典范本。

谈及历史上我国造园经验总结和理论概括的集大成者，非明末杰出造园家计成所著的《园冶》莫属。这部世界最早记述造园的不朽著作，将造园技艺与美学意蕴熔为一炉，具有极高的学术理论、实际运用和国际交流之价值。

《园冶》虽然是公认的风景园林学科的必读之书，但也是最难读懂的古书之一。许多读者希望有像《营造法式》、《清式营造则例》那样的图书面世，以实际图例来解释其内容，以便直接运用于当今的造园实践之中。于是，这本《园冶图释》便应运而生，使《园冶》这棵"古树"开出了"新花"。

《园冶图释》的主要作者吴肇钊先生系我公司的设计大师，曾长期担任公司总工程师之重要职位；另两位作者吴迪、陈艳也是我公司优秀的主创设计师和高级管理人员。吴肇钊先生师承园林界泰斗——已故的汪菊渊院士、当今的孟兆祯院士，设计作品多见于国内各地和海外，屡获殊荣。在数十年从事园林设计创作同时，吴肇钊先生潜心研究并实践《园冶》之精髓，历经多年精心打造，终于使得这一空前的力作呈现在读者眼前，可谓是《园冶》研究的一个重要里程碑，也是园林界的一项创新成果。

书中选用作者千余幅精美的手绘彩色园林图，系统地图释了《园冶》从造园规划到细部的创作理论、方法以及一些具体施工技术，成为读懂《园冶》的最佳选择。书中亦不乏我公司在海内外的工程实例，属我公司重大科研成果之一。我认为，《园冶图释》不仅可作为我国造园理论与艺术创作的教科书、参考书，它对园林规划设计的创新和发展也具有积极的现实意义并将产生深远的影响。该书承蒙中国建筑工业出版社以中、英文双语出版，相信更有利于其走向世界，并在国际上大放异彩。

藉此《园冶图释》隆重出版之际，我谨代表公司一表祝贺之意，祝贺该书得以顺利面世；二抒自豪之情，为公司有如此的优异人才和优秀成果而自豪；三致感激之忱，感谢该书作者所付出的心血和辛劳，感谢出版社高效而周全的工作。

我相信，通过阅读《园冶图释》，可以进一步理解《园冶》中"天人合一"的核心理念和宇宙观，它与当今提倡的"以人为本"、"保护环境"、"可持续发展"等观点不谋而合，为眼下方兴未艾的造园实践提供了"借古开今"的健全精义，从而使"虽由人作、宛自天开"的中国园林之花开遍中外大地。

（中外园林建设有限公司董事长王金满）

Preface 6

Landscape art is the splendid legacy of civilization of mankind. China has long been regarded as the mother of landscape architecture in the world. People from home and abroad gasp with admiration at the exquisite traditional gardens in the south and the majestic imperial gardens in the north of China. Traditional Chinese gardens are unique models to follow and imitate far and wide.

Yuan Ye written by the distinguished garden designer Ji Cheng in Ming Dynasty is in every aspect the most famous works among the notable ancient books on the conclusion of experience and theory in landscape architecture in our country. It is the first book to record the garden building techniques in the world. This book, which combined the gardening skills with the aesthetics, boasts superb value in theoretical research, practical application and international exchange.

Yuan Ye is widely seen as a must in landscape architecture and one of the hardest book to understand as well. Many readers expect *Yuan Ye* to be edited as *Building Standards of Song Dynasty (Ying Zao Fa Shi)* and *Building Rules of Qing Dynasty (Qing Dai Ying Zao Ze Li)* with illustrations employed to explain the meaning of the books so that the theory of the books can be readily applied to the current landscape practice. *Yuan Ye Illustrated* is therefore produced to meet this need. *Yuan Ye*, an ancient book about traditional Chinese landscape architecture is sure to be given a new life with the publication of *Yuan Ye Illustrated*.

Mr. Wu Zhaozhao, the main author of *Yuan Ye Illustrated*, is the design master of our company. He had assumed the post as the chief engineer of our company for many years. Mr. Wu Di and Mrs. Chen Yan are another two authors of the book. They are also the leading designers and senior executives of our company. Mr. Wu Zhaozhao, the student of late academician Mr. Wang Juyuan, and academician, Mr. Meng Zhaozhen, who are authorities on landscape architecture, has created many design works both at home and abroad and most of his design works were awarded national and international prizes. For dozens of years Mr. Wu Zhaozhao has been involved in thorough theoretical researches on the essence of *Yuan Ye* and putting his studies into practice when undertaking landscape design works. This great book is finally published thanks to Mr. Wu's efforts for years. It is not only a milestone in the research on *Yuan Ye* but also an innovation achievement in landscape architecture industry.

More than one thousand pieces of color landscape paintings drawn by the author have been selected to interpret *Yuan Ye* systematically in light of garden planning, theory and method on creation of details and related construction techniques. It will surely become the best choice for readers to understand *Yuan Ye*. The book is also one of the important scientific achievements of our company with the compilation of domestic and overseas projects finished by our company. I sincerely believe that *Yuan Ye Illustrated* can not only be taken as a textbook and reference book on garden building theory and artistic creation but also play a positive role in the development and innovation of landscape planning and design works and produce a far-reaching impact in the same respect. We are honored to have China Architeture & Building Press help us in publishing this book in Chinese and English. I am fully confident that *Yuan Ye Illustrated* in Chinese and English will be recognized and highly appreciated by people from different countries.

I would like to take this opportunity to express my sincere congratulations on the grand publishing of *Yuan Ye Illustrated*, my proud feelings on such outstanding staff and achievements of our company and my wholehearted gratitude to the authors for their hard work as well as to China Building Industry Press for their efficient and comprehensive undertaking.

I believe that one could further understand *Yuan Ye* by reading *Yuan Ye Illustrated* by referring to the harmony between man and nature as the core philosophy and cosmology, which has the same idea by coincidence with the current advocated thoughts of people-oriented, environmental protection and sustainable development. *Yuan Ye Illustrated* provides a useful tool for the practitioners in current flourishing landscape field to learn from past experience and knowledge to sharpen their professional edge. And I wish more people from home and abroad will get to know Chinese landscape architecture represented by the building principle put forward by Jicheng "The garden shall look like being shaped by nature although it is built by people".

Mr. Wang Jinman
Chairman of the Board Landscape Architecture Corporation of China

自序一

我出生于1944年，1966年毕业于北京林学院园林专业。回想大学期间，恰逢"文革"，鉴于多方原因只能充当"逍遥派"，为不遭批斗，经老师介绍鼓足勇气去中央美术学院学绘毛主席油画像，尽管亦学两年素描、水彩，与美院相比乃天壤之别，只得夜以继日奋战，油画颜料可谓成筐成筐用完。"功夫不负有心人"，经过近两年努力，油画作品也"挤入"1966年的全国美展。学校礼堂及饭厅外墙均挂上我画的巨幅毛主席及中央领导画像，可谓利用二年派性斗争时间多学了一个专业，绘画功底为后来园林设计助一臂之力。

大学毕业后十分有幸，学术上能得力于几代园林与建筑宗师的培育。早在20世纪70年代，园林界泰斗汪菊渊先生将我选入其主编《中国古代园林史》编委（是其中最年轻的），为期五年编撰江南园林章节，给学子奠定了古典园林的理论和实例研究的扎实功底。为此，20世纪80年代初江苏省建委调我赴南京参加编写《江苏园林名胜》一书，恰童雋、陈从周老先生均为顾问，在一年多的时间里，童老《江南园林志》学术的精辟与惊人的记忆令人难以忘怀，其创作图稿的精练与色彩至今仍为学子震撼和崇拜，并得其引荐求教杨廷宝泰斗。杨老十分耐心讲图并亲笔示范，其准确工整的画风一直是学子的样板。陈从周老先生文采卓绝为世人公认，其对园林的鉴赏评论亦是难以比拟的。著者主笔的石涛"片石山房"修复设计得其指导，受益匪浅。原拟山房长廊内碑刻选用石涛画作，陈老教导"片石山房"已是石涛画本再现，碑刻则应为诗书作品，可谓绝妙至极！陈老并亲自撰写碑记与碑名。全国惟一"园林大师"朱有玠先生对著者的园林设计起到了定位作用，其首次看到学生建成的园子，就赞誉颇有画意，以后每见到学生均告诫"画本再现的风范"是造园成功的准则，至使著者至今仍坚持设计先绘出画本后，再进行施工图设计以确保画本再现。恩师孙筱祥先生、孟兆祯先生不仅学术渊博，而且多才多艺，让学生吸收了各具特色的丰富营养，他们的厚爱加速了学生的成才。孟先生一直是学生园林学科的"辞源"，本书掇山章节先生均页页过目，亲笔修正，实令学生感激万分。

正当我年富力强之际，甘伟林司长、王泽民总经理将我调入中国对外园林建设总公司任总工，得以参与海外园林工程设计与主持施工，德国、美国、法国、加拿大、日本、新加坡、马耳他等以及中国香港均留下园林作品与设计。在完成中国园林在海外传播任务的同时，自觉地更翔实考证、研究祖国古典园林技艺，并坚持中国画的手法绘园林效果图，否则是难以应答各方洋学者提出的"尖攒"、溯源等学术难题，以及现场画图和举办个人画展的要求。现在看来，这些对个人学术上的成长饱含积极意义。

学子虽然具备研究江南古典园林的功底，更重要的是经历40余载设计建造古典园林的实践，诸如：石涛"片石山房"的复建，扬州瘦西湖"卷石洞天"、"白塔晴云"、"二十四桥景区"、"静香书屋"等古园林的复建等，被孟兆祯院士誉为有的作品"已达到新中国成立后全国一流的先进水平"。1989年江苏省园林优秀作品评选时，曾囊括前三名。此后，宗师们的鼓励，同行、同仁频频敦促学子完成《园冶图释》科研成果，出版社亦寄予厚望并决定中英文对照出版。众望所托，笔者只得不顾才疏学浅，以为抛砖引玉吧。

此书的编辑出版，是向传道授业宗师致谢，亦为之汇报。同时亦是让更多中外人士读懂《园冶》。由于图释《园冶》是探索的开篇，不足之处敬请同行们不吝斧正。

吴肇钊

2010年8月于岭南

Preface by Author 1

I was born in 1944 and graduated in 1966 from Beijing Forestry College in Landscape Architecture major. I recall that in my college time, in the middle of "the Great Cultural Revolution", due to many reasons I could only resort to acting as a "happy camper" who took no sides in order to avoid being harshly criticized with physical abuse. Through recommendation of my professor I summoned enough courage to take classes at Chinese Central Art Academy to learn canvas portrait oil painting of Chairman Mao. Although I had already spent two years learning sketch and water color painting, comparing with the works done by their own college majors, I could still see a significant difference. I had to spend days and nights to improve my skills, consuming buckets after buckets of oil paint. My effort paid off. After nearly two years of hard work, my art work finally entered national art exhibition. My canvas paintings of Chairman Mao and other national leaders were hanged at my college auditorium and walls outside campus canteens. So during these two years of factional struggle I wasted no time and mastered a new major and that painting skill helped me tremendously in my future landscape garden design work.

After graduating from college, academically and professionally I was very fortunate to get nurtured under generations of great masters in garden design and architecture. As early as the 1970s in the 20th century, Wang Juyuan, a leading authority in Chinese landscape architecture, invited me to join the editorial board of *Chinese Ancient Garden History*, of which he was the editor-in-chief (and I was the youngest). Five years of working on the project on Southern Chinese Garden chapter had laid a solid foundation for me in Chinese classical garden theory and practice. Because of this project, in the early 1980s, Jiangsu Construction Board appointed me to travel to Nanjing and work on compiling *Jiangsu Garden Attractions*. During that period of time, we had senior consultants Tong Jun and Chen Congzhou. During more than a year's time working with them, I was impressed by Mr. Tong's incisive scholarship on *Annals on Southern Chinese Gardens*; his concise and colorful design work still remain awe-inspiring and amaze me to this day. I was introduced by him to learn from Yang Tingbao, a leading authority on architecture. Mr. Yang was very patient in design work interpretation and he showed design examples in person by himself. His accurate and neat style has been a model for me to follow. Mr. Chen Congzhou's Chinese literary talent is well-known and unsurpassed, and his appreciation and critiques on garden art are marvelous. The restoration design of Shi Tao's "Pianshi Shanfang" led by the author had been guided by Mr. Chen and I had benefited from him greatly. Originally, Shi Tao's paintings were supposed to be the contents of the stele inscriptions inside the long corridor; however, Mr. Chen pointed out that "Pianshi Shanfang" itself was already the realization of Shi Tao's paintings and as such, the stele inscriptions should show his poetry and other writings. What a brilliant idea! Mr. Chen then wrote the record of the stele inscription and named the garden. The sole national "Garden Master" Zhu Youjie played a positioning role on the author's landscape garden design career. When he first saw a garden that I built, he complimented that it was pretty picturesque. Later whenever he saw me he had always admonished that the success of a garden design is "the realization of an art work". This made me keep making an artistic drawing before making a constructed one to ensure that the artistic version is realized. My mentors Sun Xiaoxiang and Meng Zhaozhen are not only erudite but also multi-talented. They let their students absorb all the rich nutrients of different characteristics and their deep love made their students qualified for garden design work sooner. Professor Meng is respectively regarded as a live dictionary by landscape architecture students. He personally reviewed every page on the chapter about rock and mountain building and for that I appreciate immensely.

During the prime of my life, Gan Weilin, China's Construction Ministry's director, and general manager Wang Zemin assigned me Chief Engineer's position in China's Overseas Landscape Architecture Construction Company. This gave me opportunities to participate oversea design and construction projects and I had left my work in Germany, the United States, France, Canada, Japan, Singapore and Malta and so on. While propagating Chinese garden art overseas, I consciously conducted detailed researches and studies about classic Chinese garden art and insisted on using methodologies in Chinese landscape painting in my designed garden renderings; otherwise it would be difficult to field various critical academic questions, some of them dating back to ancient times, posed by international scholars. They also requested me to make artistic renderings at the site and host personal art shows etc. Looking at those events retrospectively, I now realize that they were really beneficial for my personal academic growth.

Armed with the skills to study classic southern Chinese gardens, I had more importantly more than 40 years of experience designing and building classic gardens, such as the renovation work with Shi Tao's "Pianshi Shanfang", Shouxi Lake's "Juanshi Dongtian", "Baita Qingyun", "24 Bridge Scenic Zone", and "Jingxiang Shuwu" in Yangzhou. Academician Meng Zhaozhen even complimented some of my work as reaching the "first class status after the founding of the new China". In 1989, I had won all top three prizes in Jiangsu Province's Outstanding Landscape Garden Design. Since then, after encouragements from great garden masters and frequent urgings of my peers and colleagues, I planned to finish my research project on *Yuan Ye Illustrated*. With high hopes, the press decided to publish the book in Chinese and English bilingually. With all the expectations, I ventured on this project and wish to throw out a minnow and catch a whale.

The purpose of this publication is to show appreciation to my past mentors and professors' teachings; it is also my report to them. Another goal of this publication is to let more people, Chinese or not, to understand *Yuan Ye*. Since *Yuan Ye Illustrated* is but my first attempt of exploration, there must be some mistakes or errors, and I would invite all my peers and colleagues to correct me.

Wu Zhaozhao
August, 2010

自序二

自幼在苏州和扬州的园林长大，耳濡目染，对于一山一水，一草一木，亭台楼阁，鸟语花香有深厚的情感。此次有幸跟随吴肇钊教授著《园冶图释》一书，本身就是一次再学习、再理解和再提高的过程，希望通过本书，为中国园林的延续和普及尽一些绵薄之力。

从事园林设计工作多年，在如何处理人与自然和谐共生的问题上时常感到茫然和困惑，我们常常说要尊重自然，融入自然，然而我们的强势和对利益的追逐，使我们在面对自然时常常横刀直入，傲慢无礼。中国古典园林浓缩了中国文化精华，是多学科多专业的综合体，对场地的理解、环境的认识、意境的追求、自我价值的实现和工程细节的追求，没有相当的功力、阅历及丰富的积累和修养，不要说建造就连观赏都很难理解其中真意。中国古典园林恬淡安详，从容大度，历经千年，任时代变迁依然魅力不减，活力四射，这旺盛的生命力正是《园冶图释》一书的原动力。

现在景观设计多强调西方构图和线条，不再注重意境、峰回路转等含蓄缠绵的表达，丢掉的是自己的特色，改变的是自己的心，当文化和自我意识回归时，又感到强烈失落。古典园林正是我们的精神滋养和物质财富，放慢脚步，降低对物质的追求，让我们的心处于宁静平和的状态，你会发现古老、睿智而深厚的文化从来都没离开。

吴肇钊教授潜心研究古典园林多年，早年亦有大量修复园林获奖，时至今日依旧从事古典园林和现代景观设计工作，真正是从理论到实践，融会贯通，这才有《园冶图释》一书的推出，是毕生研究实践的总结。中国园林讲究诗情画意，《园冶图释》就是按照这一思路，将计成骈体古文原作以画的形式分解展现给读者，寓情景于书画，便于读者深入理解计成原作和古典园林的诗画意境。

兴造论：阐述造园结构妙在因地借景，得体实用；园林建筑：三分匠人，七分主人，若是园林设计则造园师作用要占九成；巧于因借，精在体宜，宜亭斯亭，宜榭斯榭，精而合宜；并以实例工程沁园和汝昌瀛洲进行论述，深入理解园林兴造。

园说：综述造园及诗情画意园林空间，由山水意境的引导和实际工程效果的表达，理解园林意境创作和寄情山水的诗话空间塑造。

相地：综述园林选址和勘察及因地制宜的造园要领，以清代瘦西湖园林群为实例，分析各类园址造园优劣；更以现存名园为例，从专业角度分析其扬长避短，因地制宜的造园手法，利于深入理解现存园林精品。

立基：以"影园"总体规划方案为图，发表论文"借长洲镂奇园——影园复建设计"为说明，孟兆桢院士的模型三方面图文并茂详述立基，以实际项目（计成营造的影园）为例更易于理解和读懂原著。

屋宇：分别从门、楼、堂、斋、房等二十二个方面选择现存具有典型性和代表性的屋宇单体，以工程图纸的方式从平面、立面、剖面详细表达，是一次全面细致的资料收集整理过程。

装折、栏杆、门窗、墙垣及铺地：以原图为基础，从现存的具有代表性的名园中收集资料，从平面、立面、剖面及节点上下工夫，力求全面、细致、准确，成为重要的设计参考资料。

掇山：从掇山的技巧，到山石结体的基本形式，到石头的组合，从掇山的优劣对比以及中国山水画追求的"三远"意境，生动形象地将掇山分解说明。同时按计成园山、厅山、楼山等分门别类，以名园为例，图文并茂，清晰明了。

峰石部分更是从美学上的相石、石的拼掇，到园林名石集锦，详尽透彻。

选石：从自然山水到认识石皴，依皴合掇，对石的赏析，事无巨细。

借景：以书画的方式理解意境创作上的借景，再从设计图上说明意境的处理手法。

总之，《园冶图释》是多年实践经验的积累和汇总，不当之处还请不吝赐教。

陈艳、吴迪于岭南孟秋

Preface by Author 2

Growing up around gardens in Suzhou and Yangzhou, we breathe with the mountains and waters, the grasses and trees, the pavilions and buildings. We hear the birds chirping and smell the flowers' fragrance, and grow deep love for them. It's very fortunate for us this time to work together with Professor Wu Zhaozhao to co-author *Yuan Ye Illustrated*. It's a process of re-learning, re-comprehending, and re-improving for us. We wish through this book we could help continue and popularize Chinese garden art.

Throughout many years of garden design work, we often feel confused or at loss on the harmonious coexistent relationship between man and nature. We often claim to respect nature and blend with nature; however, due to our might and pursuit of profit, we're frequently rude and show little care for nature. A Chinese classic garden is a condensation of Chinese culture, where it integrates multi-disciplinary fields including comprehension of the site, recognition of the environment, pursuit of certain artistic state, self-realization and pursuit of refined construction. Without well-honed skills, experiences and much accumulated trainings and accomplishments, it is difficult to understand the gist of a garden, not to mention to build one. Chinese classic gardens are tranquil and serene, calm and magnanimous. After a thousand years of vicissitude it still shows its vitality with enduring charm. This thriving force is what drives the writing of *Yuan Ye Illustrated*.

Modern landscape designs stress more on western garden design oriented physical forms and compositions and pay little attention to mentally oriented artistic conceptions. When touched upon such topics such as meandering paths, hidden scenes and other subtle finesses, designers often feel a bit bashful. This is a sign of losing one's uniqueness and true self. And when traditional culture and self-awareness return with popularity, one may feel being left out. Classical gardens are exactly our spiritual nourishments and physical assets. Slow down our pace, reduce our quest for materialism, and put our heart at a serene and tranquil state, and then we would find that the ancient, wise and deep-rooted culture has been with us all the time.

Professor Wu Zhaozhao has devoted himself to the study of classical gardens for years and has won many awards in many restorative garden designs in his early years. He still conducts classical and modern garden design projects today and is truly an expert in fully integrating theory with practice. The publication of *Yuan Ye Illustrated* is a summary of his life-time research and practice. Chinese gardens emphasize on poetic connotations and picturesque images. Along this line of thought, *Yuan Ye Illustrated* attempts to interpret Ji Cheng's ancient parallel prose texts with illustrative pictures, and it explains certain garden design circumstances with texts and pictures, helping readers to fully comprehend Ji Cheng's work and poetic and picturesque scenes in classical gardens.

On Design and Construction: It expounds that the gist of a garden structure lies in its contextual scene leveraging, appropriateness and practical considerations. Garden Architecture: Three part craftsmanship with seven part mastermind. For a garden design the role of the designer is about ninety percent. Ingenious adapting and leveraging and building a Ting or Xie wherever it is appropriate is the way to go. The chapter cites real projects Qin Yuan and Ruchang Yingzhou for examples for deeper understandings of garden design and construction.

On Garden Design: A general review of garden design, poetic and picturesque garden space construction. It explains on how to construct a landscape garden that invokes an artistic conception of a natural landscape, and presents results of real construction projects. It stresses on understanding of artistic conceptions and the creation of garden spaces that enable natural landscape bound poetic connotations.

Siting: A general review of garden siting, prospecting, and context-sensitive garden construction. It cites the garden group of Shouxi Lake of the Qing dynasty and analyzes the pros and cons of variant garden site types. It takes existing well-known gardens as examples, and performs in-depth analyses on their strengths and weaknesses, as well as their specific approach on garden building based on concrete contexts. This would help with a deep understanding of current existing garden classics.

Base: Based on "Ying Yuan" master planning plan as graphical illustrations, the published paper "Build an Exquisite Garden on a Long Isle – Comments on Ying Yuan's Restoration" as textual notes, and Academician Meng Zhaozhen's models, this chapter explains in full detail on real garden work – Ji Cheng's Ying Yuan – to help ease the understanding of the original work.

Buildings: For typical and representative individual buildings, such as Men, Lou, Tang, Zhai, and Fang etc., twenty two categories are selected for explanation purposes. They are presented as construction drawings using plans, façades, and section views, representing a thorough and detailed information collection and management process.

Decorations, Railings, Doors and Windows, and Pavings: Based on the original graphics and drawing examples from existing representative classic gardens, this chapter spends time on illustrations on plans, façades, sections and details, making effort to become a comprehensive, detailed and accurate design reference.

Rockery: From rockery construction techniques to basic forms of rockery to rock combinations, this chapter compares various characteristics of different rockery types and the "Three Distance" artistic conceptions sought after in Chinese landscape paintings, depicting a vivid illustration of rockery art. It categorizes rockery type into Yuan Shan, Ting Shan, and Lou Shan etc. using examples from well-known gardens with clear and easy-to-understand text and graphic illustrations.

Rock Selection: From natural landscape to recognition of grains and textures in rock painting, this chapter illustrates how to appreciate and select rocks on their visual properties with details at length.

Scene Leveraging: This chapter helps with the understanding of scene leveraging by way of artistic painting. Furthermore, via design drawings it explains on handlings of artistic conceptions.

In conclusion, *Yuan Ye Illustrated* is an accumulation and summary of many years of experiences and practices. If there is any errors or mistakes please feel free to send in your comments.

Chen Yan and Wu Di, at Mengqiu in Lingnan

《园冶图释》缘起

《园冶》是我国17世纪杰出的造园学家计成的不朽著作。该书成于明末崇祯四年(1631年)，为中国最早记叙造园的专著，内容包括理论、规划、设计、营造等内容。《园冶》梓行后，估计因阮大铖的原因，在清代一度被列为禁书。该书从民间书坊流传至日本，被纳入《东洋美术史》，并改名《夺天工》。1931年朱启钤等人从日本购回《园冶》残本，将其补成三卷，遽付影印，由陶兰泉收入《喜咏轩丛书》内。1933年阚铎委托日本村田治郎就日本内阁文库藏本加以校对，将图式加以绘正，并分别句读，然后由中国营造学社付印出版。1957年城市建设出版社将中国营造学社的原版进行影印重刊，1981年陈植先生的《园冶注释》由中国建筑工业出版社出版，使得《园冶》能够广泛传播。

1. 园冶图释是读懂园冶的最佳方式

《园冶注释》虽已成为风景园林学科的必读之书，但不少内容令人似懂非懂，特别是对指导造园实践作用发挥不够，许多读者希望能像《营造法式》、《清式营造则例》以实例图来解释，以便能直接运用于造园实践。

1982年笔者在南京参加编著《江苏园林名胜》一书时，陈植老先生就找到笔者希望调入南京林学院为其助手，重点完成园冶图释的科研。学科泰斗汪菊渊老先生1979年主持国家重要科研"中国古代园林史"研究，笔者被选为其中最年轻的编委，为期5年，重点编撰江南园林部分。当时汪老强调力求寻找到计成造园的实物，经多年史料稽考和实地考察，笔者只找到计成在扬州建造的"影园"遗址，并为此依据众多史料进行了影园复原的设计，发表《计成与影园兴造》论文于中国建筑工业出版社《建筑师》第23期(1982年)，文章重点是依据复原的影园实体，从诸如：选址、借景、延山引水、地貌、建筑等提炼出其造园特色，结果均符合计成《园冶》中的论述，由此可得出：园冶书中的论述均是其实践与实例的总结。为确认结论的正确，笔者将计成写屋宇的各种形式，掇山篇中的各类叠山，在江南园中都找到了相对应的实例，甚至多个实例，可见用江南园林实景来图释计成造园理论是读懂《园冶》的最佳方法，也是可行的。

笔者因具备了研究江南古典园林的功底，又经历40余载设计建造古典园林的实践，诸如：石涛"片石山房"的复建，扬州瘦西湖"卷石洞天""白塔晴云""二十四桥景区""静香书屋"等古园林的复建等。笔者亦有幸代表国家在美国华盛顿营造了"峰园""翠园"，在德国斯图加特营造了"清音园"，并获"93'国际园林博览会"大金奖，另在加拿大、新加坡、中国香港均有设计作品。几十年的实践让笔者加深了对原文的理解，同时也积累更多的素材。

2. 园冶图释意向

《园冶图释》是在理解原著忠实原著的前提下，结合现存江南古典园林实例，以笔者多年园林实践为基础，以钢笔手绘电脑上色图形式图释《园冶》的精神要旨，图文并茂、通俗易懂。它也是笔者多年园林设计工程实践经验的总结，希望对于现代人理解古典园林专著和现代景观规划设计具有借鉴意义。

《园冶》中有些文字（如"冶叙"，"题词"等）与图释无关，读者可参考陈植先生的《园冶注释》来加深对其的理解。

The Origin of *Yuan Ye Illustrated*

Yuan Ye is the monumental work of Ji Cheng, a distinguished Chinese garden designer in the 17th century. The book was completed in the 4th year of Chongzhen (1631) at the end of the Ming dynasty. It is the earliest treatise in China on garden design and construction, whose contents include theory, planning, design, and construction etc. After its printing, probably because of Ruan Dacheng, it was banned in the Qing dynasty. The book was circulated among non-government book stores and eventually ended up in Japan, where it was included in *Japanese Art History* and was renamed as *Excelling Nature*. In 1931, Zhu Qiqian and others purchased it from Japan with missing pages, and made up three volumes and rushed to photocopying, and then it was included in Tao Lanquan's *Xiyongxuan Collection*. In 1933, Kan Duo entrusted Japanese Jiro Murata to proofread it against collections in Japan's Cabinet Library, adding graphics and punctuations, and then let China's Construction Society Press to print and publish. In 1957, Urban Construction Press photocopied and re-published the version from the China Construction Society. In 1981, Mr. Chen Zhi's *Yuan Ye Commented* was published by China Architecture & Building Press, making the work disseminated widely.

1. *Yuan Ye Illustrated* is the best way to understand *Yuan Ye*

Although *Yuan Ye Commented* has already become a must-read book in landscape architecture community, some of its contents are still obscure and lacking effectiveness in its role as a guide to garden design. Many readers wish it could be like *Building Guide* or *Building Examples in the Qing Dynasty* with real graphical illustrations to facilitate garden design practice.

In 1982 when I was in Nanjing taking part in the compilation of *Jiangsu Garden Attractions*, senior researcher Chen Zhi found me and wished me to be his assistant in Nanjing Forestry College working on *Yuan Ye Illustrated* project. In 1979, when the leading authority Mr. Wang Juyuan was directing national key project "China Ancient Garden History", I was elected to be the youngest editorial board member. I had served for five years and mostly working on Southern Chinese gardens. At that time, Mr. Wang emphasized the need to find real garden sites of Ji Cheng's design and after years of historical literature research and field explorations, the author only found relics of "Ying Yuan", a garden that Ji Cheng built in Yangzhou. Based on this finding and numerous historical records, I had restored its original design and published "Ji Cheng and the Building of Ying Yuan" in *Architects* journal, volume 23, 1982, by China Architecture & Building Press. The focus of the paper is the extracted unique garden design characteristics based on the restored garden such as site selection, scene leveraging, mountain extension and water channeling, terrain design and architecture etc. The result is all compatible with what is written in *Yuan Ye*. From this we could draw conclusion that all the discourses in *Yuan Ye* are summaries of his real practices. In order to prove that my conclusion is right, the author found matches of all the architecture and rockery styles in his book to those in real Southern Chinese gardens, some even with multiple instances. This shows that using illustrations of real garden scenes to comprehend *Yuan Ye* is feasible and the best way.

The author possesses solid skills in researching Southern Chinese classic gardens and has over 40 years of classic garden construction experiences with building examples such as the renovation of Shi Tao's "Pianshi Shanfang" and Yangzhou Shouxi Lake's "Juanshi Dongtian", "Baita Qingyun", "24 Bridge Scenic Zone", and "Jingxiang Study" and other classic garden renovation projects. The author also enjoyed the privilege of representing China to build "Peak Garden" and "Verdant Garden" in Washington DC, USA, and "Qingyin Garden" in Stuttgart, Germany, and had won the grand prize in International Garden Show in 1993. The author also has design work done in Canada, Singapore, and Hong Kong, China. Decades of practice helped the author deepen the understanding of the original text and at the same time, accumulate more materials.

2. The Intention of *Yuan Ye Illustrated*

Yuan Ye Illustrated is written under the premise of faithfully understanding the original text. Based on the author's years of garden design practice and in consideration of current existing Southern Chinese garden real sites, I employed manual pen drawing coupled with computer coloration technique to illustrate the essential points of *Yuan Ye*, making it easy to understand by combining texts with graphics. It is also a summary of the author's many years of practice in landscape garden construction projects. I wish it could help contemporary generation understand this classic work and would facilitate modern landscape architecture planning and design.

Some writings in *Yuan Ye*, such as "Preface" and "Inscription" and so on, are irrelevant to the illustration and readers could reference Chen Zhi's *Yuan Ye Commented* for help with its interpretations.

《园冶》冶叙

　　余少负向禽志，苦为小草所绁。幸见放，谓此志可遂。适四方多故，而又不能违两尊人菽水，以从事逍遥游；将鸡塒、豚栅、歌戚而聚国族焉已乎？銮江地近，偶问一艇于芜园柳淀间，寓信宿，夷然乐之。乐其取佳丘壑，置诸篱落许；北垞南陂，可无易地，将嗤彼云装烟驾者汗漫耳！兹土有园，园有"冶"，"冶"之者松陵计无否，而题之冶者，吾友姑孰曹元甫也。无否人最质直，臆绝灵奇，侬气客习，对之而尽。所为诗画，甚如其人，宜乎元甫深嗜之。予因剪蓬蒿瓯脱，资营拳勺，读书鼓琴其中。胜日，鸠杖板舆，仙仙于止。予则着'五色衣'，歌紫芝曲，进觥觩为寿，忻然将终其身，甚哉，计子之能乐吾志也，亦引满以酬计子，于歌余月出，庭峰悄然时，以质元甫，元甫岂能已于言？

　　崇祯甲戌清和届期，园列敷荣，好鸟如友，遂援笔其下。

石巢 阮大铖

Preface of *Yuan Ye*, by Ruan Dacheng

　　Since I was young, I have been dreaming of secluded life in the mountains; however, I had to work as a government official involuntarily. Fortunately I was dismissed from office and put on exile to return home and I thought that my life aspirations could be realized. Unfortunately it was war time everywhere and I had duty to fulfill my filial piety by taking care of my parents and that prevented me from traveling freely. I wondered if I had to lead a boring and mundane life and living with my folks and family for the rest of my life?

　　Yizheng is not far from here, and I hired a small boat and arrived at a place between Wuyuan and Liudian and stayed there for two wonderful nights. I loved this beautiful place because although it is located in a populated area it gives you an impression of a natural landscape, and that satisfied both my love for nature and for people. Now I laugh at those who travel far distance roaming around. There are gardens here and so there is creative garden design work. The one who can write a treatise on garden design is no other than Ji Wupi, i.e. Ji Cheng, of Wujiang, and the one who gave the title of the book, *Yuan Ye*, is my friend Cao Yuanfu of Dangtu.

　　Wupi is frank and simple and extraordinarily gifted. He is not philistine or hypocritical. His hand writings and paintings are just like his own character. No wonder Yuanfu loved him.

　　Therefore, I turned a piece of side corner waste land into a garden by clearing up the weeds, creating a rockery and managing water bodies. It is used as a place for reading and playing musical instruments. During happy festival time, I would accompany my folks either walking or taking a ride, and singing and dancing in the garden. I mimicked Laolaizi by wearing "Wucaiyi", singing "Zizhiqu", toasting to celebrate birthdays for seniors. I plan to live this leisure and pleasure life style for the rest of my life. Indeed, Mr. Ji used his talent to have realized my personal interest, and I should have a full cup to toast him for my appreciation. When all the dust is set and all the singing and dancing are stopped, looking at the rising new moon against still mountain silhouette, if I ask Yuanfu about this, what can he say?

　　It is now April in Chongzhen 7th year (1634). With thriving garden scene and chirping birds, I write this preface under this beautiful scenery.

《园冶》题词

　　古人百艺，皆传之于书，独无传造园者何？曰："园有异宜，无成法，不可得而传也。"异宜奈何？简文之贵也，则华林；季伦之富也，则金谷；仲子之贫也，则止于陵片畦；此人之有异宜，贵贱贫富，勿容倒置者也。若本无崇山茂林之幽，而徒假其曲水；绝少"鹿柴""文杏"之胜，而冒托于"辋川"，不如嫫母傅粉涂朱，只益之陋乎？此又地有异宜，所当审者。是惟主人胸有丘壑，则工丽可，简率亦可。否则强为造作，仅一委之工师、陶氏，水不得潆带之情，山不领回接之势，草与木不适掩映之容，安能日涉成趣哉？所苦者，主人有丘壑矣，而意不能喻之工，工人能守，不能创，拘牵绳墨，以屈主人，不得不尽贬其丘壑以徇，岂不大可惜乎？此计无否之变化，从心不从法，为不可及；而更能指挥运斤，使顽者巧、滞者通，尤足快也。予与无否交最久，常以剩山残水，不足穷其底蕴，妄欲罗十岳为一区，驱五丁为众役，悉致琪华、瑶草、古木、仙禽，供其点缀，使大地焕然改观，是亦快事，恨无此大主人耳！然则无否能大而不能小乎？是又不然。所谓地与人俱有异宜，善于用因，莫无否若也。即予卜筑城南，芦汀柳岸之间，仅广十笏，经无否略为区画，别现灵幽。予自负少解结构，质之无否，愧如拙鸠。宇内不少名流韵士，小筑卧游，何可不问途无否？但恐未能分身四应，庶几以《园冶》一编代之。然予终恨无否之智巧不可传，而所传者只其成法，犹之乎未传也。但变而通，通已有其本，则无传，终不如有传之足述，今日之"国能"即他日之"规矩"，安知不与《考工记》并为脍炙乎？

崇祯乙亥午月朔　友弟郑元勋书于影园

Inscription of *Yuan Ye*, by Zheng Yuanxun

All the arts and crafts have books written about them and they are handed down from ancient times, except those on garden art. Why? This is because "Garden building varies depending on people, location and time. Moreover, there is no fixed rule to follow, and as a result, there could be no monograph written about it." How do we understand "vary"? The Liang dynasty's emperor Jianwen built Hualin Garden with his royal nobility style. The West Jin dynasty's magnate Shi Chong built his Jingu Fragrant Garden, while during the Warring States period, the impoverished Chen Zhongzi had to build a small vegetable garden at Yuling. This is what "vary" means. People's social economic status is not easy to change. If there is no quiet environment in verdant mountains, no good reputation as "Qushui Liushang", little beautiful scenes as "Lu Chai" or "Wen Xing", but just posing or imitating famous sites like Wang Wei's "Wangchuan Villa", it would be mere copycat of the beauty and make the unworthy place more ugly. This is what "vary" means on different physical site characteristics. As long as the designer has a complete and balanced view of the situation, a garden can be built either complex and refined or simple and natural. Otherwise, if it is built in a far-fetched manner, totally dependent on carpenter and bricklayer's work lacking orographic treatment and with little hiding and shielding effect of landscapes, how can it have the joy and pleasure of a garden living life? It is a pity that a designer has everything in his mind but just could not express it well enough to let craftsmen understand. As a consequence, craftsmen can only mechanically follow the instructions without understanding the gist and the ability to implement with flexibility; the garden designer can only resort to hopes of completing the construction and giving up on principal and original design ideas. This is a shame. When Wupi, i.e. Ji Cheng, designed his gardens, he had got his general rules in his mind and could improvise in implementation without sacrificing the principles. There is no match on this in other garden designers. Moreover, he could direct the constructions on site, possessing a high level of practical skills, and could make stubborn rocks appearing agile and queer, and stuffy space circulating physically and psychologically. This proves very pleasing. We have been staying friends for a long time and I often thought that a small scale garden design would not fully unleash his garden design talent and fancied that it would be great if we could gather all the wonderful attractions in the world including rocks, flowers and trees, and precious animals to let him design and dispose making a total improvement over existing site situation. This is very satisfying and I would hate not to have a talent like him! So Wupi can only build grand rather than small gardens? Not so. We say there could be variations to people and place. In this profession, to design a garden according to specific location and people with appropriate variations, there is no body who can match the caliber of Wupi.

Take my south city garden, for example, it is located between a flat land by the water and water bank with some willows, the area coverage of the garden is also small. With some light design work from Wupi, it immediately became spacious yet with depths. I thought I was very good at garden design, only to find out that I was at most at the amateur level comparing with his. There are numerous celebrities and well-learned people around and if they would like to build small gardens to enjoy their miniaturized natural landscape, why don't they consult Wupi? However he could not accept all the requests and work at different places simultaneously and can only rely on his Yuan Ye to do the job. Even so, I am afraid that Wupi's wisdom and skills could not be taught and passed on; what can be taught and passed on are only those fixed rules. So with this sense, there is no true transfer of what he is best at. Having said that, I think to accommodate to certain circumstances, one has to have something to count on. Although the transfer cannot be a complete one, it is better than nothing and at least it can provide something for people to follow. Wupi's attainment is nonpareil in China and his work would be a model for future generations to follow. Who can say that *Yuan Ye* would not match the stature of "Zhouli · Kaogongji" and would be extolled in future?

Written on May 1, Chongzhen 8th year (1635) at Ying Garden

《园冶》自序

不佞少以绘名，性好搜奇，最喜关仝、荆浩笔意，每宗之。游燕及楚，中岁归吴，择居润州。环润皆佳山水，润之好事者，取石巧者置竹木间为假山，予偶观之，为发一笑。或问曰："何笑？"予曰："世所闻有真斯有假，胡不假真山形，而假迎勾芒者之拳磊乎？"或曰："君能之乎？"遂偶成为"壁"，睹观者俱称："俨然佳山也"；遂播闻于远近。适晋陵方伯吴又于公闻而招之。公得基于城东，乃元朝温相故园，仅十五亩。公示予曰："斯十亩为宅，余五亩，可效司马温公'独乐'制。"予观其基形最高，而穷其源最深，乔木参天，虬枝拂地。予曰："此制不第宜掇石而高，且宜搜土而下，令乔木参差山腰，蟠根嵌石，宛若画意；依水而上，构亭台错落池面，篆壑飞廊，想出意外。"落成，公喜曰："从进而出，计步仅四百，自得谓江南之胜，惟吾独收矣。"别有小筑，片山斗室，予胸中所蕴奇，亦觉发抒略尽，益复自喜。时汪士衡中翰，延予銮江西筑，似为合志，与又于公所构，并骋南北江焉。暇草式所制，名《园牧》尔。姑孰曹元甫先生游于兹，主人偕予盘桓信宿。先生称赞不已，以为荆关之绘也，何能成于笔底？予遂出其式视先生。先生曰："斯千古未闻见者，何以云'牧'？斯乃君之开辟，改之曰'冶'可矣。"

时崇祯辛未之秋杪否道人暇于寤冶堂中题

Preface of *Yuan Ye*, by Ji Cheng

When I was young, I was known for drawing and painting. I also enjoyed traveling in the wild and seeking unique attractions. Among all the landscape paintings of past times, the most I love are Guan Tong and Jing Hao's misty landscape with a grand imposing manner. So I often learned from their works. I traveled in the north and the two lake areas. At mid-age, I returned to my native land Jiangsu and chose to settle down in Zhenjiang.

Zhenjiang is a scenic place and it is surrounded by mountains on all its four sides. Some local people piled up some oddly shaped rocks among bamboo groves and when I spotted them by chance, I laughed at it. They asked me why laughing and I said, "I was told that in the world there are real and fake stuffs. Why don't you leverage real mountains' images but faked it by making it like Spring Gods piling fist-sized rocks?" And then someone asked, "Well, can you do it?" As such, by this chance encounter, I made a steep cliff for them and all those who looked at it lauded, "It is indeed an excellent mount!" Since then, my fame on mountain making spread out. Coincidently, because of my fame I was invited by Mr. Wu Youyu, who once served as a governor in Changzhou. He purchased in the east of the city a piece of land, which used to be the old garden of Wenguo Handa of the Yuan Dynasty, occupying only fifteen Mus (Note: One Chinese Mu is about 666.67 square meters in size.). Mr. Wu told me, "Use 10 Mus to build residential houses and build a garden on the rest 5 Mus land, to imitate Sima Wen's 'self pleasure'." I observed the site and found that the terrain elevation is high but the water source is deep. There are tall trees in the air and winding tree branches touching the ground. I said, "Based on the actual situation, to build a good garden here it is not only necessary to pile up rocks to increase the height but digging down to increase the depths, to make the trees grown on the ground appear to be growing on the mountain sides to have a varied and mixed look. Rocks can be set between the exposed winding and buckling tree roots to simulate ancient landscape paintings. And build pavilions and terraces by the mountain and water to make varying heights of scenery have reflections in the water. Make deep and tortuous streams and gullies and have corridors flying over. The wonderful scenes would be unexpectedly beautiful." After the completion of the garden, Mr. Wu gladly commented, "From the garden entrance to the exit, although there are only four hundreds some steps, I am able to feel the beauty of the Southern Chinese landscape gardens. Everything beautiful is shown."

There are also other smaller undertakings. Although the construction size is small, I could fully express all the wonderful ideas in my mind and that made me very happy. During this period of time, there is a secretary Wang Shiheng who invited me to build a garden in the west of city of Yizheng. It looks like we share the same interest with that of Wu Youyu. The two gardens have since been well-known throughout China.

During my spare time in garden building, I sorted out the design schemas and presentations, and entitled it as "*Yuan Mu*", or garden pasture literally. Mr. Cao Yuanfu traveled from Dangtu to Yizheng and I accompanied him to tour my gardens. He stayed for two days and praised for my work whatever he saw. He felt that viewing my gardens is just like enjoying Jing Hao and Guan Tong's landscape paintings, and asked me whether I could write about them? So I took out my manuscripts to show him and he said, "These have never been seen or heard of before since thousands of years ago. Why is it titled (Yuan) 'Mu'? Since this is your original work, it should appropriately be entitled as 'Ye'!"

Written in late autumn in Chongzhen 4th year (1631), in my spare time at Huye Hall.

园冶图释

Yuan Ye Illustrated
——Classical Chinese Garden Explained

兴造论

On Design and Construction

世之兴造¹，专主鸠匠，独不闻三分匠、七分主人之谚乎？非主人也，能主之人也。古公输巧，陆云精艺，其人岂执斧斤者哉？若匠惟雕镂是巧，排架是精，一梁一柱，定不可移，俗以"无窍之人"呼之，甚确也。故凡造作，必先相地立基，然后定其间进，量其广狭，随曲合方，是在主者，能妙于得体合宜，未可拘率。假如基地偏缺，邻嵌何必欲求其齐，其屋架何必拘三、五间，为进多少？半间一广，自然雅称，斯所谓"主人之七分"也。第园筑之主，犹须什九，而用匠什一，何也？园林巧于"因"、"借"，精在"体"、"宜"，愈非匠作可为，亦非主人所能自主者，须求得人，当要节用。"因"者：随基势之高下，体形之端正，碍木删桠，泉流石注，互相借资；宜亭斯亭，宜榭斯榭，不妨偏径，顿置婉转，斯谓"精而合宜"者也。"借"者：园虽别内外，得景则无拘远近，晴峦耸秀，绀宇凌空，极目所至，俗则屏之，嘉则收之，不分町疃，尽为烟景，斯所谓"巧而得体"者也。体、宜、因、借，匪得其人，兼之惜费，则前工并弃，即有后起之输、云，何传于世？予亦恐浸失其源，聊绘式于后，为好事者公焉。

A construction is primarily based on craftsmanship. However, have you ever heard the expression "three part craftsmanship with seven part mastermind"? Here "master" does not mean the owner of a garden, but rather its planner or designer. In ancient times, Lu Ban had mastery craftsmanship while Lu Yun was extraordinarily skillful, but were they only craftsmen who only knew how to use their tools? If a craftsman's skill is only to know how to craft something meticulously and to follow construction instructions blindly without any creative idea, then it is more appropriate to label him as a "mindless person". Therefore as long as it is a building construction, he must first observe and study the site in order to determine how the foundation is to be laid, and then decide how wide and deep it should be. He must measure the foundation site and lay out rectangular courtyard based on the site's shape. This all depends on whether the designer could adapt to the site's unique characteristics appropriately. You cannot follow the rules blindly; neither can you improvise without constraint. If the site is irregular in shape, you can adapt your design without forcing it to be shaped regularly. There is no strict rule to dictate whether to use three or five roof truss units. Why always using a whole number of units anyway? As long as it is natural and elegant, even if it is half a unit it will be fine. This is so-called "seven part mastermind". Therefore, the role of an architect of a garden is nine out of ten while a crafter one out of ten. Why? This is because a garden design stresses on ingenious customization and mutual leveraging. It strives for proper composition with the most appropriate layout, and this is not what a crafter is able to do, neither can it be realized by the architect's subjective wishful thinking. The designer must be right and effective. What is "Yin"? It means in order to follow the natural profiles of a local terrain and preserve the square shape with completeness of a site, one may need to cut some tree branches that obstruct a view, or direct some spring water to flow over rocks, so that different views can be leveraged. Build a gazebo or pavilion where it is appropriate. A garden path might as well be naturally wiggling and appears secluded. This is what "delicate and appropriate" suggests. What is "Jie"? It means although the inside of a garden differentiates from the outside, there is no difference in what one can see at various distances. Verdant mountains against the blue sky in the distance, or dark red colored old Buddhist temple nearby, wherever one's eyes can see, as long as it is a beautiful scene, it should be included in the view. If it is a view not so tasteful, it should be blocked. A farmland or grassland can all be made into a misty landscape. This is so-called "ingenious and befitting". So in order to achieve the effect that is most befitting, appropriate, terrain-following and mutual leveraging, if there is no well-judged architect and the construction budget is tight, all previous effort would be nullified. Even if there are masters like Lu Ban or Lu Yun, with these restraints in place, great masterpiece cannot be handed down to next generations. It is my concern that this fine art of garden design and construction would get lost through future generations, so I decided to illustrate the garden design patterns for garden art enthusiasts.

---

1 兴造论是阐述园林的规划设计，园林建筑是"三分匠人，七分主人"，即建筑师的作用占七成。若是园林设计则造园师作用要占九成。园主人也非能自主。园林"巧于因借，精在体宜"，依据用地现状经营山水，栽植花木，再按"宜亭斯亭，宜榭斯榭"的原则点级不同功能建筑，要做到"精而合宜"。园虽别内外，得景则无拘远近，"嘉则收之，俗则屏之"，这就是"借景"，妙在"巧而得体"。

为理解兴造论实质，故选择一座古典园林规划设计的全过程进行图文并茂剖析。

On Design and Construction expounds on planning and design of landscape gardens. Landscape architecture emphasizes on "three part craftsmanship with seven part mastermind", that is, the role of the architect is about seventy percent. If it is a landscape garden, then the role of the garden designer is about ninety percent. The master cannot act on his own. The gist of garden design is "ingenious leveraging and naturally befitting". Based on the site's characteristics, a designer lays out the land, water, and plants, and then place different types of building where appropriate, with the purpose of being "exquisite and appropriate". Although there is internal and external difference for a garden, there is no difference in how to get the scenes, no matter how close or far away they are. "Nice views should be included; un-tasteful ones should be excluded." That is what "a borrowed scene" means. It is a wonderful idea in that it is "subtly and ingeniously befitting".

In order to help understand this chapter on design and construction, a complete design case is selected with explanations in both texts and graphics.

沁园诗情画意的创作 Qin Garden's Poetic and Picturesque Design

1. 定位

笔者承担广东清远市旅游接待中心与别墅区的园林景观设计任务，宗兴董事长陈坚先生匠心独运，提出"唐风禅境"的创作理念，其意是扬弃当前西风强劲的"外来品"，诸如法国式、意大利式、欧式等风格，强调外观带有唐代风格色彩，内部结构、空间符合现代社会需要的现代"中国极品"。为此笔者将别墅群环抱的中心绿地设计成"唐风禅境"立意的古典园林"沁园"，既与坡麓环境融为一体，又使别墅业主共享充满诗情画意的园林，达到自古至今国人对最佳住宅的定位："山居之迹于寂也，市居之迹于喧也，帷园居在季孟间耳。"园居生活既可坐享城市文明，又兼得湖山乐趣，现存实例为帝王的避暑山庄，达官显贵、富商大贾的江南园林拙政园、网师园、个园等。因尺度缩小的空间实体提供了可行、可望、可游、可居的现实条件，无需长途跋涉之苦，可得山水逍遥之功；满足业主悠哉游哉、信步山水、散情消遣的闲情逸致。"小中见大"，"以少胜多"的沁园艺术效果呈现的是立体的画、多调的曲、即兴的诗。

2. 立意

古典园林的构成，包含自然要素与人工要素互相对应对立而又统一的两个方面。

自然要素：地形、植物、动物。

人工要素：园林建筑、园林工程。

自然要素是起主导作用，决定园林景色的自然特征；人工要素主要是实用价值，满足游憩、遮荫遮雨、园居活动。二者相辅相成、相得益彰，自然要素惟有通过人工要素才能发挥作用，人工要素不但保证着与自然要素直接发生关系的游园活动，更满足园居的实用需求。故人工要素的处理以接近自然为准则，尽量顺应自然，美化自然，使其结合于自然之中，以成为与自然要素融为一体的观赏因素。人工要素要在发挥自然美的前提下，巧用建筑的匠思，因为建筑处理得再曲折有致也不能构成风景园林。所以说，古典园林创作是自然要素融为一体的观赏因素。所以说，古典园林创

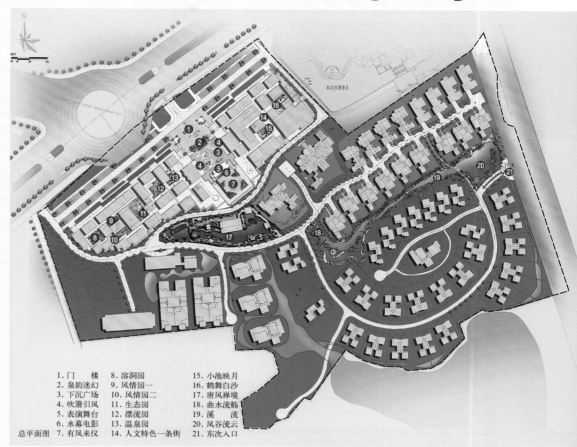

1. 门楼 8. 溶洞园 15. 小池映月
2. 泉韵迷幻 9. 风情园一 16. 鹤舞白沙
3. 下沉广场 10. 风情园二 17. 唐风禅境
4. 吹箫引凤 11. 生态园 18. 曲水流觞
5. 表演舞台 12. 漂流园 19. 溪流
6. 水幕电影 13. 温泉园 20. 凤谷流云
7. 有凤来仪 14. 人文特色一条街 21. 东次入口

总平面图

清远市旅游接待中心与别墅区总平面图
Master Plan of Qingyuan City's Vistor Center and Recreational Residence

1. Setting

The author accepted the planning and design project for Guangdong's Qingyuan City's visitor center and recreational residence. Mr. Chen Jian, the client and Chair of the Zongxing board, proposed Zen-like Tang dynasty design style, shunning current strong western exotic styles, such as French, Italian and other European ones. He emphasized it should aim for a nonpareil having a Tang dynasty appearance with its inside and spatial layout catering to modern living needs. To achieve this goal, the author designated the central open space as the site of the classic Tang dynasty style Zen garden, to mingle with the hilly surrounding, filling this recreational residential complex with a poetic and picturesque air. Since the ancient time, an ideal residence setting has been conceived: "Mountain living can be too quiet while urban living too noisy. It is only the garden living that can have equilibrium with the best of both worlds." With garden living, one can enjoy both the urban civilized convenience and the natural landscape of mountains and lakes. Existing examples include Mountain Summer Resort and the private gardens of the powerful and wealthy, such as Humble Administrator's Garden, Master of Nets Garden, and the Bamboo Garden etc. Because of the scaled down spatial layout, it is convenient for residents to walk, to watch, to wander, and to live. One can avoid the travail to enjoy the natural landscape. This satisfies the needs of clients who wish to live a life of ease and leisure, walking and wandering effortlessly in a leisurely and carefree mood. This "less is more" and "seeing big in small setting" artistic effect present us with perspective imagination, rich intonation, and improvised poetry.

2. Conception

Classic Chinese gardens have both natural and artificial fundamental components that are both opposed to each other and at the same time well-integrated.

Natural Components: Terrain, Plants and Animals.

Artificial Components: Landscape

作的是自然美与人工美统一的艺术形象。

中国园林的创作方法，园林界泰斗孙筱祥先生早有精辟论著，即三个创作境界：

（一）生境——自然美和生活美相结合；

（二）画境——原始的自然美中注入艺术美；

（三）意境——浪漫主义理想美境界（包括：美的感情、美的抱负、美的品格、美的社会）。

沁园的设计严格遵循了三个境界的创作方法。

3．创作

沁园的设计首先是考虑生境（自然美），主要是通过山水地形的创作。笔者早年著作《夺天工》中有专文探讨古典园山水理法，因地制宜是其基本原则。尽管技法各不相同，但可总结归纳为：延山引水、宛自天开，阜培窐疏、相映生辉；池中理山、尤特致意；叠石丘壑、突兀深邃四种。沁园位置处于坡麓地形，故选择"延山引水"的创作手法，即将园外山体借景入园，园内人造的平岗岩峦作为山麓的余脉处理，然后把山水引导入园形成多种水景。园林建筑既要包含玩赏实用与审美享受的双重内容，又要组织到山水艺术环境中，其布局则是"宜亭斯亭，宜榭斯榭"，均因境成。园内植物配置均需有画意、入意境，并赋予不同的品格，其作用：装点山水、衬托主题；分隔联系、含蓄景深；陈列鉴赏、景象点题；渲染色彩、突出季相；散布芬芳、招蜂引蝶；隐蔽园界、拓展空间。

画境、意境的创作是诗情画意的体现。组成沁园的八景是八幅以山水为主体的"自然写意山水画"。建筑营造和花木配置按山水风骨量体裁衣，烘云托月，以提高自然山水的艺术感染力。匾额对联则是画幅寓意的精髓，系饱含"禅"之意境的诗句。为此根据总体设计作出理念草图，然后依据草图完成全套施工图设计，再根据施工图进一步追求艺术灵感和源泉，并按清朝文人和园林鉴赏家张潮在《幽梦影》中所说的："山之光、水之声、月之色、花之香……真足以摄招魂梦，颠倒情思"来指导定稿沁园八景图。

Buildings and Constructions.

Natural components take the leading role, and they determine the natural characteristics of a landscape. Man-made or artificial components mainly have pragmatic concerns, and they satisfy recreational needs, provide shelters and room for other activities. These two components are complementary and they enhance each other. Natural components play their roles through artificial guarantees, while artificial components ensure the recreational activities that directly interact with natural components to satisfy practical recreational needs. Therefore, the layout of artificial components should be accommodating and close to nature to enhance its beauty, so that man-made features can be fitted into nature as if growing as part of it to form an integrated viewing element. The artificial element should ingeniously extend itself under the condition of full development of natural element, for without natural element in full play, no matter how clever and masterful a building is designed, it still falls short of good landscape architecture design. So the artistic image that a classic Chinese garden design is aimed for is about the integration of natural and artificial beauties.

The methodologies of classic Chinese garden design, which Mr. Sun Xiaoxiang, a leading authority, had long been expounded, are exemplified in three states of creative design:

(1) Life – the integration of natural beauty and the joy of life.

(2) Art – pristine natural beauty imbued with artistic beauty.

(3) Mind – the ideal romanticism, which includes the feeling of beauty, the aspiration of beauty, the character of beauty, and the society of beauty.

3. Design

The first consideration for Qin Garden's design is the Life state, the natural beauty. This is mainly realized by garden terrain and water body design. The earlier work "Excelling Nature" by the author is dedicated to study classic Chinese garden's reasoning and methodology on garden terrain and water body design. Context-oriented design is the key. Although there are various techniques, in general, it can be summarized as the following four types: mound extension and water flow channeling by imitating nature, banking up with earth to form a mound and clear up the low-lying land to enhance their beauty by contrast, managing a rocky mound in a pond to carry meaning, and designing of piling rocks to achieve towering yet deep spatial effect. The Qin Garden is situated on a slope; therefore, the mound extension and water channeling approach is applied here. With this approach the hilly area outside the garden can be extended to the inside. In the garden, the man-made mound is perceived as the extension of the outside hill, and the water body is treated as flowing from outside and they form various water scenes. Buildings in a garden serve two roles: to possess recreational and practical values with aesthetic concerns and to organize garden terrain with various water bodies. The overall arrangement is supposed to be "placing a gazebo or pavilion as appropriate", all depending on actual context. All garden plants are picturesque and can lead to a state of mind of high aesthetic values. They have different characteristics and their purposes are multiple: decorating the landscape and setting as foils to the main theme, dividing and connecting different areas and creating depths of scenes, displaying for appreciation and bringing out a theme, strengthening coloration to stress seasonal change, diffusing flower fragrance and attracting bees and butterflies, and hiding garden boundary and expanding garden space.

The creation of states of Art and Mind is the realization of poetic and picturesque garden design. The eight scenes that form the Qin Garden are eight realistic "natural impressionistic landscape painting". The arrangement of buildings and plant design are modeled after vigor of style of these landscape paintings. They are positioned in a way to enhance the artistic effects of the natural landscape. The traditional Chinese Bian'e (Note: a horizontal inscribed

4. 小结

中国传统园林是自然美、艺术美、意境美的综合创作,有些内容是施工图表达不出来的,故表现图入画十分重要,它能较全面真实地体现出设计者的意图。笔者曾在天津设计红学家周汝昌纪念园,并请吾师孟兆祯院士撰写碑记,先生阅看全园鸟瞰效果图时凝视片刻,即提笔赋诗一首《小瀛洲咏》:"引水穿流谋凫院,堤廊桥舍绘缠绵;以水为心邀祥云,石庭成趣耐流连",笔者的设计理念被刻画得淋漓尽致。从此以后凡传统园林设计,笔者不仅完成施工图,还附有景点效果图,并指出需按施工图检查工程质量,按效果图审定建设的艺术效果,特别是自然要素的部分。

board) and Duilian (Note: an antithetical couplet written on the left and right gate posts) are the quintessence of Chinese painting, as they are poems giving rise to Zen-like frame of mind. The master conceptual design was made based on these ideas, and it is followed by a whole set of construction drawings. Further artistic inspirations were sought after the completed construction drawings. The eight landscape paintings were created after inspirations found in Qing dynasty scholar and garden connoisseur Zhang Chao's "Shadow of Tranquil Dreams": "the light of mountains, the sound of waters, the color of the Moon, and the fragrance of flowers … are indeed sufficient to evoke soul dreams, causing infatuated meditation."

4. Conclusion

Traditional Chinese gardens are integrated creations combining natural beauty, artistic beauty, and the beauty of the mind. Some of the contents cannot be fully expressed through construction drawings, and as such a perspective representational drawing becomes crucial, because it can reflect more truly about the intentions of the designer. The author once designed in Tianjin a memorial garden for Zhou Ruchang, a well-known researcher of the classical Chinese literature "A Dream in Red Mansions", and invited my professor Meng Zhaozhen, an academician in the Chinese Academy of Engineering, to write an inscription. After staring at my perspective painting for a while, Professor Meng composed a poem "Small Yingzhou": "Channeled water courting the duck pond, gallery on embankment lingering bridges and houses; centered water body inviting auspicious clouds, and rock garden generating lasting interest." The author's design idea is portrayed completely. Since this project, as long as it is a traditional Chinese garden design, the author not only completes construction drawings, but also artistic perspective renderings, and requires using construction drawings to check on construction quality, and using perspective renderings to validate artistic effect, especially the natural element parts.

沁园设计鸟瞰效果图
Perspective View of Qin Garden

沁园设计总平面图
General Plan of Qin Garden Design

沁园八景・匾额・对联
The Eight Scenes of Qin Garden, its Bian'e and Duilian

沁园八景图：四面八方 "非必丝与竹，山水有清音"
One of the Eight Scenes of Qin Garden: Four Sides and Eight Directions – "the sound of nature excelling that of artificial instruments".

沁园八景图：沁园云境 "云生山有衣，月落潭无影"
One of the Eight Scenes of Qin Garden: The Cloud Scene – "cloud draped mountains with cover while moon falling pond without trace".

沁园八景图：莲坛禅韵"掬水月在手，弄芳香满衣"
One of the Eight Scenes of Qin Garden: Lotus Pagoda with Zen Charm – "moon in hand when holding water in palms; fragrance lingering on clothes when playing with flowers".

沁园八景图：苇舟野渡"始从芳草去，又随落花回"
One of the Eight Scenes of Qin Garden: Ferry Crossing by Reed Port – "departing from fragrant grassy patch and returning with fallen flower petals".

沁园八景图：烟渚柔波 "岛影含树影，天光透水光"
One of the Eight Scenes of Qin Garden: Misty Islet over Mild Ripples – "tree silhouettes in islet shadows and water illuminating through sky light".

沁园八景图：一朝风月 "延山引泉谋禅境，亭桥廊榭摹唐风"
One of the Eight Scenes of Qin Garden: Scene Sought After – "seeking zen-like scene by extending mountain and channeling water; imitating Tang style architecture by building pavilion bridges and hall corridors".

沁园八景图：丝竹琴台 "行到水穷处，坐看云起时"
One of the Eight Scenes of Qin Garden: Musical Instrument Playing Place – "go to the end of water; sit and watch cloud rise".

沁园八景图：紫竹苔院 "苔痕映阶绿，修篁入帘青"
One of the Eight Scenes of Qin Garden: Purple Bamboo in Mossy Yard – "green steps set off by mossy walkway and slender bamboos against verdant shade".

"小瀛洲咏"的思绪

"Small Yingzhou" – A Train of Thought, by Meng Zhaozhen (the only academician on landscape architecture in Chinese Academy of Engineering)

吴君肇钊在风景园林设计方面走中国传统的路子而且有所创新。平时交往就有"酒逢知己千杯少"之感，我不会喝酒便以言谈取代。他又擅长绘画，综合平时所学的积累加以抒发，以设计绘画表达自己的心志与情感。其中有《小瀛洲》一卷我觉得有些意思。寻觅"住世瀛壶"有些浪漫色彩。我便揣摩他的立意、相地和借景，即兴以七言诗来提炼一下设计意匠。我不会写诗，更不是诗人，但出于我也是一名风景园林设计师，中国园林要以诗画塑造空间，应该是写得出一点感慨。

地宜是水，理水之理法首先是"疏水之去由，察水之来历"。先创造水环境，再于水境中谋水院之景。扬州瘦西湖有"凫庄"前车之鉴，水院漂浮水面，朝雾薄云

Mr. Wu Zhaozhao's garden design follows traditional Chinese style yet creative. There is an old Chinese saying that goes, "when like minds meet, they can never drink too many bottles." I had the same feeling when I was with Mr. Wu. Since I cannot drink, I used conversation instead. Mr. Wu is a master of artistic drawing. Based on what he learned and accumulated, he uses his design drawings to express his ideas and emotions. One of his works, "Small Yingzhou", is particularly interesting and intriguing, as it seeks the artistic and romantic concept of "eternal life in this world". I was trying to interpret his design concept, his thought on siting and scene leveraging, and wrote an impromptu poem to extract his artistic conception. I cannot write a poem and I am even far from being a poet; however I am a landscape architect and know that classic Chinese gardens can hardly be formed without poetry and artistic elements, so I thought I could express some of my feelings.

A great design is with appropriate arrangement of water body. The gist here is that "deciding where the water should be directed and finding out where it is coming from". A physical micro-environment with water is established

汝昌瀛洲效果图

Ruchang Yingzhou Perspective View

时便产生虚无缥缈感。所以我起句"引水穿流谋凫院"。用的园林建筑是堤、廊、桥、舍、组合的效果却有中国文学讲究的"缠绵"。所谓"缠绵",再缠绵,最后也不得一句道破。对园景而言,这才耐人流连、寻味,园林意味深求。水院建筑的比重并不少,但从谋篇而言是以水为心,而不是以建筑为心,建筑布置应该是与水找关系,邀水贯穿于环境。"石令人古,水令人远。"水因映天成影而令人远,借水邀闲云便是发挥延展空间的地宜之举。人们可将美好的希望委托于云。于是,第三句"以水为心邀祥云"应心而出。人又为什么赏石甚至誓石、甘为石奴呢?石不仅坚固延年,而且不变心,所以古之誓石者要"与石为伍"。视石为人,充分体现了美学家李泽厚先生从美学对园林作的科学而生动的概括,是"人的自然化和自然的人化。"我就以此收章,"石庭成趣耐流连"作为追求不尽之意的收尾。这些话说得是否合宜,欢迎大家指正

孟兆祯(中国工程院院士)

first and then a water court scene is developed. Taking lessons from Fu Zhuang, the designer of Shouxi Lake in Yangzhou set the water court in a way as if it floats on water. With a thin layer of morning fog, it adds an illusory feeling. This is why in the poem I started with "Channeled water courting the duck pond". The garden constructions deployed have a combination of dykes, corridors, bridges, and small houses, and therefore it has achieved the "lingering" effect, often emphasized in classic Chinese literature. People seek lingering and lingering effect until it cannot be done so any more. On garden scene design, this method can effectively make people linger and ruminate, soliciting far-reaching zen-like mind. For water court design, although the weight of buildings is heavy, the focus is still water, not buildings. Therefore, the layout of buildings should focus on the relationship to the water, and inviting water channeling through the whole setup. "Rock invokes ancient feeling while water far-reaching." You get a more far-reaching feeling when there is sky reflection in the water, and inviting leisurely clouds via water proves to be an effective way of extending spaces – an appropriate measure for a garden design. People may embed their good wishes in the cloud; therefore the third sentence in my poem, "centered water body inviting auspicious clouds", was written with this thought. Then why people enjoy rocks, obsessed with it, or even be a slave of it? That is because a rock is not only sturdy and lasting forever, but also will never have a change of heart. As such, all those rock lovers in ancient times would be "in company with rocks". Viewing a rock as a human being reflects a vivid generalization of landscape aesthetics, as esthetician Li Zehou aptly puts it, "Human's naturalization and nature's humanization". Let me end with "rock garden generating lasting interest", as a conclusion implying to quest for unending meaning. Whether my statements are appropriate or not, I welcome your comments and corrections.

Meng Zhaozhen(Academician, China Academy of Engineering)

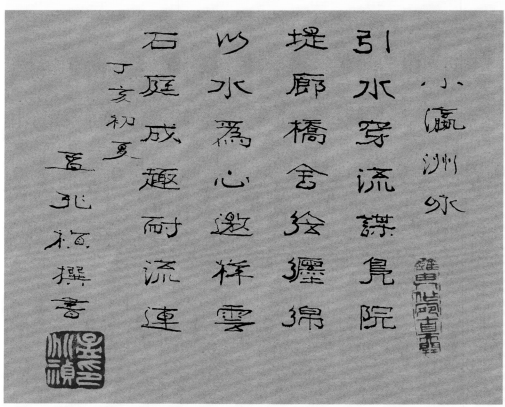

孟兆祯院士赋诗《小瀛洲咏》
Academician Meng Zhaozhen's Peom "Xiao Yingzhou"

园冶图释

Yuan Ye Illustrated
——Classical Chinese Garden Explained

园说

On Garden Design

凡结林园，无分村郭，地偏为胜，开林择剪蓬蒿；景到随机，在涧共修兰芷。径缘三益，业拟千秋，围墙隐约于萝间，架屋蜿蜒于木末。山楼凭远，纵目皆然；竹坞寻幽，醉心即是。轩楹高爽，窗户虚邻；纳千顷之汪洋，收四时之烂漫。梧阴匝地，槐荫当庭；插柳沿堤，栽梅绕屋；结茅竹里，浚一派之长源；障锦山屏，列千寻之耸翠。虽由人作，宛自天开。刹宇隐环窗，仿佛片图小李；岩峦堆劈石，参差半壁大痴。萧寺可以卜邻，梵音到耳；远峰偏宜借景，秀色堪餐。紫气青霞，鹤声送来枕上；白苹红蓼，鸥盟同结矶边。看山上箇篮舆，问水拖条枥杖；斜飞堞雉，横跨长虹；不羡摩诘辋川，何数季伦金谷。一湾仅于消夏，百亩岂为藏春，养鹿堪游，种鱼可捕。凉亭浮白，冰调竹树风生；暖阁偎红，雪煮炉铛涛沸。渴吻消尽，烦顿开除。夜雨芭蕉，似杂鲛人之泣泪；晓风杨柳，若翻蛮女之纤腰。移竹当窗，分梨为院；溶溶月色，瑟瑟风声；静扰一榻琴书，动涵半轮秋水，清气觉来几席，凡尘顿远襟怀；窗牖无拘，随宜合用；栏杆信画，因境而成。制式新番，裁除旧套；大观不足，小筑允宜。

There is no differentiation between town and country in almost all garden designs. It would be great if the site is quiet and slightly remote. Weeds need to be removed where there is a grove or forest. Since a garden leverages nearby natural landscape as appropriate, flowers like orchard or narrow leaf stoneseed should be planted at the stream or brook side. The garden paths should go along with pine, bamboo, or plum trees to lay a solid foundation. Garden walls should be intermittently hidden among vines, and buildings meandering up and down of tree branches. Looking afar from a building, one sees beautiful scenes everywhere; walking into a bamboo grove, one is obsessed with tranquility and solitude. A building should be erected high without obstruction and there should be ample space in front of its windows. This way large extent of water scenery can be included and all seasonal flower scenes can be enjoyed without obstruction. Plane tree shadow spreads on the ground and a yard is covered under ash tree shade. Poplar and willow trees are planted along a dyke while plum flowers are scattered around a building. One may build a cottage in a bamboo forest, or dredge long flowing water body. One can face a bright and colorful screen of mountain scenes, or long arrays of green landscape. Although it is made by man, it is as if created by nature. Ancient temple looming through the window scene feels like artwork created by Tang dynasty landscape painter Li Zhaodao; steep vertically cut rock formation gives an impression of a Yuan dynasty landscape painter Huang Gongwang. Close by, a Buddhist temple can be a neighbor so that a soothing chanting can be heard; farther away, gorgeous mountains can be a treat to the eyes. Looking into the distance and seeing auspicious purple and blueish cloud, one could almost hear crane's chirping at the pillow side; nearby water plants with red and white colors attract birds like gulls to hang around. If you like to enjoy the mountain, you could ride in a sedan chair, and if you prefer water scene, you could carry a walking stick. With the garden wall slanting up and the long bridge spanning the water, you do not need to envy Tang dynasty Wang Wei's Wangchuan Villa, nor do you have to match up to Jin dynasty Shi Chong's Jin Gu Fragrant Garden. A bay of water is more than you would need to spend the summer, and acres of garden are far more than you would need to enjoy the spring. You can raise some deer to tour the garden together, and stock some fish for fishing. Having a drink on a summer night in a pavilion, adding some ice to quench your thirst, you feel cool breeze blowing through the bamboo grove; sitting close to a stove on a winter day in a house, boiling snow water to make tea, it sounds like roars of waves. All your thirst is quenched and worries washed away. Raindrops on banana leaves at night sound like teardrops from a mermaid, and the slender willow branches bent by morning breeze bring imagination of a lady's waist line curve. Plant a few bamboos in front of a window, and grow some pear trees to form another yard. Misty moonlight casts wavering shadows on you books and music instruments in the study, and the bleak and howling autumn wind blows over the pond, rippling the moon reflection. Breezes blow through your tea table and bamboo mat, and that instantly brings about a feeling of worldly transcendence. In general, on garden construction, the size of windows can vary and should be more adaptable; the railing design should follow no fixed rules, as it should be decided according to specific conditions. New designs are preferred over conventional ones. It is very appropriate to build smaller garden elements this way, although it is not as sufficient for grander garden constructs.

造园地段以偏静为胜；景物因借随机；在涧畔就整理兰芷
It would be great if the site is quiet and slightly remote. Since a garden is leveraging nearby natural landscape as appropriate, flowers like orchard or narrow leaf stoneseed should be planted at the stream or brook side.

园说篇是计成以开拓的视野综述造园，其诗情画意的评述十分精彩而又形象，但很难全部用园林实景来体现。鉴于古代造园蓝本皆依据画家的画作，故从古代山水画中寻找相关画面是图释园说篇内容的唯一途径。但效果也只是辅助理解园说的基本内容而已，因为"园说"是企图通过文字的举例，来阐述形象与意境的关系，引导造园者由形象思维走意境创作的道路。其实，形象与意境的联想，一方面来自设计者的安排，一半则有赖于欣赏者的素养与情操。所以造园不是毫无旨趣地造些房屋，种些花木，而是旨在构成博大旷达的意境"寄情"。

"On Garden Design" is Ji Cheng's general broader view on garden design. His poetic discourse, although vivid and wonderful, can be very difficult to exemplify with actual physical landscape scenes. Since all the ancient garden design blueprints are based on fine artists' drawings or paintings, it must be the only way to interpret the art of garden design from ancient landscape paintings. However, the landscape paintings can only be an auxiliary way to aid the interpretation on garden design, since "On Garden Design" attempts to use examples written in scripts to relate the image to the artistic concept, leading garden designers on conceptual designs via graphical thinking. In fact, the association between graphics and artistic concept is, on the one hand, determined by the designer's work, and on the other, by the viewer's sentiment and artistic experiences and accumulations. Therefore, garden design is not to tastelessly build some buildings and plant some plants, it is to construct a broad-minded spiritual and emotional sanctuary.

园林路径顺三益而开，基业传千秋。三益指梅、竹、石，"梅寒而秀，竹瘦而寿，石丑而文"是谓"三益之友"
The garden paths should go along with three Yis (Note: Yi here means beneficial element.), that is, plum flowers, bamboos, and rocks, to lay a solid foundation for generations to come. "The coldness of a plum flower makes it beautiful. The slenderness of a bamboo tree represents longevity. An unhandsomely rock manifests its suaveness." This is the so-called "friends of three Yis".

园林中建筑要曲折，架于树梢之上；倚山楼而望远，放眼都成美景；能接纳千顷汪洋的波光，收揽四时烂漫的花信

A building should be meandering and over trees and there should be a broad view from the building. This way large extent of water scenery can be included and all seasonal flower scenes can be enjoyed without obstruction.

步竹坞而寻幽，醉心便在此中

Walking into a bamboo grove, one is obsessed with tranquility and solitude.

桐影遍地，槐荫满庭；沿堤插些杨柳，绕屋种点梅花

Plane tree shadow spreads on the ground and a yard is covered under ash tree shade. Poplar and willow trees are planted along a dyke while plum flowers are scattered around a building.

结茅屋竹林，疏瀹一派的长流

One may build a cottage in a bamboo forest, or dredge long flowing water body.

环窗隐塔影，好像是李昭道所绘的小幅山水画
Ancient temple looming through the window scene feels like artwork created by Tang dynasty landscape painter Li Zhaodao.

岩峦堆劈石，就像黄公望（号大痴道人）所画的半壁山水。图为黄公望《富春山居图》局部（亦可参见本书选石篇"黄子久山水画"）
Steep vertically cut rock formation gives an impression of a Yuan dynasty landscape painter Huang Gongwang. Shown in the picture is part of "Fuchun Mountain Dwelling" by Huang Gongwang (It can also be found in the chapter about rockeries "Landscape Painting by Huang Zijiu".)

利用障锦山屏，列千寻之耸翠的环境造园
One can leverage verdant mountain ranges and form a surrounding screen of mountain views around a garden.

园林可以和萧寺为邻，梵音到耳；远峰偏宜借景，秀媚的山色堪饱眼福
Close by, a Buddhist temple can be a neighbor so that a soothing chanting can be heard; farther away, gorgeous mountains can be a treat to the eyes.

遥看紫气青霞的景象，鹤声送来枕上，真可谓仙境兮
Looking into the distance and seeing auspicious purple and blueish cloud, one could almost hear crane's chirping at the pillow side.

白苹红蓼，鸥盟同结矶边
Nearby water plants with red and white colors attract birds like gulls to hang around.

城垣起远空，长桥跨水面
A town is up far and remote, and a long bridge crosses the water.

一湾曲水，不仅消夏；百亩幽芳，何止藏春；养鹿同游，放鱼捕钓
A bay of water is more than you would need to spend the summer, and acres of garden are far more than you would need to enjoy the spring. You can raise some deer to tour the garden together, and stock some fish for fishing.

听夜雨打芭蕉，像夹杂鲛人的泪点
Raindrops on banana leaves at night sound like teardrops from a mermaid.

是晓风吹杨柳，又像舞翻蛮女腰肢
Slender willow branches bent by morning breeze bring imagination of a lady's waist line curve.

凉亭小酌，竹树风生
Cooling off with a drink in a pavilion and enjoying a cool breeze through a bamboo grove.

移竹当窗，分梨为院
Plant a few bamboos in front of a window, and grow some pear trees to form another yard.

月色溶溶，影扰一榻琴书
Misty moonlight casts wavering shadows on you books and music instruments in the study.

风声瑟瑟，掠过池面，波漾半轮秋月
Bleak and howling autumn wind blows over the pond, rippling the moon reflection.

窗户无拘，随宜合用；栏杆信画，因境而成

The size of windows can vary and should be more adaptable; the railing design should follow no fixed rules, as it should be decided according to specific conditions.

制式新番，裁除旧套

New designs are preferred over conventional ones.

景到随机，因境而成（香港粉岭公园）

A garden is leveraging nearby natural landscape as appropriate, and a scene is formed in context. (Fenling Park, Hong Kong.)

景到随机，因境而成（德国法兰克福春华园）

A garden is leveraging nearby natural landscape as appropriate, and a scene is formed in context. (Chunhua Garden in Frankfurt, Germany.)

景到随机，因境而成（德国春华园）

A garden is leveraging nearby natural landscape as appropriate, and a scene is formed in context. (Chunhua Garden, Germany.)

景到随机，因境而成（新加坡蕴秀园）

A garden is leveraging nearby natural landscape as appropriate, and a scene is formed in context. (Yunxiu Garden, Singapore.)

格式求新颖，俗套必屏除；称大观虽不足，建小筑正合宜。这是笔者在设计小型园林时严格遵循的原则。

园林界每十年在德国举行一次的国际园林博览会，相当于体育界的奥林匹克运动会。其方法是由参展国向当届园林博览会组委会提出申请，经批准取得参展资格，后可派专家前往园林博览会展出现场考察，并作出参展设计方案，经园林博览会全权总代表签字批准后，再办理施工人员进场施工手续，施工期一年，博览会展期一年。在展览期间进行三次评选，评出金奖，得金奖的国家可在博览会期间举行一周的升旗仪式及文化艺术、商业活动。

笔者有幸于1991年代表国家前往德国斯图加特市（1993年国际园林博览会展地）考察现场，并结合实地完成博览会全权总代表批准的中国参展园的方案。1992年清音园现场施工亦笔者与商自福先生总其事。值得庆幸的是：清音园荣获1993年国际园林博览会大金奖，又获德国政府授予荣誉奖章，并拆建保留于斯图加特市生态景观带的山顶上，成为该市园林景观的闪光点之一。

"New designs are preferred over conventional ones. It is very appropriate to build smaller garden elements this way, although it is not as sufficient for grander garden constructs." This is exactly what I follow strictly when I design smaller gardens.

Every ten years in landscape architecture community, there is an international garden exposition held in Germany. It is equivalent to the Olympic games in sports. Here are the procedures. Garden contest applicants submit their applications to the expo organizing committee, and once their applications are approved, they can send experts to the expo site for visits, field work, and initial designs. After the expo's plenipotentiary's approval, the construction workers' paperworks are processed. The construction work lasts for a year. The exposition runs for the same length. Evaluation and selection run three times during the expo and a final gold medal is selected. The country winning the gold medal enjoys one week of national flag raising ceremony as well as related art, cultural, and commercial events.

The author is privileged to represent China to visit the garden expo site in Stuttgart, Germany, and after the garden expo plenipotentiary's approval, completed the Chinese garden design – for the 1993 International Garden Expo. In 1992, at the Qingyin Garden construction site, I worked together with the construction lead Mr. Shang Zifu. To our great pleasure, Qingyin Garden garnered the International Garden Expo's grand golden medal and the honorary medal awarded by the German government. After the expo, it was relocated to a hilltop in Stuttgart's ecological landscape zone, becoming a highlight of the city's landscape scene.

荣获：1993年德国国际园林博览会大金奖
The International Garden Expo's grand golden medal, Germany, 1993.

荣获：1993年德国政府授予荣誉奖章
The honorary medal awarded by the German government, 1993.

1993年国际园林博览会总平面图（德国－斯图加特市）
1993 International Garden Expo Master Plan, Stuttgart, Germany

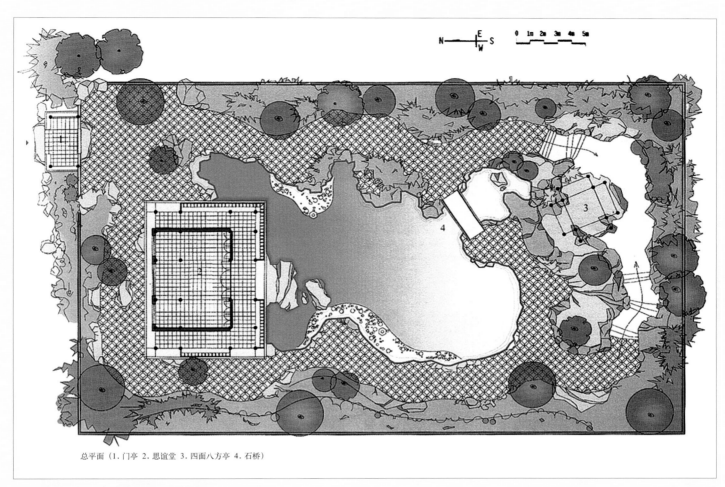

总平面（1. 门亭 2. 思谊堂 3. 四面八方亭 4. 石桥）

清音园平面图
Plan of Qingyin Garden

临水的思谊堂，对联："万松时洒翠，一涧自流云"
Siyi Tang by the pond. The couplets go, "Ten thousand pine trees cast verdant shadows; one single stream flows like a cloud".

清音园门亭
Entrance Pavilion of Qingyin Garden

入门迎石峰
The greeting peak at the entrance

屹立鹰崖假山上的四面八方亭；联句："非必丝与竹，山水有清音"
The square pavilion erected on the eagle cliff rockery. The couplets go, "No need for musical instruments; there are natural music from mountain and water".

园冶图释

Yuan Ye Illustrated
——Classical Chinese Garden Explained

一 相地

1 Siting

园基不拘方向，地势自有高低；涉门成趣，得景随形，或傍山林，欲通河沼。探奇近郭，远来往之通衢；选胜落村，藉参差之深树。村庄眺野，城市便家。新筑易乎开基，只可栽杨移竹；旧园妙于翻造，自然古木繁花。如方如圆，似偏似曲；如长弯而环璧，似偏阔以铺云。高方欲就亭台，低凹可开池沼；卜筑贵从水面，立基先究源头，疏源之去由，察水之来历。临溪越地，虚阁堪支；夹巷借天，浮廊可度。倘嵌他人之胜，有一线相通，非为间绝，借景偏宜；若对邻氏之花，才几分消息，可以招呼，收春无尽。架桥通隔水，别馆堪图；聚石垒围墙，居山可拟。多年树木，碍筑檐垣；让一步可以立根，斫数桠不妨封顶。斯谓雕栋飞楹构易，荫槐挺玉成难。相地合宜，构园得体。

A garden site is not confined by orientations and the elevations can be high or low. Natural point of interest should be there immediately after the entrance, and the scene should be formed by naturally following the terrain, either to be integrated with the mountains or connected with the rivers. For garden explorations close to towns, it is better to stay away from major thoroughfare; to build a garden in rural area, one should take advantage of trees of different densities and heights. For a countryside garden, it needs to have a great view; for an urban garden, the convenience at home is mostly treasured. For new constructions, it is easy to lay the foundation, one can only transplant poplar and bamboo trees; for older garden renovations, it is natural to use ancient trees and flowers. The shape of a garden can be square or round, elongated or curved, as appropriate. A long and curved site should be made into a circular jasper shape; an open slope should be made into a stacked cloud shape. A pavilion should be erected at a high spot to make it more prominent; a pond should be dug at a lower spot to enhance the depth. A well-situated building should be close to water, with a clear knowledge of the water source and unobstructed water outlet. Over a brook or stream, a flying corridor is more suitable to be built; under a narrow lane with skyline light, a house corridor is more appropriate. As long as there is an outside scene connected through a vista, that scene should not be blocked; if one can have a peek at a neighbor's plants and flowers, even if it is exposed very slightly, it is sufficient to evoke feelings of lush spring. Bridging a disconnected water, one can reach another building; piling rocks to build a wall, it may look like a small hill. If there is an old tree getting in the way for construction, it might be good to yield a little so that the tree can live, or cut a few branches on top to let it grow. This is because it is much easier to build artificial buildings than to grow well-grown trees. In general, with proper siting, the garden built would naturally be decent and appropriate.

"相地篇"是论述园林的选址和查勘。不同的园基其创作手法各异，故列举众多不同园基各自的造园要领，实质上是因地制宜的规划原则和意境创作的设计方法如何贯彻到各种不同类型的实际地形与功能要求中去的概述。为能准确而又较具象读懂相地的内容，笔者选择清代康乾年间扬州瘦西湖园林群的实例来对照图释。

扬州有近2500年建城史，西汉始创，六朝初兴，隋唐已有"园林多是宅"之称，明清尤其清康熙、乾隆都曾六次南巡，驻跸扬州。富商大贾和官吏雅士争宠于皇室，园主更以能一邀"御赏"为荣，故"三十里楼台"应运而生。园

"Siting" is about the exploration and selection of a garden site. Since there are various garden siting methods, citing different examples actually outlines how to integrate artistic conception and as-appropriate planning and design method with actual siting and functional use. In order to accurately and more specifically understand the core idea of "Siting", the authors selected Shouxi Lake garden in Yangzhou in Qianlong's reigning period of the Qing Dynasty and use its graphics for explanation.

The city of Yangzhou has a long history of nearly 2500 years. It was first built in Western Han Dynasty, and became sizable in the Six Dynasty period. It was well known for "most gardens as residence" in the Sui and Tang dynasty. In the Qing Dynasty, especially during Kangxi and Qianlong's reign, it was visited for six times by the emperors and served as a station point. Officials and wealthy merchants fought for getting royal visits at their home gardens as they felt much privileged and honored if they could land one. As a result, there came a saying as "miles of edifices". Garden buildings were thriving and in

林建造争奇斗妍，标新立异，景点数以百计。清人刘大观曰："杭州以湖山胜，苏州以市肆胜，扬州以园亭胜。三者鼎峙，不可轩轾。"以瘦西湖而言，记载为："十里林亭通画舫""一路楼台直到山"。《扬州画舫录》中，谢溶生序文中曰："增假山而作陇，家家住青翠城堙；开止水以为渠，处处是烟波楼阁。"《水窗春呓》中盛赞："扬州园林之胜，甲于天下"。

1979年适逢故宫博物院举行"历代园林古画"展览，江苏省建委特派笔者赴京审定并撰写造园精艺等江南园林部分的内容，恰展品中有宫廷画家绘扬州瘦西湖园林图三十幅，借此机会，经博物院同意笔者将审核稿费换三十幅园林图的拍摄与复印权。下面所选的画幅均为笔者当年手工临画复制品，现补添色彩，并依据计成相地篇的内容选择相应的园林实景画幅进行图释与说明。

（注：图幅中"白塔晴云"、"卷石洞天"、"春台祝寿"、"望春楼"、"静香书屋"已荡然无存，现在实景是笔者20世纪80年代亲自设计并主持施工复建的。）

constant contest with each other among hundreds of scenes, each striving for being unique and creative. The Qing Dynasty's Liu Daguan said, "Hangzhou excels for its mountains and lakes, Suzhou for its urban markets and Yangzhou for its gardens. It is hard to tell which one is better than others." Regarding Shouxi Lake, the record goes, "miles of garden gazebos leading to boat houses" and "all the edifices ending at the mountain". In "Yangzhou Boat House", Xie Rongsheng writes in the preface, "by building artificial rockery to simulate real hills, every home seems to be living in verdant mountains; by channeling still water to form flowing water system, it seems to have buildings over fog shrouded water everywhere." It is highly praised in "Shui Chuang Chun Yi", "Yangzhou gardens tops all others." Indeed.

In 1979, when the "Historical Chinese Landscape Painting" exhibition was held in the National Palace Museum, Jiangsu Province Construction Council sent the author to Beijing to write about fine garden art and other Southern Chinese gardens. Coincidently among exhibits there were a royal painter's thirty landscape paintings of Yangzhou's Shouxi Lake garden. Through this opportunity and under the permission of the museum, the author traded the royalties earned for the copy right of these paintings. The following selected paintings are all those that the author copied manually. I then added some colors and I will interpret them according to Ji Cheng's "Siting" on some of the garden scenes in the paintings.

(Note: In these paintings, some of them, such as "white pagoda against blue sky and clouds", "scrolling rocks and cavernous sky", "terrace for birthday celebration in spring", "spring herald building", and "quiet and fragrant studio", do not exist anymore. The current scenes are what the author led to have restored in the 1980s.)

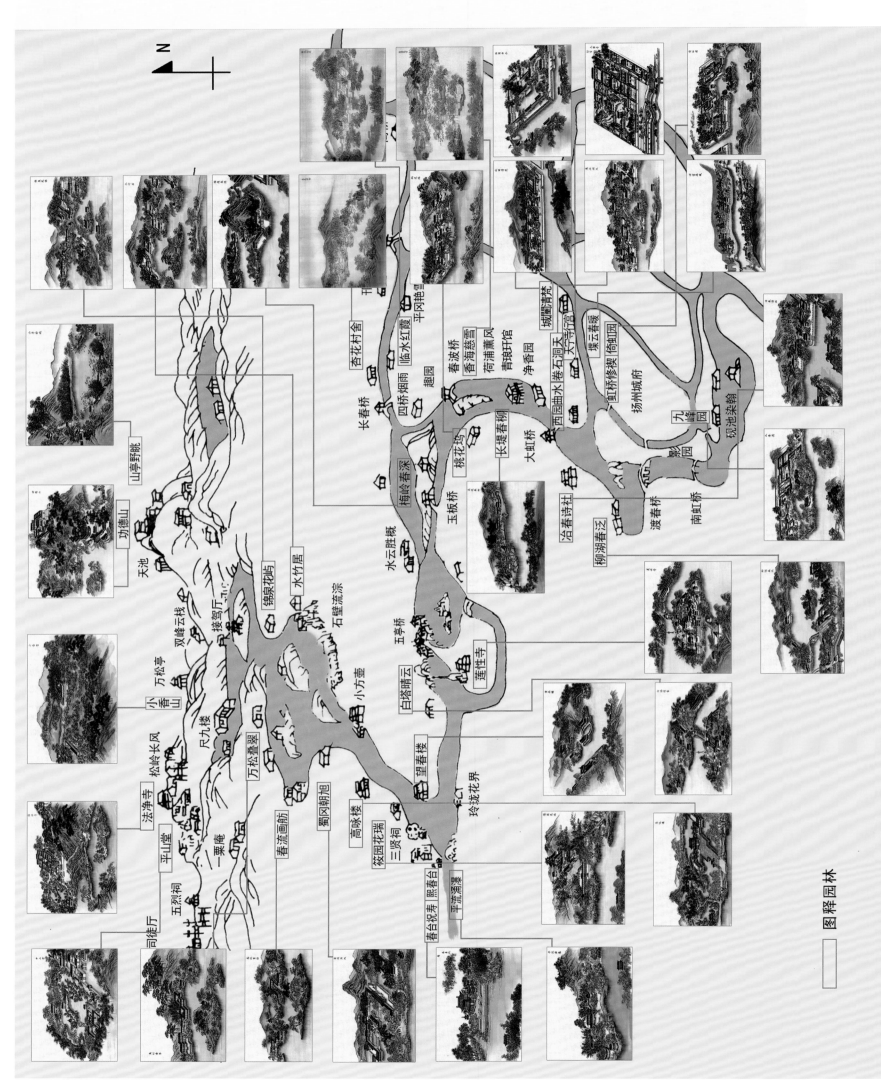

清代乾隆年间瘦西湖园林名胜图
Garden attraction map of Shouxi Lake, Qianlong's reign, Qing Dynasty

园基不拘方向,地势自有高低;涉门成趣,得景随形,

A garden site is not confined by orientations and the elevations can be up and down. Natural point of interest should be there immediately after the entrance,

"平山堂"位于蜀冈上,宋·欧阳修守扬州,以南徐(长江南)诸山拱立环向,望与槛平,因名平山堂。皇帝赐御书:"怡情"、"平山堂"、"贤守清风"额

"Pingshan Tang" was on top of Mound Shu. When Ouyang Xiu of the Song dynasty governed Yangzhou, he noticed that the surrounding mountains south of the Yangtze River were about the same height of the railings, so he named it "Pingshan Tang" (Note: mountain leveling hall). The emperor granted the names on horizontally inscribed board as "Yi Qing", "Pingshan Tang", and "Xianshou Qingfeng".

或傍山林,

either to be integrated with the mountains,

"山亭野眺"在功德山半山上,建南楼"深竹厅",临水为屋"菱荷深处",选址极佳

"Shanting Yetiao" (Note: wild outlook in mountains) was built at the hill slope of Mount Gongde. A southern building is "Shenzhu Ting" (Note: deep bamboo hall) and a water front one "Jihe Shenchu" (Note: deep among lotus and water chestnut). Both building sites are excellent.

"西园曲水"水自北而东折,若半壁,依水曲折以置亭馆,有"新月楼","觞咏楼"、"拂柳亭"诸景
The water at "Xiyuan Qushui" (Note: west garden and meandering water) comes from the north and turns to the east along a seemingly half wall. There are a few scenery spots such as "Xinyue Building", "Shangyong Building", and "Fuliu Pavilion" etc.

欲通河沼,
connected with rivers,

"城闉清梵"临河面城,旧为舍利禅院,圣祖仁皇帝赐:"大智光",乾隆南巡赐名:"慧因寺",其旁为别苑。乾隆四十八年(1783年),知府衔罗琦又加修葺,有"香悟"、"涵光"、"楼鹤"等亭,其西为药园
"Chengyin Qingfan" faces a river and city. It used to be a Buddhist temple for holding Buddhist bone relics. It got its name "Huiyin Temple" during the royal visit of Qianlong to south China. The emperor Kangxi granted the name "Dazhiguang". A garden was at its side. In the 48th year of Qianlong's reign(1783), Magistrate Luo Yi made some renovations, adding pavilions such as "Xiang Wu", "Han Guang", and "Lou He" etc. A medicinal garden was at its west side.

探奇近郭,
garden explorations close to towns,

> 远来往之通衢，
>
> to stay away from major thoroughfare

"柳湖春汛"在通泗门外，避开往来大道，南部建"渡春桥"，桥之西北，临池台榭，掩映参差；芦荻荷花，一望无际。湖心垒石为山，建亭其上

"Liuhu Chunxun" is located outside of Tongsi Gate, away from major thoroughfares. To its south, "Duchun Bridge" was built. To the northwest of the bridge some buildings by the pond were built among trees. It enjoys a broad view over lotus and reeds. A pavilion was built on top of rockeries at the center of the lake.

> 选胜落村，藉参差之深树。
>
> to build a garden in countryside, one should take advantage of trees of different densities and heights.

"小香雪"即"十亩梅园"，在今万松岭内。修水为塘，旁筑草屋竹桥，制极清雅，上赐名"小香雪居"。御制诗云："竹里寻幽径，梅间卜野居"。"浣花杜甫宅，闻说此同诸"

"Xiao Xiang Xue", that is, "Ten Mus Plum Garden", is located in today's Wansongling (Note: ten thousand pine ridge). The original water bodies were made into a pond; delicate and elegant thatch house and bamboo bridges were built. It got its royally granted name "Xiao Xiangxue Ju" and a royal poem that goes, "Seeking quiet paths in the bamboo grove and dwelling a wild house among plum trees." Another one goes, "Washing flowers in a house like that of Du Fu's and it is said that here it is really like so".

"九峰园"旧称"砚池染翰"。前临"砚池",旁距"古渡桥",老树千章,四面围绕,世为汪氏别业。即用主事加捐道(衔)汪长馨屡加修葺。得太湖石九(尊)于江南,殊形异状,各有名肖。乾隆二十七年(1762年)蒙赐御书"九峰园"额,并"雨后兰芽犹带润,风前梅朵始敷荣"一联,又"名园依绿水,野竹上青霄"一联。乾隆三十年(1765年),蒙赐"纵目轩窗饶野趣,遣怀梅柳入诗情"一联

"Nine Peak Garden" used to be called "Yanchi Ranhan". It faces the "Yanchi" in front and sits by "Gudu Bridge" (Note: old ferry bridge). There's an abundance of old trees around and it had been Mr. Wang's villa. The government official Wang Changxin had numerous renovations. From southern China, it has got nine Taihu rocks of various shapes and labels. In the 27th year in Qianlong's reign (1762) it got a royal inscription board as "Nine Peak Garden" and a couplet that goes, "Orchid buds are still moistened after the rain; plum flowers start flourishing under the wind." Another one goes, "The famous garden sits by the water; the wild bamboos go up in the sky." In Qianlong's 30th year (1765), another royal couplet was granted, "Wild pleasures come through the windows to meet the eyes; poetic feelings from the plum and willow trees harbor in my heart".

村庄眺野,

For a countryside garden, it needs to have a great view,

"杏花村舍"在长春桥东北村庄,园主构竹蓠茅舍于杏花深处而得名(摘自《扬州园林甲天下》)

"Xinghua Cunshe" (Note: apricot blossom village) is a village located to the northeast of Changchun Bridge. It got its name because the owner built thatched houses with bamboo fences deep in the apricot groves. (Excerpt from *"Yangzhou Gardens Rule"*)

城市便家。
for an urban garden, the convenience at home is mostly treasured.

"堞云春暖"在北门外城隅,傍城临水,便于居家。且屋宇参差,竹树蓊郁,大有濠濮间想
"Dieyun Chunnuan" is located outside the north gate. It was close to the city and by water, so it was a good site for residence. Houses at various elevations and among thick trees made you feel in the wild.

"锦泉花屿"立意是充分利用水中岛屿与溪涧水泉,其植物景观反映出开基之时以栽柳移竹为主
Design concept of "Jinquan Huayu" is to fully take advantage of islets, streams and springs. Its plant design reflects its initial implementation using transplanted willows and bamboos.

新筑易乎开基,只可栽杨移竹; | For new constructions, it is easy to lay the foundation, one can only transplant poplar and bamboo trees.

(此句与下句对偶,故其意思是:新造园林容易按自己的意思规划建造,但为了早日成景,不得不多选用竹子和杨柳之类的速生树木。)

(This is in good comparison with the following: For newer garden constructions, it is easier to plan and design according to one's original idea; however, in order to form a landscape earlier, one has to use bamboos, poplar, and willow trees because they grow faster.)

"长堤春柳"由虹桥而北,沿岸皆高柳,拂天踠地,凝望如织,候选同知黄为蒲沿堤东向为园

"Changdi Chunliu" extends from the Rainbow Bridge to the north. Along the shore, with tall willows that reach the sky and touch the ground, it looks like a dense knitting work. A garden is at the east end of the dyke.

旧园妙于翻造,自然古木繁花。

for older garden renovations, it is natural to use ancient trees and flowers.

"冶春诗社"王士正(禛)赋《冶春词》即此地。冶春,本酒家楼,后为候选州同王士铭园亭,捐知府衔田毓瑞重构。临河有"香影楼",又有"冶春楼"、"怀仙馆"、"秋思山房"、"云构"、"鸥谱"二亭,旧园翻造,古木繁花

"Yechun Shishe" is where Wang Shizheng (Zhen) wrote "Yechun Ci". Yechun was a restaurant and became a garden pavilion by Wang Shiming before it was renovated by Tian Yurui. By the river there are "Xiangying Lou", and then "Yechun Lou", "Huaixian Guan", "Qiusi Shanfang", "Yungou", "Oupu" etc. The old garden was renovated and the old trees mixed with luxuriant flowers.

"筱园花瑞"园为编修程梦星别墅，后汪廷璋等辟其西，广数十亩为芍药田。(花)有并头三萼者，因作"瑞芍亭"以纪胜，亭之北有"仰止楼"，修竹万竿，缘猗可爱。楼之东有丛桂数十本。建亭三楹，名曰"旧雨"。有古藤覆荫，花时烂漫，如张锦幄。折而西，有"梅岭小亭"踞其上

"Xiaoyuan Huarui" was a villa of Cheng Mengxing. Then Wang Tingzhang expanded its west side to dozens of Mu as a peony field. Since the flower has three calyxes, a pavilion named "Ruishao Ting" was built. North of the pavilion there is "Yangzhi Lou", surrounded with thousands of lovely slender bamboos. East of the building there are dozens of sweet tea olive trees. There is a three-unit pavilion, named "Jiuyu" (Note: old rain), and it is covered with old vines and luxuriant flowers. To the west, there is a "Meiling Xiaoting" (Note: plum ridge small pavilion) on top.

如方如圆，

（从这句起，讲建筑布局。）

The shape of a garden can be square or round

(From this one on, it starts to discuss the layouts of buildings.)

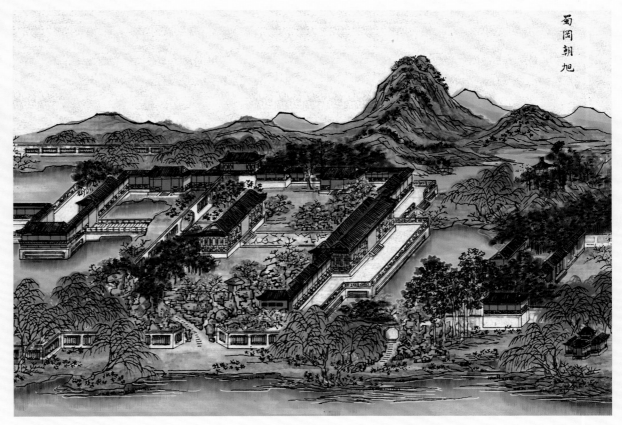

"蜀冈朝旭"为李志勋所构，张绪增重建。自堤北折登山，有亭曰"指顾三山"，亭后东折而下为"射圃"、"竹楼"、"迎晖亭"。左近蜀冈"西山朝来，致有爽气者"也。此园相形度势，均为方形构图，中规中矩

"Shugang Chaoxu" was built by Li Zhixun and renovated by Zhang Xuzeng. Along the embankment and turning north, one sees a pavilion named "Zhigu Sanshan". From behind the pavilion turning east and going down, you see "Shefu", "Zhulou" (Note: bamboo building), and "Yinghui Ting" (Note: sunlight greeting pavilion). The close-by Shugang shows "the cool at dawn in the west mountain". This garden fits the context well; its square form follows the rule strictly.

似偏似曲；
elongated or curved, as appropriate.

"春流画舫"充分利用偏曲自由的园基建造，记载："四面垂簾，波纹动荡为织。"别有"清荫堂"、"涵清阁"、"绿云亭"诸景，梅花深处置屋曰："傲寒春晓"

"Chunliu Huafang" takes full advantage of leaning and free forms of the foundation for its construction. According to record, "It seems to have hanging curtains on all four sides, with the ripples as the weaving pattern." There are also "Qingying Tang", "Hangqing Ge", "Lvyun Ting" and other scenes. Deep in the plum grove there is a house that is named as "Aohan Chunxiao". (Note: enduring coldness and heralding spring)

"万松叠翠"为奉宸苑衔吴禧祖构。园基如玉环般的长弯，面对蜀冈，冈上万松森立，滴露飘花，近落衣袖

"Wansong Diecui" was built by Wu Xizu. The foundation looks like a jade ring. It faces Shugang, which is endowed with a pine forest. Flowers flourish and dews drop on you when you are really close.

如长弯而环璧，	A long and curved site should be made into a circular jasper shape,
（山水之间，地基多狭长，故园林建筑常带状布置。）	(In a natural landscape, the site is mostly elongated; therefore the buildings are mostly designed in long and linear form.)

似偏阔以铺云

（平地造园，易于铺展，花团锦簇，状若云霞。）

an open slope should be made into a stacked cloud shape.

(To build a garden on a flat ground, it is easy to expand. It should look like flowers and clouds.)

"高咏楼"相传宋·苏轼题《西江月》词于此。按察使衔李志勋所构，候选道张绪增重建。乾隆二十七年（1762年），皇上临幸，赐御书"高咏楼"额，并"山堂返照留闲憩，画阁开窗纳景光"一联。楼之南为露台，过桥为"旷如亭"。园内佳树参差，山皆黄石垒之。前植古梅，后临方池，水木清华，延览不尽

"Gaoyong Lou" is where it is said that the Song dynasty's poet Su Shi wrote his "Xijiang Yue". It was built by Li Zhixun and was renovated by Zhang Xuzeng. Qianlong paid a royal visit in his 27th year (1762) and wrote the horizontal tablet "Gaoyong Lou" and the poetic couplet "Leisure kept via reflection of the mountain hall; Scene contained through the opening windows of the atelier". South of the building is the open terrace and across the bridge is "Kuangru Ting" (Note: open pavilion). Beautiful trees scatter in the garden and all rockeries were constructed using yellow stones. There are old plum trees in front and square-shaped pond behind. Quiet and beautiful plants fill the garden.

高方欲就亭台，低凹可开池沼。

A pavilion should be erected at a high spot to make it more prominent; a pond should be dug at a lower spot to enhance the depth.

"梅岭春深"俗称"小金山"，清时保障河自北而来与迎恩河会，二水涟漪，回绕山麓。此景为瘦西湖中心区构思焦点，境仿镇江金山"移来金山半点何惜乎小"

"Meiling Chunshen", commonly known as "Xiao Jinshan" (Note: little gold mountain), is where Baozhang River from the north converged with Ying'en River in the Qing dynasty. The joint rivers meander around the mountain foot. This scene is the central focus point of the Shouxi Lake scenery zone, and it imitates Jinshan of Zhenjiang. "Who cares if it's a bit small if we could move part of Jinshan?"

功德山

"功德山"为蜀冈最高处，山体如流云般舒展。山上恭备坐起，乾隆三十年（1765年）御书"天池"匾额，并"渌水入澄照，青山犹古姿"一联。山西为"双峰云栈"，"听泉楼"跨池上

"Gongde Shan" (Note: mountain of merits and virtues) is the highest point at Shugang. The mountain body is as fluid as flowing clouds. There are daily living facilities on the mound. In his 30th year (1765) reign Qianlong wrote the royal tablet "Tian Chi" (Note: heavenly pond) and a couplet "Green water reflecting clear pictures; Verdant mountains showing ancient posture". "Shuangfeng Yunzhan" (Note: double peaks with dwelling in the cloud) is to be west of the mountain and "Tingquan Lou" (Note: spring listening building) is over the pond.

平流涌瀑

> 卜筑贵从水面，立基先究源头。
> A well-situated building should be close to water and one needs to study the water's source.

"平流涌瀑"是借助西郊天然水源建造，水源处建亭，水由亭下，前过石桥，循山麓，穿竹径，数折后入湖。修廊邃室，补置花木竹石，与熙春台相映带

"Pingliu Yongpu" (Note: still current and gushing waterfall) was built leveraging the natural water source in the western suburb. The pavilion was built at the source of the water, which flows down from the pavilion, passing the stone bridge, along the mountain foot going through the bamboo grove, and finally turning back into the lake. Long corridors and rooms in the deep, coupled with added flowers, bamboos, and rocks, echo well with Xichun Tai.

疏源之去由，察水之来历。
with a clear knowledge of the water source and unobstructed water outlet.

"望春楼"选址水北来折东的转折处，引水入园，凿石为池，并架左右二桥，湾环如月。隔湖与熙春台相对，互为借景
"Wangchun Lou"'s (Note: building for anticipating spring) water source comes from the north and turns to east. Water was introduced into the garden and rock work was done to form the pond, which is coupled with the left and right bridges that resemble a crescent moon. It forms an opposing scene as it faces Xichun Tai across the lake.

"香海慈云"乃布政使卫江春于溪浦之北置之栅池，内涵碧沼，广盈数亩。水上架"海云龛"，奉大士像。临池南建"来熏堂"，堂之东构廊桥"春泼桥"，以连陆地（摘自《扬州园林甲天下》）
"Xianghai Ciyun" is that Wei Jiangchun built north of Xipu. It has a large span of water body of several Mus. There is "Haiyun Kan", with persons of great virtues worshiped there. "Laixun Tang" was built close to the south of the pond, and a corridor bridge "Chunbo Qiao" was built east of the hall to connect with the land (from *Yangzhou Gardens Rule*).

临溪越池，虚阁堪支；	Over a brook or stream, a flying corridor is more suitable to be built
（"越池"，不通，难解。1957年城建版作"越地"，则"临溪越地"可解为遇到小溪要跨过去，可在上面支"虚阁"。虚阁，即无窗，此处可作廊桥解。）	("Flying over a pond" is hard to understand. In 1957, there was an Urban Construction Press edition referring it as "flying over land", which can be understood as "leaping over when a brook is encountered". A "void construct", which can be interpreted as a windowless covered bridge, can be built crossing over it.)

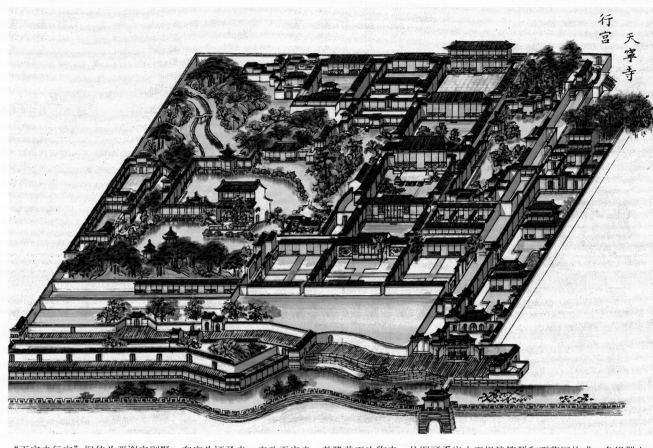

夹巷借天，浮廊可度。

Under a narrow lane with skyline light, a house corridor is more appropriate.

"天宁寺行宫"旧传为晋谢安别墅，在唐为证圣寺，宋改天宁寺。乾隆曾五次临幸。从图可看出由五组建筑群和西花园构成。各组群之间留有夹巷，均凌空接以浮廊，使之相对独立然又连为一体

"Tianningsi Xinggong" was said to be Jin dynasty's Xie An's villa. It was Zhengsheng Temple in the Tang dynasty and Tianning Temple in the Song dynasty. Qianlong paid royal visits for five times there. From the illustration, we can see that it consists of five building complexes and a west garden. There're narrow alley ways between the complexes and they're all in the form of flying corridor, forming a relative independent yet connected network.

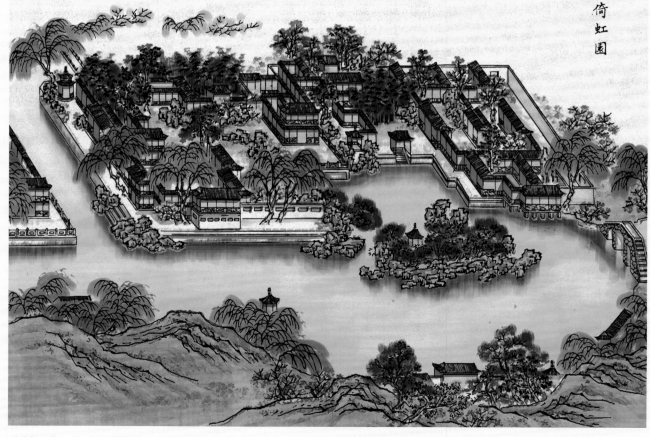

倘嵌他人之胜，有一线相通，非为间绝，借景偏宜。

As long as there is an outside scene visible through a vista, that scene should not be blocked.

"倚虹园"一称"虹桥修禊"，乾隆二十七年（1762年）临幸，赐御书"倚虹园"匾额。园傍城西濠，三面临河，三面借景：东为"堞云春暖"、北为"西园曲水"、西则"冶春诗社"

"Yihong Yuan" is also referred to as "Hongqiao Xiuqi". Qianlong paid a royal visit in his 27th year (1762) and wrote a royal horizontal tablet "Yihong Yuan". The garden is by the west moat and faces the river in three sides and leverages scenes in three sides, too. To the east is "Dieyun Chunnuan", north "Xiyuan Qushui", and west "Yechun Shishe".

"法净寺"即大明寺。位于蜀冈，平山堂，芳圃，双峰云栈，功德山等景均历历在目

"Fajing Si", also referred to as Daming Temple, is at Shugang. All the scenery spots such as Pingshan Tang, Fangpu, Shuangfeng Yunzhan, Mount Gongde are all visible here.

若对邻氏之花，才几分消息，可以招呼，收春无尽。

If one can have a peek at a neighbor's plants and flowers, even if it is exposed very slightly, it is sufficient to evoke feelings of lush spring.

"临水红霞"乃园主周柟于此遍植桃花，与高柳相间。每春深花发，烂若锦绮故名。其东侧邻里"平冈艳雪"植红梅数百本，香艳袭人。其北隔水"杏花村舍"、"邗上农桑"均为田园农家风光（摘自《扬州园林甲天下》）

"Linshui Hongxia" (Note: water front rosy clouds) was owned by Zhou Nan, who planted peaches everywhere, interspersed with tall willow trees. It got its name because of the splendid flower booming in deep spring. Hundreds of red plum trees with brilliant flower colors were planted in its east neighbor "Pinggang Yanxue". Over water to the north there are "Xinghua Cunshe" and "Hanshang Nongsang", which are all idyllic rural landscapes. (from "*Yangzhou Gardens Rule*")

架桥通隔水，别馆堪图；

Bridging a disconnected water, one can reach another building.

"莲性寺"旧名法海寺，圣祖仁皇帝赐今名。寺四面环水，中有白塔，有"夕阳双寺楼"、"云山阁"。门前为法海桥，寺后为莲花桥（五亭桥），过桥则是"白塔晴云"，桥东湖上为"凫庄"，皆入画

"Lianxing Si" used to be called Fahai Temple, and the current name was given by emperor Kangxi. The temple is surrounded by water on all its four sides. There is a white pagoda in the center and there are "Xiyang Shuangsi Lou" and "Yunshan Ge". In front of the gate it is Fahai Bridge and behind the temple it is Lotus Bridge (Five Pavilion Bridge). Across the bridge it is "Baita Qingyun", and to the east of the bridge it is "Fu Zhuang". They are both picturesque scenes.

聚石垒围墙，居山可拟。

Piling rocks to build a wall making it look like a small hill.

"水竹居"奉宸苑衔徐士业园。园前面河，后依石壁，可谓拟山而居，寄情山水。水中沙屿可通者曰"小方壶"。并石而起者，为"花潭竹屿"、"蒔玉居"、"静香书屋"、"妍清室"、"阆风堂"，最后为"曲室"。乾隆三十年（1765年）赐名"水竹居"，又赐"水色清依榻，竹声凉入窗"一联

"Shuizhu Ju" (Note: water and bamboo dwelling) was Xu Shiye's garden, which faces a river in front and against a rock wall behind. This is a living by an imitated mountain and a spirit dwelling in nature. The reachable sand islet in water is referred to as "Xiao Fanghu". Those erected along the rocks are "Huatan Zhuyu", "Shiyu Ju", "Jingxiang Shuwu", "Yanqing Shi", "Langfeng Tang", and finally, "Qu Shi". Qianlong 30th year (1765) a royally given name "Shuizhu Ju" (Note: water and bamboo dwelling) was granted, along with a couplet that goes "The look of water cleanses onto the beds; the sound of bamboo chills into the windows".

卷石洞天

"卷石洞天"奉宸苑卿衔洪徵治垒石为山，玲珑窈窕，丘壑天然。有"群玉山堂"、"夕阳红半楼"、"委宛山房"、"薜萝水榭"、"契秋阁"诸景

"Juanshi Dongtian" was built by Hong Zhengzhi by piling exquisite and graceful rocks that resemble nature. There are scenes such as "Qunyu Shantang", "Xiyanghong Banlou", "Wanwei Shanfang", "Biluo Shuixie", and "Qiqiu Ge".

> 多年树木，碍筑檐垣；让一步可以立根，斫数桠不妨封顶，斯谓雕栋飞楹构易，荫槐挺玉成难。
>
> If there is an old tree getting in the way for construction, it might be good to yield a little so that the tree can live, or cut a few branches on top to let it grow. This is because it is much easier to build artificial buildings than to have well-grown trees.

桃花坞

"桃花坞"道衔前嘉兴通判黄为荃建，候选州同郑之汇重修。旧有"蒸霞堂"、"澄鲜阁"、"纵目亭"、"中小亭"诸胜。今增置长廊曲槛，均避让古木繁花，望如错绣

"Taohua Wu" (Note: peach cove) was built by Huang Weiquan, and renovated by Zheng Zhihui. There were "Zhengxia Tang", "Chengxian Ge", "Zongmu Ting", "Zhongxiao Ting" several scenery spots. Nowadays there are long corridors with curved railings to give way to old trees and flowers.

> 相地合宜，构园得体。
>
> In general, with proper siting, the garden built would naturally be decent and appropriate.

白塔晴云

"白塔晴云"系按察使衔程扬宗、州同吴辅椿先后营造，乾隆四十四年（1779年）候选道张霞重修。此园相地十分讨巧，巧于因借，对岸与莲性寺、白塔相对，并通过瘦西湖有代表性的建筑五亭桥与之连接贯通，可谓达到事半功倍的艺术效果

"Baita Qingyun" was built by Cheng Yangzong first and then Wu Fuchun. Zhang Xia renovated it in Qianlong 44th year (1779). The garden boasts in ingenious design, as it leverages surrounding scenes smartly. It faces Lianxing Temple and the white pagoda across the river, and connects to them via the Shouxi Lake's representative architecture Five Pavilion Bridge. This way it doubles its artistic effect with half the effort.

"春台祝寿"系乾隆二十二年（1757年）奉宸苑卿衔汪廷璋起"熙春台"。记载中："飞甍丹槛，高出云表。"台地二层，颇具"横可跃马，纵可方轨"的气势。且为北、西二水交汇处。对景望春楼、五亭桥、白塔，浑厚端庄，可称得上"相地合宜，构园得体"的佳作
"Chuntai Zhushou" was built in Qianlong 22nd year (1757). Wang Tingzhang built "Xichun Tai". According to historical record, its "flying roof ridge's way high above the cloud". There're two layers of terraces having an air of being able to "run horses or chariots". Here it is where the water converges from west and north. The opposing scenes consist of Wangchun Lou, Five Pavilion Bridge, and the White Pagoda. It is vigorous yet dignified, and deserved to be regarded as a masterpiece of "naturally decent and appropriate".

"相地"一章是园址的选择、踏勘、评价，地形与造型构思的统一，以及各类型园址的布置处理。以现代来看：是园址选定和总平面规划设计的范畴。该章节实质上是介绍了如何将因地制宜的规划原则和意境创作的设计方法贯彻到各种不同类型的实际地形与功能要求中去的概略途径。

"相地"一章将园址基地分为六大类型：山林地、城市地、村庄地、郊野地、傍宅地、江湖地。根据实践经验，分节论述了每一类地形的优劣，如何在功能上或意境创作上发挥其优点、克服其缺点，比起前代这方面的论述系统得多、详尽得多。尽管如此，也仅仅是文字上的综述。为方便理解文中内容，同时指导园林师造园实践，现选择现存江南古典名园实例，用图文并茂的形式展现，以方便学习、研究与借鉴，达到"借古开今"的旨意。

The chapter "Siting" is about a garden site's selection, exploration, and evaluation. It is about the conception of form in accordance with the natural physical terrain, as well as various layouts of different garden types. From a modern perspective, it belongs to the general master planning category. In essence, this chapter discusses in general how to integrate the as-appropriate planning principle with artistic conception in different physical settings taking functional use into consideration.

"Siting" divides garden sites into six broad categories: hilly forest land, urban land, village land, suburban land, residence side land, and river and lake land. Based on field experiences, it is discussed in the chapter the advantages and disadvantages of each different type, and how to take advantage of its strength while overcoming its shortcomings on functionality and artistic conception. It is a much more systematic and elaborate discourse comparing to prior work. Even so, it is a discourse based on texts only. In order to understand and interpret the contents and guide landscape architects' practices, we hand-picked some Southern Chinese classical landscape gardens and added some illustrative graphics to show by examples. This is to facilitate better learning, studying and referencing, taking the wisdoms and lessons from the past and applying them to our current work.

（一）山林地

园地惟山林最胜，有高有凹，有曲有深，有峻而悬，有平而坦，自成天然之趣，不烦人事之工。入奥疏源，就低凿水，搜土开其穴麓，培山接以房廊。杂树参天，楼阁碍云霞而出没；繁花覆地，亭台突池沼而参差。绝涧安其梁，飞岩假其栈；闲闲即景，寂寂探春。好鸟要朋，群麋偕侣。槛逗几番花信，门湾一带溪流，竹里通幽，松寮隐僻。送涛声而郁郁，起鹤舞而翩翩。阶前自扫云，岭上谁锄月。千峦环翠，万壑流青。欲藉陶舆，何缘谢屐。

(1) Hilly Forest Land

Hilly forest land proves to be the best kind for garden design. This is because a terrain has high and low elevations, with its various forms and depths. It can be either a steep cliff or broad flat land. The fine natural features are already there for you without much travail. For a hidden place with some dredging, a low-lying terrain can be easily made into a pond; digging to make some caves at foothill and piling dirt to make some mounds that can be used as bases for buildings. Tall mixed trees with high-rise buildings mingle with skies and clouds; flowers of various kinds spread everywhere and pavilions and building decks of various shapes are in and out over the pond. Where there is no path over water, there is a bridge, and where there is flying rock with a cliff, there is a plank road. The scene is everywhere whenever you are at ease; the spring scene is always there whenever you feel quiet and alone. Beautiful birds can be your pals and joyful moose can be your sightseeing buddies. Smell of spring blossom; sight of brook outside the garden. Lone cottage in bamboo groves; thatched house in pine forest. Surging sound of wind blowing through pine leaves; dances by elegant waterfowls. When one lives in such a place and enjoys reclusiveness, who cares what is going on elsewhere? With verdant mountains and valleys with lush landscaping, one can enjoy nature right here without the travail of traveling afar.

无锡寄畅园 Jichang Garden in Wuxi

无锡寄畅园是著名的江南古典园林之一。始建于明正德年间（1506～1521年），原属秦姓私园。初名"凤谷行窝"。背山临水，地僻景幽，饶有山水林木之雅。清初，秦氏后裔又请当代著名造园家张涟、张鉽叔侄重行布置，疏泉立石，园景益胜。既具有现存一般江南古典名园所共有的委曲宛转、妙造自然的特点，又具有苍凉廓落、古朴清旷的独特风格，因而别树一帜，久享盛誉。清帝乾隆南巡后，曾仿照此园于北京清漪园（即今颐和园）内建造了"惠山园"，后改名"谐趣园"。

寄畅园的风景布局与空间结构，是结合园内地形和周围环境，根据其东西狭窄，南北纵长，西枕惠山山麓，地势西高东低的特点，因高培土，就低凿池，创造了与园基纵长方向相平行的水池和假山。全园以水池为构图中心，池东是一带临水亭榭，池西用假山呼应，组成空间。

水池"锦汇漪"，广仅三亩，然而平波潋滟，形成了园中开朗明净的空间。由于周围山石、建筑、绿化的点缀配置，勾勒出曲折窈窕的水面轮廓。尤以池东的临水建筑"知鱼槛"，和西岸伸向池心的石矶"鹤步滩"相对夹峙，形成了水面中心对景，增加了风景层次。水池北段，由于七星桥和廊桥分隔，又将池水分成两个不

Wuxi's Jichang Garden is one of the most well-known Southern Chinese classical gardens. It was first built in Zhengde period in the Ming Dynasty (1506-1521) and was a private garden of an owner whose family name is Qin. Its original name is "Feng Gu Xing Wo", literally meaning "Travel Lodge in Phoenix Valley". It faces water and has mountains at its back. It is blessed with a remote and quiet setting with lush natural hilly forest. At the beginning of the Qing Dynasty, the descendants of the Qin family invited famous garden designer Zhang Lian and his nephew Zhang Shi to redesign on the garden's rock and water scenes and made much improvement. It has both Southern Chinese garden's veiled implicitness in delicate treatment of nature and a tint of bleak, broad stroke simple crudeness. Hence its uniqueness and lasting fame. After Qing Dynasty Emperor Qianlong's Southern China tour, a copied garden "Huishan Garden" was built in Beijing's Qingyi Garden – today's Summer Palace – and it was later renamed as "Xiequ Garden", literally meaning "garden of harmonious interests".

The landscape and spatial structure of Jichang Garden shows its respect for its natural setting. It takes both its site terrain and outer natural context into consideration. Its site is narrow east-west wise but elongated from south to north. On the west it is the foothills of Mount Hui and as such the elevation is higher on the west and it slopes down to the east. Adding fill on higher elevations and digging for pond on lower ones, a format of pond and rock parallel to the elongated side is formed. The pond is the focus of the whole garden's composition; on its east side there are a couple waterfront pavilions, and its west side is echoed with artificial rockery to frame the space.

The pond "Jin Hui Yi" is only three Mus in size; however, its water body appears smooth and wide, forming the garden's clear and open space. The surrounding rockeries, buildings and vegetation punctuate and define the ponds perimeter. The pond is flanked by the east's waterfront building "Zhi Yu Jian" and the west's pond center pointing rocky "He Bu Tan". This forms the pond center's facing scenes, enriching scene layers. The northern part of the pond is

同情趣的小水面，显得深邃不尽，玩味无穷。七星桥平卧波面，池水轻拍，倒影如画，更显得水面幽深宁静。在此更可看到远处锡山塔景的借景，耸立在树梢檐角，从而显得余意不尽。如在东岸临水屈曲的亭廊、郁盘亭和知鱼槛小憩，则可观赏池西的曲岸回沙、假山疏林和园外的惠山远景，晴峰明媚，似在咫尺之间。亭旁老树横偃斜披，如苍龙探首波面，古拙入画。

池西假山，与水池基本平行。假山处于惠山之麓，前迎锡山晴峰，后延惠山远岫。假山造型，模拟惠山九峰连绵逶迤之状，并把它作为惠山的余脉布置，堆成平岗小坡的形式，使之与惠山雄浑自然的气势相应。假山高度一般三至五米，与水池比例相称，与锡、惠两山尺度亦属得宜，既能掩映连络锡、惠两山的自然山色，将假山与自然山峦融为一体；又能与池中倒影相映成趣，滉漾入画。

假山构筑材料土多石少，石料全部采用黄石，与惠山的土质石理统一。大斧劈皴，浑厚嶙峋，因而显得假山与真山脉络相连，体势相称。且石脉分明，坡脚停匀，没有人工斧凿的痕迹，成为从园东眺望惠山远景的绝妙前景。

假山内部安排的岩壑涧泉——八音涧，涧道盘曲，林壑幽深。置身其中，但见奇岩夹径，怪石峥嵘，高林蔽日，浓荫满地，空间紧凑曲折，殊有前不知其所穷，后不知其所止之感，与山外开朗清旷的风景形成强烈对比。八音涧的设计，是利用流经墙外的二泉伏流，伴随着涧道曲径，引导到假山中来。依据地形倾斜，顺势导流，创造了曲涧、澄潭、飞瀑、流泉等诸般水景。既有蓄聚渟泓的泉潭，又有迂回轻泻的曲涧。悬空挂注的流泉，产生了铿锵的泉鸣，潆回辽绕的曲涧，又产生了琤琮的水声。加上整个涧流，或浮石面，或伏石罅，或傍山涯，或流谷底，不但呈现了忽断忽续、忽隐忽现、忽急忽缓、忽聚忽散的不同水景，而且产生不同音响。水音与岩壑共鸣，不啻八音齐奏。《画筌》云："山本静，水流则动。"假山更因八音涧的布置，增加了生动灵活之趣，堪为现存江南古典园林中叠山理水的杰出范例。

由于寄畅园的山水掌握了惠山九峰连绵逶迤、岗峦起伏的形象，和江南水乡重洲浅渚，湖港交叉的特点，把假山当作大

divided by the Seven Star Bridge and the corridor bridge, further dividing the pond into two sub-areas with distinct interests; this creates depths of scenes with unending interest. The Seven Star Bridge is flat over the water surface. With water lightly touching its underside and picturesque reflections, it creates an atmosphere of serenity and tranquility. One could spot the reflection of the pagoda over the tree canopies and building eaves on the distant Mount Xi, further enhancing the feeling of unending ruminations. If you are at the east bank's waterfront zigzagged corridor Yupan Pavilion and Zhiyu Jian for a brief rest, you could enjoy the pond west bank's curved shore, rockery, sparse grove and the distance scenery of Mount Hui outside the garden. When it is clear and shining, it feels as if it is just close by. The old tree by the pavilion slants against slope, as if an old dragon dipping into the water, forming an aged and unadorned picture frame.

The rockery west of the pond parallels to the pond. It is located at the foothills of Mount Hui, with its front facing Mount Xi and its back extending to Mount Hui. The modeling of the rockery is after Mount Hui's nine peaks' meandering pattern, and acts as an extension of Mount Hui, to form a hilly but gentle slope, compatible with Mount Hui's natural but forceful momentum. A typical height for a rockery is about 3-5 meters, matching the pond's scale, and at the same time, appropriate with that of Mount Xi and Mount Hui. This way it can both set its background off Mount Xi and Hui and connect itself with them, integrating them as a whole, and simultaneously forming integrated reflections in the pond, creating an impressionistic picture with rippling images.

The rockery has less rock than earth, with its rock consisting of pyrite or sericite, consistent with the soil type and texture of Mount Hui. The rich texture of vigorous cutting effect makes this man-made rockery seemingly extending the mountain's moving trend, fitting well in the context. It is so well connected and punctuated, without the feeling of human intervention that it acts as a perfect foreground looking towards Mount Hui from the east of the garden.

Inside of the rockeries, there is a stream called "Eight Tone Gully" meandering. Coupled with dense forest, it feels deep and quiet. Being inside of the gully, you are flanked by rocks of different shapes. Tall trees with shady and winding gully path forms a compact and ever-changing spatial sequence, making you feel uncertain about its starting and ending places, contrasting sharply against what you see and feel in the outer much broader and open landscape space. The design of "Eight Tone Gully" takes advantage of two underground streams passing by outside the garden. They are directed by the winding gully path to the rockeries inside the garden. It makes the best use of the sloping terrain, creating winding streams, a clear pond, a water fall, and a flowing spring several water scenes. It has both still, static, and reserving pond, and moving, dynamic, and flowing gully stream. The hanging and dropping water stream creates crisp sound; the water flowing in the meandering and deep gully added another deeper acoustic effect. The stream water sometimes floats on top of flat rocks, or cuts through rock cracks; it sometimes flows by a cliff, or meanders through a gully. Visual wise, it is sometimes continuous or broken, sometimes visible or non-visible, sometimes rapid or slow, and sometimes converging or diverting, showing a variety of scenes. Acoustic wise, due to different rock and space formations it goes through, it makes distinct sounds. That is why it is referred to as the "Eight Tone Gully". It is described in "Hua Quan", "Originally a mountain appears to be static; however, because of flowing water, it appears to be dynamic." The rockery, because of the design and installation of "Eight Tone Gully", appears to be more lively and

山余脉布置，把水池当作大水缩影处理，因而平岗小坡，高低有致；曲岸回沙，婉转相迷，达到了"虽由人作，宛自天开"的境界。

vibrant. This is indeed a great example of "although man-made, it appears to be natural" classic garden in Southern China, rich in tradition of rockery and water handling.

The design of Jichang Garden shows great mastery of mimicking and extending the moving trend and innate charm of Mount Hui, with its nine peaks in their rolling and meandering posture and composition. In line with Yantze River Delta area's abundance of water bodies laced with rivers and lakes, setting the mountains off as background and the rockery as a natural extension of its ridges, using the pond as a miniature of a larger water body, the garden employs variable elevated and rolling terrains that embraces meandering and varying streams and pond, achieving the ultimate state of "natural appearance out of man-made construction".

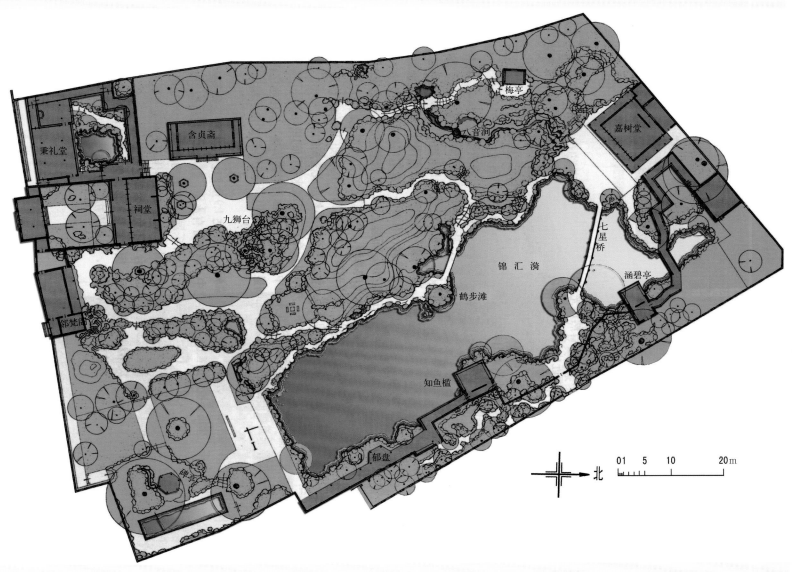

无锡寄畅园平面图
Plan of Jichang Garden, Wuxi

苏州拥翠山庄 Yongcui Villa of Suzhou

拥翠山庄为苏州近郊典型之山庄园林，位于虎丘二山门内，憨憨古泉西侧之山坡，面积1亩有余。

清光绪十年（1884年）春，朱庭修与僧云闲寻访得虎丘古憨憨泉于试剑石右。有巨石载其上，汲饮甘洌。同游者洪文卿、彭南屏、文小坡皆大喜，踊跃为谋，众集

Yongcui (Note: embracing verdant green) villa is a typical Suzhou suburban garden. It is located inside the second gate of Hu Qiu (Note: Mount Tiger) on the slope west of Hanhan old spring, occupying over one Mu (Note: approximately 1/15 of one hectare).

In the spring of Guangxu 10th year in the Qing Dynasty (1884), Zhu Tingxiu and Monk Yunxian found Hanhan spring by a Hu Qiu's sword testing stone. The spring flows on top of a huge rock and the spring water tastes chill and wonderful. Their companies Hong Wenqing, Peng Nanping, and Wen Xiaopo were jubilant and came up enthusiastically with various

钱数万，于泉旁笼隙地，围短垣，随地势高下错屋十余楹，有抱瓮轩、问泉亭、灵澜精舍、送青簃诸胜，杂植梅、柳、蕉、竹数百本。凭垣而眺，四山瀚蔚，人景皆绿，故名拥翠山庄。光绪十一年（1885年）正月落成，杨显为记。

民国13年（1924年），于灵澜精舍之北，原陆公祠址重建送青簃。拥翠山庄建造年代较晚。苏州解放后，于1959年全面整修。"文化大革命"期间，庄内陈设匾对遭破坏，长期闭园。1983年恢复之，今全园保存完好。

全园呈纵长方形。又巧于利用虎丘自然地形，依山势遂分四层，逐层升高，布局自然，为苏州自然格式之台地园林。园门居南，石级数十，门楣书庄名。两侧之壁嵌有"龙、虎、豹、熊"行、草书大字石刻四方，传为咸丰八年（1858年）桂林陶茂森所书刻，由他处移置。入庄门为抱瓮轩，面阔三间，位居园之首层。轩东花窗粉墙相围，墙外即古憨憨泉，为石栏四围之古井，轩后有边门可通。拾级而上，二层置问泉亭，亭敞三面，东南面泉，北壁置"庐山瀑布"挂屏及石碑两方。亭西堆叠湖石假山，峰石起伏，蹬道蜿蜒，配植花木，有白皮松、紫薇、黄杨、石榴，自然有致，与园外之葱茏林木，融为一体。亭之西侧上方，后增构月驾轩，乃取郦道元《水经注》"峰驻月驾"句意。壁间嵌嘉庆元年（1796年）钱大昕隶书"海涌峰"石碑。此碑为光绪十三年于虎丘山麓发现，得以保存。再上一层为园之主构灵澜精舍，三楹主厅踞北面南，前置平台，俯览园景，翠木茂密，石径盘转，深得山林之趣。东侧突出围墙，更有平台，可纵观虎丘山麓景色，仰眺云岩古塔。精舍之北，为送青簃小庭院，两翼庑廊相接，为民国年间重建。回廊之壁嵌有"拥翠山庄记"书条石。综观全园，结合地形，高下开合，错落有致，园中无水，邻借古泉，近览远眺，别具一格。

拥翠山庄有堂构7处，匾额6块，对联5副，石刻、书条石16块，古树名木银杏、桧柏等4株。

ideas. They raised among themselves tens of thousands to have started building the garden. By the spring and farm fields, they built short walls and following the terrain, some buildings in sequence. There are Baoweng Xuan, Wenquan Pavilion, Linglan Jingshe, Songqing Studio etc. with plum, willow, banana, and bamboo trees spreading around. Looking into the distance against the wall, one sees vibrant greens everywhere, and that is why it is called Yongcui Villa. It was completed in January of Guangxu 11th year(1885), and it was recorded by Yang Xian.

In the 13th year of the Republic of China (1924), at the north of Linglan Jingshe, Songqing Yi was rebuilt at the site of Lugong Temple. Yongcui Villa was built a bit later. After the liberation of Suzhou, it was renovated completely in 1959. During the "Great Cultural Revolution", the furniture and script tablets were destroyed, and it remained as closed for a long time. It was restored in 1983, and now all garden is preserved in good condition.

The shape of the garden is rectangular. It takes advantage of Hu Qiu's natural terrain, and has four terraced levels with natural layout. It is a terraced Suzhou garden on natural format. Tens of steps lead to the garden entrance in the south, with the garden name written on the lintel. Four large cursive or semi-cursive characters, "Dragon, Tiger, Leopard, and Bear", are inlaid on the stone in each side. It is said that they were engraved by Tao Maosen in Guilin in Xianfeng's 8th year (1858) and moved here. The entrance is Baoweng Xuan. With three units, it is located at the ground floor level. To the east of the building there are surrounding painted decorative walls, outside of which is Hanhan old Spring, which is an ancient well surrounded by stone fence. There is a side door connecting to the back of the building. Going up the stairs on the second floor one sees Wenquan Pavilion, with three faces. On its southeast it faces the spring. On its north side, there are "Lushan Falls" hanging panel and two stone tablets. Lake rockeries are erected on the west of the pavilion. With its peaks up and down, in-between paths meandering, coupled with ornamental plants such as bungeana, crape myrtle, boxwood and pomegranate, it is so natural that it blends well with the luxuriant vegetation outside the garden. On the higher ground on the west of the garden, a newer addition, Yuejia Xuan, was added, and its title was derived from "Fengzhu Yuejia" in Li Daoyuan's work "Shui Jing Zhu". On the wall there is an engraved stone tablet with Qian Daxin's clerical script style "Hai Yong Feng", written in Jiaqing's 1st year (1796). This stone tablet was found and preserved in Guangxu 11th year at the foothills of Hu Qiu. One story up is where the garden's main building, Linglan Jingshe, is located. This three-unit building faces south, with a platform in front, overlooking the garden scene. With thick vegetation and winding stone paths, it gets the gist of natural mountainous landscape. On the east there is an outgoing fence with a terrace where one can enjoy Hu Qiu's mountainous views and look up at the Yunyan ancient pagoda. North of Linglan Jingshe is a small courtyard named Songqing Yi. It has two side corridors connected to it, and it was re-built during the years of Republic of China. On the side wall of the corridor engraved a stone tablet with "Yongcui Villa History". In general, the garden is well fitted into the natural terrain with closed and open spaces up and down forming a balanced patchwork. It does not have a water body in the garden; however it leverages the nearby ancient pagoda offering both near and far away scenes and is uniquely characteristic.

Yongcui Villa has seven buildings, six horizontal inscribed boards, five character couplets, sixteen engraved stone tablets, and four ancient and precious tress such as *Ginkgo biloba* and *Sabina cypress*.

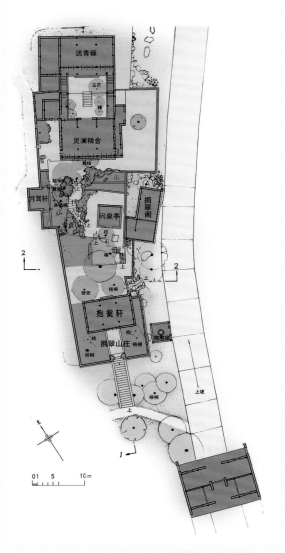

苏州拥翠山庄平面图
Plan of Yongcui Villa, Suzhou

扬州蜀冈名胜　Attractions in Shugang, Yangzhou

扬州市西北郊，有一道蜿蜒起伏的山冈，称为"蜀冈"。相传它联通西蜀，是昆仑山北岭山脉由四川、湖北、河南、安徽延展到江苏六合、仪征境内，再向东到扬州西北，形成了这道石块少、黏土多的丘陵地带。蜀冈昔日是邗城、汉城、唐城和宋大城的所在地，也是扬州历代的名胜区，屡经兴废。现蜀冈突起的东峰功德山、中峰平山名胜保存较好，两峰对峙，颇为壮观。岭上万松竞翠，苍郁欲滴，殿堂庙宇，隐现其中，蔚然如画。

蜀冈名胜包括大明寺、鉴真纪念堂、平山堂、芳圃四部分，是历史悠久，声名卓著的游览胜地。

大明寺

始建于5世纪六朝刘宋年间，宋孝武帝因以大明为纪年，故名大明寺，至今已有一千五百多年。隋仁寿元年（601年），

In the northwest of city of Yangzhou, there are some rolling hills called "Shugang". It is said that it has roots in the west of Sichuan and connections to the north ridges of Kunlun Mountain via Sichuan, Hubei, Henan, Anhui, extending to Liuhe and Yizheng in Jiangsu, and further east to the northwest of Yangzhou. It is a hilly area with more clay and less rocks. Shugang used to be Han City, Han City, Tang City and Songda City's locations, and it is an attraction area in different historical times. It had seen its times of rise and fall. Now the east protruding peak Gongde Mountain and the middle peak Pingshan attractions are preserved well. The two peaks face each other, forming a spectacular scene. The buildings and temples hidden among verdant pine trees on the ridges form a picturesque scene.

Daming Temple

It was first built in the fifth century in the Liuchao Dynasty during Liusong years. The emperor Songxiaowu used Daming as his annals year and thus naming the temple Daming Temple. Now it has over a thousand five hundred years of history. In the first year of Renshou in Sui Dynasty (601), an edict was issued from the imperial court to build a pagoda for people to worship a Buddha's relics. The pagoda of nine stories was built on top of the hill. Since the completion of construction, Daming Temple was once called Qiling Temple. Many poets in Tang dynasty, such as Li Bai, Liu Yuxi, Bai Juyi, Gao

朝廷下诏建造供奉舍利之用的栖灵塔一座，建于冈上，高达九层。此塔建成后，曾一度称大明寺为栖灵寺。唐代诗人李白、刘禹锡、白居易、高适、刘长卿等都为此写过诗篇。如李白登塔诗云：

宝塔凌苍苍，登攀览四荒。

顶高元气合，算出海云长。

塔毁于唐会昌三年（843年），寺名复为大明寺。清乾隆三十年（1765年），高宗南巡时，忌讳"大明"二字，改名并敕题"法净寺"。现仍名大明寺。

入寺门，为天王殿，四大金刚分列两侧，神态威严。过天王殿，穿松柏甬道即为大雄宝殿，殿内各色佛像人物神采各异，栩栩如生。古大明寺誉满中外的更重要原因，是由于鉴真和尚住持过，并从这里扬帆东渡，为缔结中日友好写下了灿烂的第一章。

鉴真纪念堂

从大雄宝殿向东，便是鉴真纪念堂了。纪念堂包括正殿、碑亭和回廊，共占地2540平方米，其中正殿为345平方米，碑亭为58平方米，回廊为262平方米。这组建筑是我国著名建筑学家梁思成教授生前的最后作品，其线条刚健、结构工整，充分体现出我国唐代雄浑朴实的建筑风格。纪念堂的南面还有门厅，以及有关鉴真东渡事迹和史料的陈列室。

鉴真是中日两国人民文化交流的先驱者，他是扬州江阳县（今扬州市）人，俗姓淳于，十四岁随父出家扬州大云寺，后在长安、洛阳等地游学，专攻佛教的律学兼及他说。中年以后住持扬州大明寺，传戒讲律，兴建寺塔，是淮南地区有威望的宗教首领。鉴真在从事宗教活动的同时，还从事很多社会实践，并与群众接触，使他取得多方面的文化成就。成为一个有丰富学识和才能的和尚，有条件承担中日文化交流的使命。在唐开元、天宝时日本国派遣使臣的活动达到了高潮，使臣除与中国修好外，便是把中国的文化、学术输入日本，并带来了大批留学生和留学僧。长期在中国的日僧荣睿、普照就是随日本天平四年（唐开元二十年，公元732年）派出的遣唐使丹比治广成来到中国的，两人在洛阳、长安等地研究佛学十年，闻得鉴真是律学的大师，便于天宝元年（公元742年）南下来到扬州，邀请鉴真赴日传法。鉴真欣然同意，带同弟子、匠作多人，启

Shi, and Liu Changqing etc. composed poems for the temple. For instance, Li Bai wrote poem Deng Ta:

Oh, grayish pagoda; I see distant desolate lands after climbing atop.

My vitality consummates at the pinnacle, marking foggy cloud spread.

The pagoda was destroyed in Tang dynasty's Huichang 3rd year (843), and the temple's name was restored as Daming Temple. When Emperor Gaozong had his southern tour in Qing Dynasty's Qianlong 30th year (1765), he treated the name of Daming as a toboo, and so he changed it to "Fajing Temple". Now it is still named as Daming Temple.

After entering the temple's gate, it is Tianwang's Hall, with Buddha's four great warrior attendants standing at each side, with majestic air. Passing the hall, through a corridor of pine and cypress trees, it is Daxiong Baodian, the main hall. Inside of it there are various characters with lifelike vivid expressions. The more important reason that the old Daming Temple is so well-known is that the famous monk Jianzhen had once been the abbot of the temple and departed from here to sail east to Japan, paving the way for Sino-Japanese friendship.

Jianzhen Memorial Hall

Jianzhen Memorial Hall is located to the east of Daxiong Baodian. The memorial consists of a main hall, a monument pavilion, and a corridor, occupying 2540 square meters, with the main hall having 345 square meters, the monument pavilion 58 and corridor 262 square meters. This building complex was the last design work of Liang Sicheng, our country's famous architect. Its energetic outline, tidy and neat structure, fully reflects the forceful and simple architectural style of the Tang dynasty. South of the memorial hall, there is an entrance hall, as well as a showroom displaying history and historical artifacts of the monk's eastward trip to Japan.

Jianzhen is the pioneer of cultural exchanges between China and Japan. His hometown is Jiangyang County, today's Yangzhou City. His secular family name is Chunyu. When he was fourteen years old, he became a monk, following his father at Dayun Temple in Yangzhou. Later he had study tours to Changan, Luoyang and other places, focusing his study on Buddhism laws and principles and others. After middle-age he became the abbot of Daming Temple in Yangzhou, performing initiation ceremonies, teaching Buddhist laws, building temples and pagodas, earning religious prestige and authority in Huainan area. While engaging religious activities, Jianzhen was also active in social works, getting in touch with the general public, achieving multiple accomplishments in many areas and becoming a well-learned and capable monk, qualified for the mission of cultural exchanges between China and Japan. In Kaiyuan and Tianbao years in the Tang dynasty, it reached a peak point for Japan to send envoys to China. The duty of the Japanese envoys was to maintain friendly relationship with China, as well as introducing Chinese culture and learning. Along with the envoys, there was large number of students and student monks. For example, two of the monks who lived in China for a long period of time, Rongrui and Puzhao, came to China along with Tajibi, the Japanese enjoy sent in Tianping fourth year of Japan (Kaiyuan 20th year in the Tang dynasty, 732). The two monks spent ten years studying Buddhism in Luoyang and Changan. When they got to know about the great Buddhist law master Jianzhen, they traveled down south to Yangzhou in the first year of Tianbao of the Tang dynasty (742), and invited Jianzhen to Japan to preach Buddhism. Jianzhen gladly accepted their invitation, assembled Buddhist disciples and different trade artisans and got ready

程东渡。出生入死，历尽艰苦，经过五次失败，经受了牢狱之灾，风涛之险，疫病之困。先后十年，牺牲三十六人，日僧荣睿亦病逝广东端州（今肇庆），鉴真本人也因途中劳累过度而双目失明。但他东渡的决心毫不动摇，"为是法事也，何惜生命"，终于在天宝十二年（753年），第六次出发渡海成功，以六十六岁的高龄到达日本。在日本人民的密切合作下，他传播戒律，兴建佛寺，辛勤不懈地工作了十年，无保留地传授了我国的佛教经典，以及医药、雕刻、绘画、书法，建筑、印刷术等唐代文化。鉴真在奈良建造的唐招提寺，他弟子在寺内建造的金堂，特别是鉴真的塑像至今仍保存完好，被日本政府和人民称为国宝，每年开放三天，供人瞻仰。

公元763年五月初六，七十六岁高龄的鉴真和尚圆寂于日本唐招提寺内。据传说，鉴真死后三日，头部仍有余热，因而久久不能入殓，这说明鉴真在日本人们心目中的地位和人们对他的怀念之情。鉴真的骨塔，至今仍在日本唐招提寺后的松林中。

1963年，当鉴真和尚圆寂1200周年之际，中日两国佛教、文学艺术、医药各界人士，为纪念这位对中日文化交流作出不朽功勋的人物，在扬州古大明寺内举行了盛大的纪念活动。郭沫若副委员长，还写下了赞美鉴真和尚功绩的诗篇：

鉴真盲目航东海，一片精诚照太清。
舍己为人传道艺，唐风洋溢奈良城。

现在，鉴真纪念堂的碑亭内，横立一块汉白玉碑石，正面是郭沫若题："唐鉴真大和尚纪念碑"，背面是我国佛教协会会长赵朴初所撰纪念鉴真和尚圆寂1200周年的碑文。正殿四周有记载鉴真足迹的彩色壁画四帧，正中供奉鉴真的塑像，系楠木雕成，再现了鉴真的形神，其渊深的学识，坚强的毅力，慈善的面容，令人肃然起敬。

平山堂

大雄宝殿向西，便是有名的平山堂，系宋代文学家欧阳修于庆历八年（1048年）在扬州任太守时营造。因为坐在厅堂之中，江南诸山历历在目，拱揖槛前，若可攀跻，并与堂平，故得平山堂之名。平山堂是专为士大夫、文人吟诗作赋的场所，从欧阳修的《朝中措》一词中可看出：

to sail east to Japan. They risked their lives and endured much hardship, and after five failed attempts to cross the East China Sea to reach Japan, experiencing jail time, treacherous maritime weather, epidemic diseases, ten years of unremitting effort with thirty six lost lives, they finally reached Japan on their sixth attempt in Tianbao 12nd year (753). During the process, the Japanese monk Rongrui died of disease in Duanzhou (now Zhaoqing) in Guangdong. Jianzhen also went blind after overwork and when he finally reached Japan, he was already sixty six years old. He had unwavering beliefs in preaching Buddhism in Japan as he said, "to work for Buddhist causes we would sacrifice our lives." With the close cooperation of Japanese people, he disseminated Buddhist precepts, built Buddhist temples, worked hard unremittingly for ten years, taught Buddhist classics without reservations, along with medicine, sculpture, painting, calligraphy, architecture and printing technology and culture of the Tang dynasty. The Tangzhaodi Temple that he built in Nalo, the golden hall that his disciples built inside, and especially his statue, are all preserved well as of today. They are treated as the national treasure by the Japanese government and people, and is open for three days per year for the public to pay their respects. On May 6th, 763, the monk Jianzhen passed away in Tangzhaodi Temple at the age of seventy six. It is said that three days after he passed away, there was still warmth in his head. As a result, people would rather wait to delay the burial. This shows that he was well-respected in people's heart and he was missed tremendously by general public. The pagoda that keeps his remains is still at the pine grove behind the temple.

In 1963, when it was 1200 anniversary of Jianzhen's passing, representatives from both Japan and China's Buddhist, literary and artistic, medicine fields held a grand memorial ceremony in the ancient Yangzhou's Daming Temple, to memorize this outstanding figure who contributed so much in cultural exchanges between the two countries. Mr Guo Moruo wrote a poem in praise of Jianzhen:

Sailing east despite of blindness, Jianzhen's sincerity shines.

Preaching Buddhism and promoting culture and technology exchanges, the Tang style permeates in the city of Nalo.

At present, in the monument pavilion in Jianzhen memorial hall, there is a white marble stele, on which there is an inscription by Guo Moruo, "The Great Monk Jianzhen's Monument". At its back, it is an inscription commemorating Jianzhen's 1200 year's passing, written by China's national Buddhism Association Chairman Zhao Puchu. Surrounding the main hall, there are four color murals documenting Jianzhen's footprint. His statue made of phoebe wood is situated in the center of the hall for people to worship. The statue vividly portrays his character and spirit, a deep learning pundit, a person of perseverance, charitable and benevolent, drawing awe and reverence.

Pingshan Hall

West of Daxiong Baodian, the main hall, is the well-known Pingshan Hall. It was built in Qingli 8th year (1048) by Ouyang Xiu, the famous writer in the Song dynasty, when he served as the head of Yangzhou's prefecture. While sitting in the center of the hall, one can see vividly outside mountains, as if you could climb right up from the hall that seems to be at the same level at the mountain base, so it is called Pingshan – level mountain – Hall. It is especially a venue prepared for literati to write poems and articles, as can be understood in Ouyang Xiu's "Chao Zhong Cuo":

Leaning against the railing, I see the mountains looming.

平山栏槛倚晴空，山色有无中，手种堂前杨柳，别来几度春风。文章太守，挥毫万字，一饮千钟。行乐直须年少。樽前看取衰翁。

平山堂屡经战火，然而复废又兴。现堂系清代方子箴修建。堂上佳联甚多："过江诸山到此堂下，太守之宴与众宾欢"。又有联云："晓起凭栏，六代青山都到眼；晚来对酒，二分明月正当头"。韵味颇耐咀嚼。

平山堂后与谷林堂、欧阳祠相连。谷林堂是苏东坡为纪念他的老师欧阳修，于元祐七年（1092年）兴建的，集诗句："深谷下窈窕，高林合扶疏"为堂名。堂前悬挂人们爱诵的联句："万松时洒翠，一涧自流云"。堂后的欧阳祠，建于清光绪五年（1879年）。堂中镶嵌欧阳修的石刻像，刀法遒劲，在光影的作用下远看是白胡子，近看为黑胡子，往往引起游人的兴趣。

在平山堂之东，有平远楼相对峙，此楼充分利用地形的高差而筑，实为三层，从外看去只有两层。据说平远楼之名是取画家郭熙《山水训》中："自近山而望远山谓之平远，平远之意冲融而缥缈"之意命名的。楼前院中有青石巨碑一座，刻"印心石屋"四个大字，为清道光帝亲笔。

芳圃

又名西园，系富商汪应庚1736年重筑，亦为乾隆南巡之御花园。江南园林多为平地挖池堆山而建，芳圃则开山凿池构筑。按起伏山地，运用"石令人古、水令人远"的绘画原理，将黄石散点、堆叠得聚散不一，错落有致，曲折凸凹，俨若天成，水体则大有萦回曲折、时隐时现、忽收忽放、开合随景的境界，加之曲枝枯木，附藤横竖，鸟鸣蝉噪，宛若野趣横生的山林，显得幽静古朴，使游玩者几欲跃入深山大泽。

园的东侧，南北分别建有康熙、乾隆碑亭，御制石刻陈列其中。黄石丛中的天下第五泉，欧阳修在《大明寺泉水记》中称："此井为水之美者也"，是合乎实际的。

I planted willow trees in front of the hall, enjoying spring breeze.

Writing prolifically at will as the head of the prefecture I also take pleasure in drinking.

I know that one should seek pleasure while he's young but I'm already old unfortunately.

Pingshan Hall has seen many times of war, and has been restored several times and thriving. The current building was built by Fang Zijian in the Qing dynasty. There are many excellent couplets written in the hall, such as, "Mountains passing by this hall; Feast by head of prefecture enjoyed by guests", and "Rising in the morning and leaning against the railings I see verdant mountains through ages; Having a drink at night, I enjoy the bright moon hanging in the sky". They are very savory and meaningful to read.

Pingshan Hall is connected in its back with Gulin Hall and Ouyang Temple. The former was built by Su Dongpo in Yuanyou 7th year (1092) to commemorate his teacher Ouyang Xiu. It uses a poetic expression as the title for the hall, "deep valley with tall trees". In front of the hall, hangs a couplet that people love to read aloud, "numerous pine trees casting verdant shadows; one stream meandering through like a cloud". The Ouyang Temple behind the hall was built in Guangxu 5th year (1879) in the Qing dynasty. In the hall, inlaid and carved stone portrait of Ouyang Xiu. The carving stroke is full of emotions. With the lighting effect it looks like a white beard if looking from afar, but a black one if looking from close by, an interesting attraction for tourists.

East of Pingshan Hall, faces Pingyuan Building. This building takes advantage of the elevation difference at the site. It actually has three stories but appears to have only two. It is said that the building's name was taken from the landscape painting artist Guo Xi's "Shan Shui Xun", "It is referred to as Pingyuan (Note: flat and far) if you look from a nearby mountain toward a faraway mountain. It means in touch but misty." In the courtyard in front of the building, there is a large bluestone monument, on which four large characters, "Seal Center Stone House", are carved. It is a personal writing by the Qing dynasty's Daoguang emperor.

Fang Pu (Note: Fragrant Garden)

It is also named as West Garden, re-built by the wealthy merchant Wang Yinggeng in 1736. It is the royal garden for Emperor Qianlong's southern tour. Most southern Chinese gardens use a flat land to dig pond and pile up dirt to make artificial hills; however, at Fang Pu, mountains are dug up to make pond. On rolling hills, applying the landscape painting rule "rocks make one feel ancient and water makes one feel far and remote", scattering yellow stones around, some are clustering while others not. The rockery forms interesting patchwork as if formed by nature. The water body also has changing shorelines, some parts are visible while others hidden. Some are open while others closed. With live and dead tree branches going through, birds chirping and cicadas singing, it is as natural as a wild forest. It is imbued with quiet and ancient air, making tourists felt like being in big mountains with water in large expanse.

East of the garden, there are monument pavilions of Emperor Kangxi in the south, and that of Emperor Qianlong in the north. The royal stone carving exhibits are displayed inside. The fifth spring in the world, or in China, is among yellow stone groves. The statement "this well water is the beauty of waters", as noted in Ouyang Xiu's "Daming Temple Spring", seems to be realistic.

1. 牌楼　　2. 天王殿　　3. 大雄宝殿　　4. 平远楼　　5. 晴空阁
6. 报本堂　　7. 悟轩　　8. 平山堂　　9. 谷林堂　　10. 欧阳祠
11. 西园入口　12. 御碑亭　13. 第五泉亭　14. 第五泉　15. 四方亭
16. 天下第五泉　17. 鉴真和尚纪念堂

扬州蜀冈名胜平面图
Plan of Shugang Attractions, Yangzhou

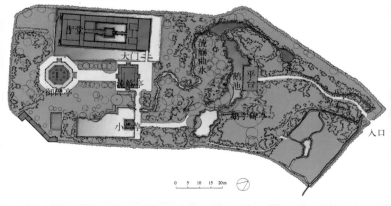

绍兴兰亭平面图
Plan of Orchid Pavilion, Shaoxing

绍兴兰亭鸟瞰图
Brid's Eye View of Orchid Pavilion, Shaoxing

绍兴兰亭　Shaoxing Orchid Pavilion

兰亭在绍兴市南郊13公里处的兰渚山下。据《越绝书》记载，越王勾践曾在此种兰渚田。至汉代，这里是一座驿亭，名兰亭。东晋永和九年（353年）三月三日，王羲之、谢安、孙绰等人，在此"修禊"，行曲水流觞之饮，并由各人赋诗以志这次聚会，并由王羲之为诸人诗集作序，这就是著名的《兰亭集序》。此序不仅文章极美，而且书法有极高的艺术价值，历来被视为我国书法艺术的瑰宝。兰亭也因这次的禊饮赋诗及《序》而闻名于后世，被尊为我国书法艺术的圣地。

宋代兰亭在兰溪江南岸山坡上，明嘉靖二十七年（1548年）迁于江北岸现址。现在的兰亭建筑物都是清代重建的，并经后世屡加修葺。

整个兰亭景区位于平地上，周围为水田。基地南北进深约200余米，东西宽约80米，入口在北端。进门经一段曲折的竹径到达鹅池，相传王羲之爱鹅，故以之名此池。池旁三角亭内碑上大书"鹅池"二字。池南为土山，山上林木茂密，将兰亭主景部分隐蔽于土山之后，起到"障景"的作用，不使产生一览无余之弊。由鹅池碑亭旁屈曲前进，到达"小兰亭"碑亭。此亭作盝顶方亭，式样较为别致。经此亭折而右，就是兰亭主题景区——曲水及流觞亭。当年王羲之等人修禊之处早已不可考，这是后世所作象征性的曲水流觞场所。曲水流觞盛行于六朝至唐宋，是文人雅集的一种形式。唐宋以后都在石上刻曲折的水槽，上覆亭子，称为流杯（觞）亭，众人各据曲水一方，羽觞随水而流，停于何人位前就应赋诗、饮酒，文人以此相娱，格调极高。明清时仍有这种风气，北京中南海、故宫乾隆花园都有这种流杯亭遗例，滁州琅琊山醉翁亭西侧也有一例。但兰亭所建流觞亭为一纪念亭，其式样作四面厅式，亭内不作曲水流觞之举，故与一般流杯亭不同。流觞亭北有一座八角重檐横尖亭，亭内有康熙手书《兰亭集序》碑，碑高6.8米，亭高12.5米，这庞然大物，似有喧宾夺主之嫌。碑亭东侧为王羲之祠，俗称"右军祠"。因王羲之在东晋曾官至右军将军，故也自称王右军（见《晋书·王羲之传》）。祠在水池之中，祠内又是水池，内外有水

Orchid Pavilion is located 13 kilometers south of Shaoxing City at the foothill of Mount Lanzhu. According to "Yuejueshu", the Yue Kingdom's king Goujian once planted orchid here. In the Han dynasty, it was named Orchid Pavilion and was an inn or rest area for mailmen. In Yonghe 9th year in the East Jin dynasty (353), on March 3, Wang Xizhi, Xie An, Sun Chuo and others, performed a religious ceremony here and played a game of composing poems with wine cups floating on a winding stream. The guests invited Wang Xizhi to write a preface for their poem collections and the preface became the well-known "Orchid Pavilion Collection Preface". This preface shows exemplar beautiful style as well as great values of Chinese calligraphy and remains as a treasure of Chinese calligraphic art. Orchid Pavilion is revered as the "Mecca" of Chinese calligraphic art because of this event and preface.

In the Song dynasty, the pavilion is located on the slope at the south bank of Lanxi River. In Jiajing 27th year in the Ming dynasty it was relocated to the current location at the north bank. The current pavilion building was built in the Qing dynasty, and it was renovated several times later.

The whole scenic area is on a flat land, surrounded by rice paddy field. The area is about 200 meters from north to south and 80 meters east-west wise. The entrance is at the north. After entering the gate and passing a winding path, you arrive at the Goose Pond. It is said that Wang Xizhi loved geese, and so the pond was named. Two large characters, "Goose Pond", were written on the stele in the nearby triangular pavilion. South of the pond is a hill, where the vegetation is thick, shielding part of the pavilion scene. This "scene blocking" treatment is to avoid the situation where everything can be seen at a glance. Along the winding path by the Orchid Pavilion, you can reach a stele pavilion called Orchid Pavilion Minor. The roof form is called "Luding", with unique style. Passing this pavilion and turning right, one sees the main theme area, Qushui Liushang, or winding stream with floating cup pavilion. The exact location of the event cannot be found and this place is just a symbolic one commemorating that event. Qushui Liushang had been very popular from the Six Dynasty all the way to the Tang and Song dynasties, as it was an elegant way for literati to meet each other. After the Tang and Song dynasties, winding troughs were carved in large stone tablet, on which a pavilion was built, and it was referred to as flowing cup pavilion. Everybody stood at a curve, floating cups flowed along the trough. If the cup stopped in front of somebody, he would need to write a poem and drink the wine. This was an entertainment among literati and was regarded as very stylish and of good taste. It still prevailed in the Ming and Qing dynasties. For example, there are such relics in Beijing's Zhongnanhai and Qianlong Garden in the Forbidden City. There is also an example at the west side of Drunken Man Pavilion in Chuzhou's Langya Mountain. However, the Liushang Pavilion built at Orchid Pavilion area is a memorial pavilion, with four-side hall style and with no Qushui Liushang inside, different than common flowing cup pavilion. North of the flowing cup pavilion, there is an octagonal based double-eaved pavilion, in which there is an Emperor Kangxi hand-written Orchid Pavilion Collection Preface stele, measuring 6.8 meters in height. The height of the pavilion is 12.5 meters, a bit too high, causing unnecessary distractions. East of the

夹持，可称是此祠一大特色。

兰亭布局曲折，竹树森郁，环境气氛极佳。曲水流觞利用兰溪水引至鹅池，经流杯渠而流至北面诸池再泄于江之下游，处理十分成功。若当天朗气清、惠凤和畅之时在此行修禊之会，也可一展我国古代文化之风雅情趣。可惜流觞亭及御碑亭二者均以巨大体量排列于小兰亭与右军祠之间，布局既刻板，建筑本身也缺少意趣，实为美中不足。

stele pavilion is a temple for Wang Xizhi, commonly called "Youjun Ci". This is because he was promoted to a general's position in the right army, and he also called himself as Wang Youjun (see "Jin Book: Wang Xizhi autobiography"). The temple is located inside a pond and there is also a pond inside the temple. It is flanked on both sides by waters, showing its uniqueness.

Orchid Pavilion's layout is tortuous and the bamboo grove is thick, creating a perfect atmosphere. Qushui Liushang redirects Lanxi stream water into the Goose Pond, passing through Liubei Channel to Zhu pond and exiting to downstream. This is a very successful handling. When the sky is clear and the breeze is easy, if religious ceremonies are performed here, China's ancient elegance of taste can be displayed here. Unfortunately, the Liushang Pavilion and the Royal Stele Pavilion's sizable bodies are located between the small Orchid Pavilion and Youjun Temple. The layout is rigid and the architecture also tends to be boring. This is something less than perfect.

（二）城市地

市井不可园也；如园之，必向幽偏可筑，邻虽近俗，门掩无哗。开径逶迤，竹木遥飞叠雉；临濠蜒蜿，柴荆横引长虹。院广堪梧，堤湾宜柳；别难成墅，兹易为林。架屋随基，浚水坚之石麓；安亭得景，莳花笑以春风。虚阁荫桐，清池涵月。洗出千家烟雨，移将四壁图书。素入镜中飞练，青来郭外环屏。芍药宜栏，蔷薇未架；不妨凭石，最厌编屏；未久重修，安垂不朽？片山多致，寸石生情，窗虚蕉影玲珑，岩曲松根盘礴。足征市隐，犹胜巢居，能为闹处寻幽，胡舍近方图远；得闲即诣，随兴携游。

(2) Urban Land

It is not appropriate to build gardens in noisy urban areas. If one has to be built, a quiet place would be a preferred location. Although hustling and bustling outside, it can be quiet after doors are closed. There could be stretches of paths with twists and turns and sloped garden walls looms through bamboo groves. A naturally shaped pond can be dug and a bridge can be built inside a gate. Phoenix trees can be planted if a courtyard is spacious; willow trees can be planted along winding river embankment. A piece of woodland would be appropriate where a building is hard to build. The construction of buildings should follow the characteristics of the garden site. Stone embankment should be used when it is close to water. A placement of a pavilion is both for a scene creation by itself and for providing a place to enjoy other scenes from there. Flowers planted would be enjoyable in spring breeze. Landscape buildings are hidden under the phoenix trees, with their shades covering the bookshelves inside. A bright moon is reflected in clear and clean pond and everything seems to be bathed with serene moonlight. Waterfalls look like white silk in the mirror and green mountains feel like a verdant veil surrounding the town. Peonies should be protected by some railings, which may use base rocks to build. Roses do not have to grow on trellises and trying to knitting them into a flower screen is a no-no. If one does not prune his plants, how can the plants be colorful and thriving? A wide mountain landscape scene is delightful but even a tiny piece of micro landscape can be of much interest. An exquisite banana tree shadow can sift through an unlatched window, while a deep-rooted pine or cypress tree can embed its creeping roots between cracks of rocks. With such a nice setting, it is proven that one can achieve quiet among a hustle-and-bustle urban environment; it can be even better than living in real wild mountains. If one can be in a serene and quiet state in a busy and noisy town, why bother to travel far to find such a place? One can get to this state of mind at leisure time and when it is in the right mood, friends can be invited over to share the same joy and experience.

苏州留园平面图
Liu Yuan's Plan, Suzhou

苏州留园 Suzhou Liu Yuan (Note: Lingering Garden)

留园在苏州园林中，不但以规模大著称，而且还以该园东部建筑群内部空间的庭园造景和流动的空间序列布局，具有独特的风格和杰出的艺术成就，而久负盛名，被列入世界文化遗产。

一进留园大门，是个比较宽敞的前厅，自右侧进入窄长的通道，经过三折以后，进入一个面向天井的敞厅，天井中有山石花木小景装点，随后以一个半遮半敞的小空间与入园的序幕"古木交柯"相接。这段行程原是沉闷的高墙夹弄，匠师们巧妙地采用空间大小和光线明暗的对比手法，即虚与实、放与收、明与暗相交替，以空间反复的节奏变化，来打破近乎漫长通道的单调沉闷感觉。在窄廊的两侧，还忽左忽右不断安排了采光的天井，在天井中布置富于画意的山石花木小景，使人工机械的建筑通道增添生意。

穿过重重通道进入"古木交柯"，由暗而明，由窄而阔。迎面漏窗一排，光

Liu Yuan in Suzhou, among all gardens in Suzhou, is well known not only for its large size but also for its internal garden design and flowing spatial sequence in its eastern building complex. Its unique style and artistic achievement earn its place in Chinese garden history and the world cultural heritage registration.

Once entering the gate, one finds a spacious hall in front. Entering a narrow passage way on the right, after three turns, you see an open hall facing an enclosed patio, in which there are scenes decorated by plants and rockeries, and it connects with the prelude of the garden, "Gumu Jiaoke" (Note: intertwining old woods) with a semi-open small space. Originally, this passage would be a much depressing alley, sandwiched by tall walls; however, the garden design masters smartly took advantage of changing sizes of spaces and varying lighting brightness, applying solid and void, tightening up and letting go, using bright and dark interchangeably, and with repetitive spatial change rhythms, to have successfully broken the long passage's monotonous depressing feeling. At both sides of the narrow alley, the designers placed random natural lighting open ceiling windows, in which picturesque mini plant and rockery scenes are used as real vignettes, introducing a lively atmosphere in an otherwise dull and mechanic artificial building passage.

After passages and passages one enters "Gumu Jiaoke", with a changing scene from dark to bright, and a space narrow to broad. Facing

影迷离。透过窗花，山姿水态隐约可见，这就是留园的序幕。回首南顾，清风丽日溢于小庭。雪白的粉墙衬托着一株虬曲苍劲的古木。整个空间，疏朗明净，古朴自然。从西面两个八角形窗洞中，透出"绿荫"之外的山池庭院。由窗旁小门进入绿荫区，可见一更小的天井，与"古木交柯"的天井虽一墙之隔，敞门相通，可是情趣截然不同。巧石修竹使天井显得幽雅秀丽，140平方米的空间内，显示了极为丰富的层次。

由绿荫穿过镂花木隔窗，几经曲折，来到"明瑟楼"，山光水色，滉漾夺目，这是留园西部山水园的起景。再西行，便是中部主厅"涵碧山房"。盛夏赏荷，此处最佳。由房前平台北眺，四面山林起伏，可亭翼然，绿树婆娑，浓荫蔽日。"小蓬莱"、"濠濮亭"，也一一在望。西北侧是连绵起伏的假山，石峰屹立，间以溪谷，池岸陡峭，野趣横生。这是山水园的主景所在。

当游人饱览山水景色之后来到"曲溪楼"，可以看到这一带巧妙地安排了环环相扣的空间，造成层层加深的气氛。此楼长十米，宽三米，虽极瘦长，但因其两端均有更窄长的空间与之形成对比，两端出入口又偏在西边，使室内有停留回旋的余地，所以入楼后并无穿行过道的感觉。自曲溪楼向右转入东部内院的西楼，室内地平稍稍抬高，使竖向发生变化。过西楼，是"清风池馆"，以水榭形式向西敞开，并与对岸"闻木樨香轩"为对景，眼前一片湖光山色。最后通过简洁的小过厅进入"五峰仙馆"，更对比出客厅的富丽堂皇，高深宏敞。该厅又名"楠木厅"，陈设布置精雅大方。厅前院里奇峰屏立，花木交柯；厅堂后院回廊花径，逶迤多姿，不愧为江南厅堂建筑的典型。

自五峰仙馆穿过短短一段曲折的暗廊来到"揖峰轩"，藤竹茂密，繁花似锦，一带曲廊回旋，真是安静闲适，深邃无尽。揖峰轩内有精巧的红木装饰。轩内中央粉墙上深色的窗框，使人顿觉窗外景物被构成一张张竹石尺幅画，挂在墙上。揖峰轩对面"石林小屋"隐于树丛湖石之后，绕过东廊到达这里，诱人小憩。环顾三面粉墙，窗外蕉叶、竹影随风弄姿。透过绿丛可见月洞小门，门后仍然绿荫蔽日。在这浓郁的绿色笼罩下，真不知"庭院深深，深几许"。出亭，沿廊至"鹤所"，只见

a visitor, is a wall of lattice windows, creating an interesting pattern of light and shadow. Through the window grills, you can see the garden's hill and water scene looming beyond. This is the prelude of Liu Yuan. Turning back and looking south, you see a small courtyard shining under the bright sun with light breeze. An intertwining old tree is set off against a white painted wall. The whole space feels simple, clear, ancient and natural. Through the two octagonal shaped lattice windows in the west, one sees another landscape garden beyond the current verdant yard. Walking through a small entrance by the windows into the green shady area, you see an even smaller open ceiling patio. Even it is separated from the "Gumu Jiaoke" patio by only one wall, with a connecting door, the interests are quite different. Nice looking rocks combined with slender bamboos make the open patio elegant and beautiful, and rich layers of scenes pack the 140 square meters of space.

Through the greens and the wooden lattice windows, after several turns, you arrive at "Mingse Lou". You are attracted by its landscape scenes of hill, rock and water and it serves as the starting scene of Liu Yuan's west landscape garden. Moving further west, is the central main hall "Hanbi Shanfang". It is the ideal place to enjoy the lotus flower in mid-summer. Looking north from the terrace in front of the building, one sees rolling hills with pavilions immersing among lush trees. "Xiao Penglai" and "Haopu Pavilion" can both be seen from here. On the northwest, is the rolling rockeries, with peaks and valleys and steep cliffs full of wild interests. This is the center scene of the rock-water garden.

After fully enjoying the rock and water garden, visitors reach "Quxi Lou" (Note: tortuous stream building). Here one can see well-arranged interlocking spaces, sensing the depths of the spatial sequence. The building is 10 meters long and 3 meters wide. Although it is extremely narrow and elongated, since at both ends there are even narrower spaces to contrast and the exits are located at the west side, leaving rooms for maneuvering, one does not feel a sense of passage way after entering the building. From Quxi Lou turning right and entering a convex room in the west building in the east inner yard, you step on an elevated ground level and this makes some changes vertically. Passing the west building is "Qingfeng Chiguan" (Note: water hall with clear breeze). It is a water pavilion opening to the west, facing the opposite bank's "Wen Muxi Xiang Xuan", creating a scene of lakes and mountains. Finally, walking through the simple passage hall and entering "Wufeng Xianguan" (Note: five peaks Shenxian hall) make the parlor's grandeur and spatial expanse more prominent. The building is also named as "Nanmu Ting" (Note: Phoebe Hall) with fine and generous arrangement of furniture. Its front yard has much rockeries and plants while the backyard is adorned with corridors and flower paths. This meandering yet splendid scene represents a typical Southern Chinese garden architecture design.

From Wufeng Xianguan and passing through a tortuous hidden corridor, you arrive at "Yifeng Xuan" (Note: peak salutation hall). Here there are thick vines and bamboos and clusters of colorful flowers. With a curved corridor, one feels quiet and relaxed with endless serenity. There are delicate redwood ornaments. Inside on the central wall there are dark colored window frames that make you feel that the outside rock and bamboo landscapes appearing to be landscape paintings on the wall. Facing Yifeng Xuan is "Shilin Xiaowu" (Note: rock grove small house) hidden among trees and behind the lake rockeries.

西墙空窗外，五峰仙馆前院翠竹潇洒，湖石挺拔，东墙上则是细密的菱花窗，疏密、虚实组合得宜，令人惬意。

自揖峰轩东行，为"林泉耆硕之馆"，即"鸳鸯厅"，精美雅丽，被誉为我国古典厅堂建筑的精品。鸳鸯厅正面北向，径对著名的留园三峰；冠云峰雄峙居中，朵云峰、岫云峰屏立左右。冠云峰高达三丈，为江南园林中最高大的一块湖石。三峰之下，假山石散点成趣，花草点缀其间，景色宜人。三峰四周建有"浣云沼"、"冠云亭"、"冠云楼"、"冠云台"、"伫云庵"等建筑。"冠云楼"上有匾额题"仙苑停云"四字，以示此处似神仙宫苑。

留园共占地五十余亩，分中、东、西、北四个景区。中部为山水园，明沽清幽，峰峦回抱；东部为建筑庭园，楼、馆、轩、斋，曲院回廊；西部是自然景色，北部为田园风光。四景区建筑大部分以曲廊联系，廊长700米，依势曲折，通幽度壑，使园景堂奥纵深，变幻无穷，堪称我国造园艺术佳作。

留园始建于明代嘉靖年间（1522～1566年），为封建官僚徐泰时的私园，时称东园。另有西园，就是现在的戒幢律寺。清嘉庆年间（1796～1820年）为刘蓉峰所有，更名寒碧山庄。因其地处花步里，又名花步小筑，人称刘园。光绪时（1875～1908年）归盛康（旭人）所得，重加扩建。后取刘、留同音，改名留园。抗日战争时期，留园曾遭敌伪严重破坏。新中国成立以后，经过大力整修，使这座名园获得新生，全园面貌焕然一新。为全国重点文物保护单位。

Bypassing the east corridor to get here, you feel it alludes you to take a brief rest. Looking at the painted walls on three sides, you see banana tree leaves outside the windows and shadowy bamboo groves. A small moon gate can be seen through the grove and it is still lush green behind the gate. With such thick vegetation, you really could not tell how deep ahead the courtyard is. Exiting the pavilion, walking along the corridor to reach "He Suo" (Note: crane lodge) you can see through the open windows on the west wall the outside verdant bamboo groves in front of Wufeng Xianguan. With tall and straight lake rockeries and fine and dense lattice windows on the east wall, a feeling of perfect combination of dense and sparse, solid and void emerges, which is very comforting and satisfying.

Traveling east from Yifeng Xuan, we see "Linquan Qishuo Zhiguan", that is, "Yuanyang Ting" (Note: hall of mandarin duck). Its beauty, elegance and fine quality earn its high praise as a masterpiece of Chinese garden architecture. Facing direct north from the hall is the well-known Three Peaks of Liu Yuan. The center one Guanyun Peak is flanked by Ruiyun Peak and Xiuyun Peak at its left and right. Guanyun Peak is as high as three Zhangs (Note: Zhang is an old Chinese measuring unit and it's about ten feet long.) and the tallest lake rockery among southern Chinese gardens. At the base of the three peaks, there are scattered rockeries with various plants, forming a pleasant combination. Surrounding the three peaks there are garden buildings such as "Huanyun Zhao", "Guanyun Ting", "Guanyun Lou", "Guanyun Tai", "Zhuyun An". At "Guanyun Lou" there is a horizontal inscribed board with "Xian Yuan Ting Yun" four characters, showing here being almost like Shenxian's (Note: celestial immortal being) fairy palace.

The whole Liu Yuan occupies more than 50 Mus and is divided into central, east, west, and north four scenic zones. The central one is a rock and water garden, which is clear, bright, and quiet, with rockeries surrounding the area. The east zone is full of garden architecture, with Lou, Guan, Zhai, curved yard and corridors. The west zone is full of natural scenes and the north rural and pastoral. The connections between the four zones are mostly done by curved corridors with lengths at 700 meters. The corridor is tortuous and meandering according to the terrain, leading to quiet places and often crossing gullies. With infinite change in sight, it is a great masterpiece of Chinese garden art.

Liu Yuan was first built in Jiajing's years (1522-1566) in the Ming dynasty. It is the private garden of the feudal bureaucrat Xu Taishi and was named the East Garden. There was a West Garden and is what is now Jiezhuanglu temple. In Jiaqing's years (1796-1819) of the Qing dynasty it was owned by Liu Rongfeng and was renamed as Hanbi Shanzhuang, or winter verdant mountain villa. Because it is located at Huabuli, it was also called Huabu Xiaozhu, or Liu Garden. In Emperor Guangxu's time (1875-1908) it changed hand to Shengkang (Xuren) and was rebuilt and expanded. After that, since Liu and Liu pronounce the same, it was renamed as Liu Yuan. During the anti-Japanese war, it was heavily damaged by the puppet army. Since the liberation and after much renovation, the garden received a facelift and has remained as a key national heritage conservation property.

苏州网师园中部鸟瞰图
Bird's Eye View of the Middle Section of Wangshi Yuan, Suzhou

苏州网师园 Wangshi Yuan (Master of Nets Garden) in Suzhou

网师园始建于南宋，原为北宋末年（公元12世纪初）花石纲发运使、侍郎、扬州人史正志"万卷堂"故址，号称"渔隐"。后被其子卖给丁氏。清乾隆间，宋宗元（悫庭）购得其地，建造园林，为退隐、养亲之所。宋氏欣赏史正志"渔隐"的含意，又因园近王思巷，故取其谐音，称网师园（网师是渔翁的意思，仍寓"渔隐"之意）。后来，园主多次易手，园名亦多次更迭，如"瞿园"、"蘧园"、"逸园"。1958年，由苏州市园林处整修开放。现为世界文化遗产。

网师园在苏州城东南角，"负郭临流，树木丛蔚，颇有半村半郭之趣"。全园分中轴对称的住宅部分和沿水院周边布置园林建筑的花园部分。从阔家头巷正门进入，通过轿厅到"清能早达"大厅，原为园主会见宾客和办事的地方。其后为供起居用的"撷秀楼"，楼前的砖雕门极为工细。这一组建筑和院落是网师园的正居部分。

轿厅西侧有一小门，门上有砖刻"网师小筑"四字，为花园入口。进门是小山

Wangshi Yuan was first built in the South Song dynasty. Originally, in the North Song dynasty (12th century), it was the property of Yangzhou local Shi Zhengzhi, who is a royal carrier and assistant minister. It was named as "Wan Juan Tang" (Note: the hall of ten thousand volumes) and claimed to be a hermitage in the ancient Chinese fisherman's complex. It was later sold to a man whose family name is Ding. In Qianlong's time in the Qing dynasty, Song Zongyuan (Queting) bought the land and built a garden for the purposes of his retirement use and taking care of his parents. Song appreciated the hermitage implication and also because its proximity to Wangsi Lane, to be homophonic, he named it Wangshi Yuan, which still has a strong connotation of a hermit fisherman. After that, the ownership had been changed multiple times, and the garden name was also changed as "Qu Yuan", "Qu Yuan", and "Yi Yuan". In 1958, it was rebuilt by Suzhou Garden Administration and opened to public. Now it is a site of the world cultural heritage.

Wangshi Yuan is located at the southeast corner of the Suzhou city. Facing a river and leaning against a city, with abundance of vegetation, it is an attraction that is between urban and rural settings. The whole garden is composed of a centrally symmetric residence part and a water courtyard centered garden consisted of garden architecture. Entering from the Kuojiatou Lane at the main entrance, passing a sedan hall, you see a large hall of "Qingneng Zaoda". It is originally a place for the owner to meet and conduct business with his guests. After that, is "Xiexiu Lou" (Note: beauty capture building) as a living room. The brick engraving in front of the building

丛桂轩，轩南墙下，用湖石叠成小山，种植桂花，轩北是黄石叠成的云冈假山。此轩四面出廊，循廊西行，进入"蹈和馆"，馆南有门通琴室，馆北沿廊到樵风径，是园子的主景——水院。

网师园的水面是以水院的形式来布置的，周围以建筑形成一个极为生动的闭合空间。有春、夏、秋、冬四季的特色。水院南为"濯缨水阁"，取"沧浪之水清兮，可以濯吾缨；沧浪之水浊兮，可以濯吾足"的诗意。面水临崖，夏季荫凉，沿水院西岸，渡高低起伏的走廊至"月到风来亭"，建于地势高处，是中秋赏月的佳处。出亭向北是"潭西渔隐"，附近有石矶与石板折桥锁住一湾清水，树木参差，覆盖池上，别有一种意境。其旁是"看松读画轩"，轩前的花台，所种罗汉松、古柏和白皮松等都是数百年前物，在冬季赏景最为合适。看松读画轩的左边有走廊通"竹外一枝轩"，小轩开敞，临水，竹影婆娑。其左有黑松斜出水上，有以少胜多、画龙点睛之妙。水院东面是"射鸭廊"和"半亭"，池边用黄石叠成假山石组，上攀紫藤，在其南面，墙面上攀以木香，是欣赏春景之处。

从潭西渔隐处的小门向西，便是"殿春簃"及其所属院落。殿春指芍药花，簃是竹屋。院中原为芍药圃，广植各种芍药，以赏芍药为主。其旁有小书房，甚为安静。院落的南部有假山和涵碧泉（树根井），泉上有"冷泉亭"，亭中有大英石一块，敲之清脆如磬，相传是明代文人唐寅旧居中遗物。

"竹外一枝轩"北边"集虚斋"，原为读书处。"集虚斋"东"五峰书屋"，为藏书处。

网师园面积仅八亩左右，但是整个布局似断似续，处处贯通，有行回不尽之致。水院周围建筑布局紧凑，比例适度，位置恰当，造型轻巧，舒适柔美，在苏州诸园林中，有独到之处。水院中，竹外一枝轩面对小山丛桂轩，一临水，一离岸；看松读画轩面对濯缨水阁，一离岸，一临水，形成两组动人的对景。为了加强自然气氛，又叠了云冈假山，在看松读画轩前筑了花台，作为离岸建筑的前景。这是两条习惯上的主轴线。而月到风来亭对着射鸭廊、半亭，成为一条东、西向的副轴线。在东西向副轴线两边的布局，是极好的不对称均衡构图。所以，网师园水院的构图，是苏州古典园林中不对称均衡构图佳例。

is very refined in details. This building complex and the courtyard is Wangshi Yuan's main residence area.

On the west of the sedan hall there is a small door, on which there is a brick engraving "Wangshi Xiaozhu" (Note: fisherman's small lodge) four characters. It serves as the entrance of the garden. After entering the entrance, you meet with Xiaoshan Conggui Xuan. Along its south wall, there are small hills piled up by lake rockeries with Osmanthus planted around. North of the building is Yungan rockery made of sericite or other yellow colored rocks. This building has corridors extended on all its four sides. Walking along a corridor to the west, you would enter "Daohe Guan". There is a door at the south side of this building and it leads to Tongqin Shi. North of the building along the corridor one can reach Zhuifeng Jing, which is the main scene of the garden – a water courtyard.

The arrangement of water at Wangshi Yuan follows the form of a water courtyard, which is surrounded by garden buildings, creating a very vivid and lively enclosed space. There are spring, summer, autumn and winter four seasonal characteristics. South of the water courtyard is "Zhuoying Shuige", which is derived from the ancient poem "When the water of Cang Lang is clear, it can be used to wash the red tassels on my hat; however, when it is muddy, it can be used to wash my feet." It is by the water front facing a cliff and is a shady cool place in summer. Along the west bank in the water courtyard, one walks on the corridor with varying elevations to "Yuedao Fenglai Ting" (Note: moon and wind pavilion), which was built on a higher place above the water, creating a perfect spot to appreciate the moon on mid-autumn festival. Out of the pavilion to the north, is "Tanxi Yuyin", with nearby stepping stones and flat zigzag-shaped bridge locking in a pond of clear water. Trees of various heights hang over the pond creating another kind of artistic feeling. By its side is the "Kansong Duhua Xuan", or pine watching and picture reading hall. At the flower bed in front of the building, there are Podocarpus, ancient cypress, and Bungeana trees, dating back to hundreds of years ago. It is best to appreciate these scenes in winter. On its left, there is a corridor leading to "Zhuwai Yizhi Xuan", which is a small open and water front pavilion, with bamboo trees planted around. Again on its left there is a lodgepole pine growing slantly out of water, completing with a final touch with a "less is more" effect. East of the water courtyard is "Sheya Lang" (Note: duck shooting corridor) and "Ban Ting" (Note: half pavilion). At the water bank, sericite or other yellow colored rocks form the rockeries, on which Wisteria climbs. South of it on the wall there are woody plants and this is a place for people to enjoy spring scenes at near distance.

To the west from the small door at Tanxi Yuyin, is "Dian Chun Yi" and its subordinate courtyards. Dian Chun means peony flower and Yi means bamboo cottage. Originally, the courtyard is a peony garden, with various varieties of the plant, and the purpose of the yard is to enjoy peonies. A small study is by its side, and it is very quiet. South of the courtyard there is rockeries and Hanbi Spring (tree root well), above which there is "Leng Quan Ting" (Note: cold spring pavilion). In the pavilion there is a big cristobalite. It sounds like a chime when it is knocked on. It is said to be a relics from the old house of Tang Yin, a Ming dynasty literati.

To the north of "Zhuwai Yizhi Xuan" is "Jixu Zhai", originally a reading place. East of it is "Wufeng Shuwu" (Note: five peaks study), a book collection place.

Wangshi Yuan only covers about eight Mus (Note:approximately 8/15 of one hectare); however its general layout shows seemingly on-and-off partial connections but all are actually connected, forming an indefinite loop. The garden buildings surrounding the water courtyard are tightly composed with proper scales and positions, light in weight with mild and soothing beauty. It is unique among all Suzhou gardens. In the water courtyard, Zhuwai Yizhi Xuan faces Xiaoshan Conggui Xuan, with one close to water and the other keeping a distance from the shore. For Kansong Duhua Xuan and Zhuoying Shuige, one is away from the water and the other at the water. This forms two pairs of interesting opposing scenes. To enhance the natural atmosphere, Yungang Rockery was added, and in front of Kansong Duhua Xuan a flower terrace was built, acting as a foreground scene for buildings that keep a distance from the shore. These are two customary central axes. Complementarily, Yuedao Fenglai Ting faces Sheya Lang and Ban Ting, forming a secondary east-west oriented axis. Arranging architecture and garden artifacts around this secondary axis is an excellent asymmetrical balancing composition act. Therefore, the composition of the water courtyard at Wangshi Garden proves to be a model example of asymmetrical layout among Suzhou classical gardens.

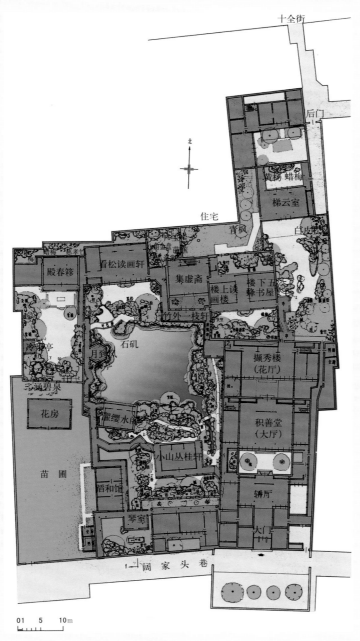

苏州网师园平面图
Plan of Wangshi Yuan, Suzhou

南京瞻园 Nanjing Zhan Yuan

南京瞻园，原为明初中山王徐达的西花园，距今已有六百年历史。清乾隆南巡时，曾驻跸于此，并题名瞻园。乾隆回京后，还命人在北郊长春园中仿瞻园形式建造了如园，足见瞻园园制之精。1853年太平天国定都天京，这里先后为东王杨秀清、夏官副丞相赖汉英的王府、邸园。天京失陷后，遭到清军破坏。同治四年（1865年）、光绪二十九年（1893年）曾两次重修。新中国成立前又被国民党特务机关占为杂院，荒芜不堪。1960年在刘敦桢教授主持下开始整建，迄1966年，建成目前所见的面貌。

瞻园是著名的假山园，全园面积仅八亩，假山就占三点七亩。自然式的山水构成园的地形骨干，结构得体，造景有法，

Zhan Yuan of Nanjing, originally as Zhongshan Governor Xu Da's West Garden in the early Ming dynasty, boasts six hundred years of history. During his southern tour, the Qing Emperor Qianlong once stayed here, and named it Zhan Yuan. After he returned to Beijing, he ordered to have built Ru Yuan in Changchun Yuan zone in the northern suburb, modeling after Zhan Yuan. This shows its exquisite quality. In 1853, Taiping Heavenly Kingdom, or Taiping Rebellion, chose Tianjing here as their capital. Their East King Yang Xiuqing and Summer Officer and Vice Minister Lai Hanying used here as their residence mansion consecutively. After the fall of Tianjing, it was destroyed by the Qing army. It was rebuilt twice in Tongzhi 4th year(1865) and Guangxu 29th year(1893). Right before the Liberation in 1949 it was occupied by the Nationalist Party's Intelligence Agency as a tenement and went into a state of disrepair. In 1960 Professor Liu Dunzhen took the lead to start a renovation and had been working on it till 1966, to have the current appearance.

Zhan Yuan is a well-known rockery garden. The whole garden

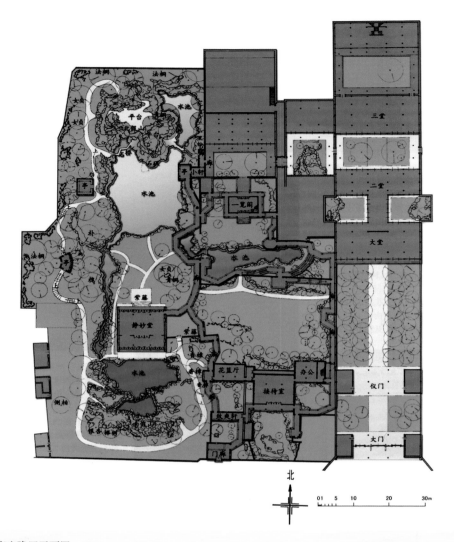

南京瞻园平面图
Plan of Zhan Yuan, Nanjing

山水之间相辅相成，山水与建筑、园路、植物之间又相互融汇，浑然一体。

主体建筑静妙堂，系面临水池的鸳鸯厅，把全园分成南小北大两个空间，各成环游路线，成功地弥补了南北空间狭长的缺陷。南部空间视野近，北部空间视野远，北寂而南喧。全园南北两个水池。南部水池较小，紧接静妙堂南沿，原为扇形水面，修建时改为略呈葫芦形的自然山池，近建筑一面大而南端收小，著名的南假山便矗立在小水池南。北部空间的水池比较开阔，东临边廊，北濒石矶，西连石壁，南接草坪，曲折而富于变化。修建时把水池东北端向北延伸西转，曲水芷源，峡石壁立，更添幽静、深邃的情趣。园中一溪清流，蜿蜒如带，南北二水池即以溪水相连，有聚有分。水居南而山坐北，隔水望山，相映成趣。南北两个性格鲜明的空间亦因此相互联系、渗透。造园者还巧妙地运用假山、建筑，进一步分割更小的空间，使游人远观有势，近看有质。布局合理，细部处理

only occupies eight Mus, with its rockery occupying three seventh. The main terrain of the garden is based on natural landscape. It has a proper structural composition with appropriate treatment of scene creation. The land/hill and water components are complementary and the composition of rock, hill, water, architecture, garden paths, and plants is seamlessly natural as a whole.

The main architecture Jingmiao Tang is a dual building facing a pond and it divides the whole garden into a smaller southern and larger northern two spatial entities, each forming their own looping visiting paths, and successfully correcting the initial defect of too large an extent along north-south axis. The scope of vision in the south is close while that in the north more distant. It is quiet in the north but noisier in the south. There are two ponds: one in the south and the other north. The one in the south is smaller and it is right at the south edge of Jingmiao Tang. Originally a fan shaped pond, during renovation it was re-shaped as natural gourd shaped. The part closing to the building is larger and it is smaller in the southern part, which terminates at the Southern Rockery. In the northern part, the pond is more open. On its east bank there is a side corridor and to its north is the rockery. Stone wall is at its west and to the south, where it meets the lawn. The pond shoreline is tortuous and diversified. When it was under construction, the northeastern tip of the pond was extended to the north and then west, seemingly seeking its source. With flanking rock at the side mimicking a gorge, it further enhances the deep and quiet feeling in the area. The one stream throughout the garden feels like a winding and meandering band connecting the two ponds in the north and the south, sometimes diverting while other times converting. Water in the south and hill in the north, looking at the hill across the water, it is mutually fun and interesting. The two distinctive spaces in the north and south are therefore linked and interconnected. The designer also cleverly made use of rockeries and building structures to further divide spaces into smaller parts so that visitors can see the general structure from afar while still enjoying the texture at close range. The layout is very reasonable and the treatment of details very exquisite. The design style is natural and poised, yet there are unique and clever surprises here and there. A fusion of personal feeling and garden setting can be achieved at this landscape garden that is felt to be made almost by nature.

There are quite a few stone artworks in the garden, some are relics of the Song Emperor Huizong's

精巧，于平正中出奇巧。情景交融，宛若天成。

瞻园石器甚多，有些还是宋徽宗花石纲遗物。著称者有仙人、倚云、友松诸石。亭亭玉立，窈窕多姿，为江南园林山石之珍品。仙人峰置于南门后的庭间，最佳一面正对入口，前有落地漏窗作框景，从暗窥明，衬以浓郁的木香，俨然条幅画卷，用以作为入口的对景和障景，十分恰当。

步入回廊，曲折前行，一步一景，涉足成趣。过玉兰院、海棠院，倚云峰置于精巧雅致的花篮厅前东南隅的桂花丛中。山石坐落的位置，恰为几条视线的交点。其余一些特点山石和散点山石分布在土山、建筑近旁，有的拼石成峰，玲珑小巧，发挥了山石小品"因简易从，尤特致意"的作用。出回廊向西，便是花木葱茏的南假山了。

南假山气势雄浑，山峰峭拔，洞壑幽深，"一峰之竖，有太华千仞之意"。假山上伸下缩，形成蟹爪形的大山岫，钳住水面。岫内暗处，仿自然石灰石溶蚀景观，悬垂了几块钟乳石，造成实中有虚，虚中有实，层次丰富，主次分明的山水景观。悬瀑泻潭，汀石出水，钟乳倒悬，渗水滴落，湿生植物杂布山岫间。苍岩壁立，绿树交映，岩花绚丽，虚谷生凉，俨然真山。山岫东侧又连深邃的洞龛，水池伸入洞中可贴壁穿行而上，游人至此，如入画中，俯视溪涧，幽趣自生。崇岩环列，直下如削，乳泉玲琮，如鼓琴瑟。

修建时曾将静妙堂南屋檐降低了一些，使游人从室内望南假山不致穷见山顶，而见两重轮廓，两重峰峦，绝壁、洞龛更显峭拔幽深。南假山水池东北有明代古树二株：紫藤盘根错节，女贞翠绿丰满。另有牡丹、樱花、红枫等点缀于晴翠之中，鸟鸣蝉噪，金鱼嬉游，泉水潺潺，更衬托出南部空间亦喧亦秀的特色。

北假山坐落在北部空间的西面和北端。西为土山，北为石山。土山有散点湖石，石山包石不见土。两面环山，东抱曲廊，夹水池于山前。池南草坪倾向水面，绿茵如毯，柳绿枫红。普渗泉静静涌出，水面清澈无澜，宛若明镜。蓝天白云，花树亭石，倒映其间，形成倒山入池、水弄山影的动人景观。北部石山独立端严，自持稳重，东南山脚跌落成熨斗形石矶平伸水面，

Huashigang, or royalty items. Famous ones include Xianren, Yiyun, and a few Yousong rocks. They are slim and graceful, and are treasures of southern Chinese gardens. Xianren Peak is placed in the open hall behind the south gate. The best one faces the entrance, with a standing lattice window as its frame. Looking at a brighter object from a dark place, with rich wood fragrance, one seems to enjoy a scrolled painting, and when it is used as the entrance's opposing and barrier scene object, it is very appropriate and indeed suits the situation perfectly.

Entering the winding corridor and walking forward in zigzag pattern, one sees varying scenes at each step, and every one of them has its own interest point. After passing Magnolia Yard and Begonia Yard, we see Yiyun Peak placed in the Osmanthus grove at the delicate and ingenious southeast corner of Flower Basket Hall. The location of the rock is at a joint place where several vistas meet. Moreover, some characteristic rocks are spread and scattered out at the dirt hill and around buildings, and some are made into small and exquisite peaks, applying and extending small rockery's role of "being simple to implement yet effective in mimicking real mountains". Exiting the corridor to the west, we see the verdant south rockery.

The south rockery exhibits forceful posture in general, tall and steep in its peaks, and deep and quiet in its lower caves. "A symbolic peak can represent a real one that is thousands of feet high". The rockery expands on top but shrinks at bottom, forming a crab leg like mountain rock, clamping on the water. In the darker area down under, natural limestone erosive landscape is emulated, with a few stalactites hanging down, mixing real with unreal, creating rich layers of views and setting off a dominating landscape scene from the rest. There are falling water into a pool, stepping stones out of water, stalactites hanging down with water dripping, and wetland plants growing on mossy rocks. Verdant rock with green trees, colorful rock flowers by empty and cool valleys, it just feels like a real mountain. East of the rockery connects a deep cave, to which the pond extends. Visitors can squeeze in the cave and climb up. When they are here, it feels as if they are entering a live picture. Looking down there is a brook and interest generates in the quiet. The rocks are surrounded and upright as if cut by axes; spring water flows and makes musical sound as if musical instruments playing.

When it was under construction, the south eave of Jingmiao Tang was lowered a bit so that when people see from here toward the south rockery they would not see the top, just the side silhouettes. With double profile with peaks and caves, it appears to be steeper, deeper, and quieter. In the northeast of the pond by the south rockery there are two ancient trees from the Ming dynasty: the wisteria tree is deep-rooted while the privet is buxomly green. There are also peony, cherry, and maple trees dotting red colors among the verdant background. Birds chirping, cicadas stridulating, gold fish frolicking, and spring bubbling. All these make the southern part busy yet pretty.

The north rockery is situated at the west and north end of the northern part. It is a dirt mount in the west and a rock one in the north. There are some scattered lake rocks in the dirt mount and for the rock mount there is no sight of dirt. It is surrounded by these mounts at two sides and embraced by a curved corridor in the east, with the pond pressed from both sides in front of the mounts. The green carpet-like lawn in front of the pond inclines towards the water with red

于低平中见层次，丰富了岸线的变化。石山西面向南延伸为陡直的石壁驳岸，水池北端伸入山坳，使人感到水源好像出自山间奥处。紧贴水面的石平桥，曲折于水池之北，既沟通了东西游览路线，又因曲桥分隔，使水面的形态和层次都增添变化。平桥附近旧有泉眼，为观泉佳处。石山体量虽大而中空，山中有瞻石、伏虎、三猿诸洞潜藏。山道盘行复直，似塞又通。山西低谷盘旋，和山道立体交叉。自谷望山，山更高远；自山俯谷，幽深莫测。沿山径，山石玲珑峭拔，峰回路转，步换景异。山顶原有六角亭一座，修建中为了遮挡园北墙外高层建筑，改亭为峭壁，峭壁前为平台，形成全园新的制高点。登临一望，山前景色历历在目。充分体现了古典园林起、结、开、合的艺术手法。

"妙境静观殊有味，良游重继又何年"。瞻园虽小，山水卓著，清风自生，翠烟自留，园制之精，驰誉中外，游人每至，流连忘返，往往尚未离去即生重游之想。

maples blending with green willows. The water from Pushen Spring flows out quietly and the water surface is so clear and clean that is reflects the skylight like a mirror. White clouds against clear blue sky, flower trees against rocks by the pavilion. All these scenes are all reflected in the water, showing a moving garden scene. The rocky hill in the northern part is solemn, staid and independent. The southeast foothill falls to a lower level like an iron sticking out horizontally to the water, showing layers at low height, adding variations along the bank. The rocky hill's west side extends to the south and rises up to form a steep cliff revetment. The northern part of the pond stretches into a col giving a sense of water source. The flat stone bridge close to the water surface meanders in the northern part of the pond. It connects the east and west parts and at the same time, enriching the form and sense of layered scene variations. Close to the flat bridge there is a spring, which is an excellent spot for enjoyment. Although the rocky hill is sizable, its center is void. Inside there are Toad Rock, Crouching Tiger, and Triple Ape and other caves hidden. The hill path is winding at first but then straightened out, as it looks clogged at the beginning but then open. Low valleys are in the west of the hill, forming three dimensional overpasses. Watching the mound from the valley you feel the distance; overlooking the valley from the mound top you feel it is unfathomably deep. Walking along the path, you feel the rocks are exquisite and steep, with changing views at each turn. There used to be a hexagon-shaped pavilion at the hilltop; however during the renovation, in order to shield high-rise buildings outside the northern wall, some steep rocks are employed instead. There is a terrace in front of the cliff, forming a commanding height in the whole garden. Standing there to take a look, everything is in sight. This fully reflects the classic garden's opening vs. concluding, and opening vs. closing artistic treatments.

"It is uniquely fascinating to appreciate the quiet wonderland and I wonder how long I would need to wait before I can enjoy it again?" Although it is small, Zhan Yuan is blessed with an outstanding rockery water garden. With clear wind and lingering fog, its exquisiteness is renowned. Tourists love to linger around a bit more, often pondering about when they could re-visit even before their leave the garden.

南京煦园 Nanjing Xu Yuan

南京煦园，也称西花园，地处南京市长江路东段原"总统府"西侧。建于道光年间，距今已有一百八十多年的历史。太平天国天王洪秀全、辛亥革命领袖孙中山先生等，都曾在此园游览和居住过。现在全园面积虽仅二十余亩，但这里仍然是清水碧潭，苍松翠竹，楼台亭阁，假山奇石，天然如画。

该园以水为主。水体南北走向，平面如长形花瓶。在造园手法上，为了突破单一狭长的水体，故用舫、阁把水体自然分割成各自独立又互相联系的三个部分。有

Xu Yuan in Nanjing, also known as West Garden, is located to the west of the Jiansu Province People's Government building in the east section of Yangtze River Road in the city of Nanjing. It was built in the Qing dynasty's Daoguang emperor's reigning period and has a history of over one hundred and fifty years. Both Hong Xiuquan, the Taiping Rebellion leader, and Sun Zhongshan, the Xinhai Revolution leader, visited and resided here. Although the size of the garden is merely twenty Mus, it is natural and picturesque with clear water pond, green bamboos and pine trees, beautiful mound, rockeries and landscape architecture. The garden is water-oriented. The alignment of the water body is north-south, forming a long vase shape. In order to break the monotonous elongated feeling, the water area is broken down into three sections, separated by a boat-shaped smaller building and a larger hall. The three parts are self-dependent yet connected. There are features that are close together and features that spread

分有聚，虽分实聚，使中部形成较开阔的水面；南舫北阁遥相呼应，东榭西楼隔岸相望，花间隐榭，水际安亭，为水体最精彩部分。平视南北两端水源莫及，自然形成大中有小，小中见大，虚实相映，层次丰富的境界。水体四周小径通幽，山石、古树、榭、亭、阁互为对景、借景，水中倒影如绘，使瓶中天地，别开生面。

瓶口有"漪澜阁"屹立水中，左右有小桥可渡。前后门窗开朗，阁中窗明几净，既可遥望外景，又可促膝品茗。屋顶为瓷质鱼脊歇山，门上方用木质彩色雄狮装饰，似笑迎贵宾，此皆为明代遗物。阁前平台昔时供赏月、演戏、听曲，乐声、水声共鸣，人影、月影齐舞，可谓佳境。依阁南望，水天相应，有一石舫相对。此舫为清代建筑，分前后两舱，卷棚屋顶，造型精巧，形象逼真，每当清风拂袖，波光粼粼，舫若徐行，神态怡然。由左右石制跳板登舟，迎面可见匾额上有"不系舟"三字，此乃乾隆皇帝南巡时亲书，故后人亦称煦园为"不系舟"。登舟北瞻，穿水遥对漪澜阁；凭栏近观，池水清澈，稍有莲花点缀，锦鳞游泳，聚集跳跃，引人入胜。

落舟登东岸，可见一碑亭，碑上刻有唐代诗人张继的《枫桥夜泊》诗：

月落乌啼霜满天，江枫渔火对愁眠；

姑苏城外寒山寺，夜半钟声到客船。

此碑为清代学者俞樾所书，据说是从苏州寒山寺移来。岸边有忘飞阁，临水而建。前面水榭瓷脊歇山，后层屋顶为卷棚式，上下建筑参差错落，变化自然。榭三面临水，周围有"美人靠"供游人坐息。榭有独立围墙，外围回廊，内院栽有桂花、芭蕉、修竹等，别有天地。对岸有"夕佳楼"，为双顶重檐，卷棚歇山，分上下两层。上有凉台，下设回廊，楼旁还有古老银杏和山石点缀，每当夕阳西斜，阳光满楼，乃吟诗作画佳处。楼西依围墙建有半方亭，亭中陈列名家手迹石刻。亭旁有阶可登凉台，迎面有太平天国时代的彩色团龙砖刻，龙墙起伏，上嵌鳞状瓦当，如龙飞舞，造型生动而有气魄，强烈地反映了太平天国时期的造型艺术特色。转折西行，来到伟大革命先行者孙中山先生办公楼，陈列着孙中山先生任临时大总统时的办公遗物。

由此东行，回廊曲折，花木丛生，四季景色应时而异，可谓城市山林。沿廊北行有古建筑五楹，造型雅致，内设各种工

apart. It appears spreading apart in the middle section but it actually converges, forming a broad open space over water. The boat building in the south and the hall in the north correspond to each other while the buildings in the east and west banks work in concert by facing each other. Pavilions are built among flowers and by the water, and become the most wonderful part on the water body. The water sources in the south and north ends cannot be seen. It is naturally formed to have small scenes in large ones and to see big from small. It has reached a state where physical scenes are blended with the imaginary and it enjoys with rich layers of varied scene depths. The garden paths around the pond lead to quiet spots. Rocks, old trees, pavilions and pagodas pose opposing scenes and leverage each other. The reflections in the water make picturesque scenes. The miniaturized natural landscape, or the world in a bottle as it is called, is very refreshing.

At the mouth of the bottle, there is "Yilan Ge" in the water, and there are small bridges reaching it. The front and back doors and windows are wide and clear, and the furniture inside the building are sparkling clean. Here one can either look into scenes in the distance or sip and taste tea in the cozy indoor space. The roof top is made of porcelain in herringbone shape in its section, and on top of the door it is a color ornament made of wood in the shape of a lion. It has a smiling gesture, intending to welcome VIPs. This is a relic from the Ming dynasty. The terrace in front of the hall used to be a place for people to enjoy the moon, opera and music. When the human figures are blended with that of the moon shadow among the ambient sound of water, everything seemed to be so wonderful and mesmerizing. Looking to the south against the hall, the water blends with the sky and there is a stone boat-shaped building at the opposite end. This building is one built in the Qing dynasty, which is divided into front and back two sections. The Juanpeng roof (Note: rolling roof) is quaint and stylish with a lifelike look. One cannot help but feel groovy and at ease in this boat-like architecture surrounded by the shimmering water rippling in light breeze. Boarding on the boat house from the side stone slate bridges one can see "Bu Xi Zhou" (Note: untethered boat) three characters on its horizontal inscribed board. This is an Emperor Qianlong's own hand writing done when he was in his southern tour. As such, Xu Yuan is also referred to as "Bu Xi Zhou" by posterity. One sees in the north Yilan Hall over water once on board of the boat house. Looking down by the railings close by, you can enjoy koi fish frolicking in the clear water around lotus flowers. This is indeed a fascinating scene.

Off the boat house to the east bank, there is a pavilion with a stele, or an upright stone tablet, on which the well-known poem "Feng Qiao Ye Bo" (Note: "Mooring by Maple Bridge at Night") written by the Tang dynasty poet Zhang Ji is inscribed:

Moon set, crows caw, frost fills the sky

River maples, fishing fires, drowsing in sorrow

Outside Gusu city, the Cold Mountain Temple

At the midnight bell, arrives the visitors' boat

(Note: The poem was translated by Yunte Huang)

The calligraphy was that of Yu Yue, a scholar in the Qing dynasty. It is said that this stone tablet was moved from Suzhou's Cold Mountain Temple. There is a Wanfei Ge built by the bank. The front water hall's roof style is Xieshan and the back part Juanpeng. Their height variations in the upper and lower buildings are naturally positioned. The building faces water on three sides and there are "Meirenkao" (Note: beauty recliners), or comfortable garden chairs for people to rest. There is a garden wall by itself connecting to a semi-surrounding corridor. Inside the yard, there are osmanthus, banana, and bamboo trees, and it forms its own space. Facing it on the opposite bank, there is "Xijia

艺、美术、书法、碑帖。最北一幢平房原为孙中山先生住处。最南一楹为"桐音馆"，介于南、北两叠山之间。四周有青桐等四时花木。馆前为南叠山。山路小径，曲折迂回，造型奇特，松柏相依，有六角亭屹立其中。亭分两层，下层山石掩映。近看亭似在幽深山壑，远观则亭耸山巅。亭角起翘，建筑艺术上称"嫩戗发戗"，各角下设云撑，出柱榫头呈鱼龙状，上配回纹挂落，为太平天国时期常用之建筑艺术。此山西麓有"方胜亭"一座，其亭系两方亭重叠，亭顶与底座似合非合，浑成一体。亭下有山石相衬，挺拔秀丽，形影相依，故亦称鸳鸯亭。馆后有北假山。山中有洞，洞洞相通，高低起伏，峰回路转，犹如迷

Lou" (Note: sunset building) and it has double eaves with Juanpeng Xieshan, and has upper and lower two stories, with a veranda on top and corridors at bottom. By the building there are ancient Gingko trees with scattered rockeries. At sunset time the sun light fills the room and that makes it a perfect place for reading poetry and making artistic paintings. West of the building against the wall there is a semi-square pavilion, in which some stone inscribed famous calligraphy works are displayed. By the pavilion there are steps leading to a veranda, which faces a color clustered dragon brick engraving done in Taiping Heavenly Kingdom's time.

The dragon wall forms a wavy motion with scale-like tiles on top simulating dragon scales. It feels like a flying dragon with vivid gesture with powerful force, effectively reflecting the plastic arts in Taiping Heavenly Kingdom's time. Turning west, we come at an office building used by the great revolutionary forerunner Sun Zhongshan. There displayed some historic relics used by Sun Zhongshan when he was the interim President of China.

Walking from here to the east, through tortuous corridors, we see various garden plants showing different colors depending on the season. This can be described as urban forest. Walking along the corridor to the north, one sees a five-unit building with an elegant style. Inside of the building there are handcrafts, art work, calligraphy, and rubbings from stone inscriptions. A northern most bungalow used to be Sun Zhongshan's residence. The most southern unit is "Tongyin Guan", located between the south and north rockeries. It is surrounded by Phoenix trees and other plants. South of it is the south rockery, through which a garden path winds with turns and changes. A hexagonal pavilion is erected among pine and cypress trees with a unique composition. The pavilion is a two-storied one and a rockery is partially integrated at its floor level. Looking from close distance you feel the pavilion is at a deep valley but seeing afar it feels positioned at the hill top. The raised corner of the pavilion is referred to as "Nenqiang Faqiang", a tenon or mortise treatment, in architecture. At each corner there is a raised support and the tip of the tenon is shaped like fish or dragon with some veins as decorations. This architectural detail was often used in Taiping Heavenly Kingdom's time. West of this rockery there is another pavilion called "Fangsheng Ting", and it is a double layered one with its top and base seemingly integrated into one body. Since it has rockeries at its base with an upright beautiful but inter-dependent

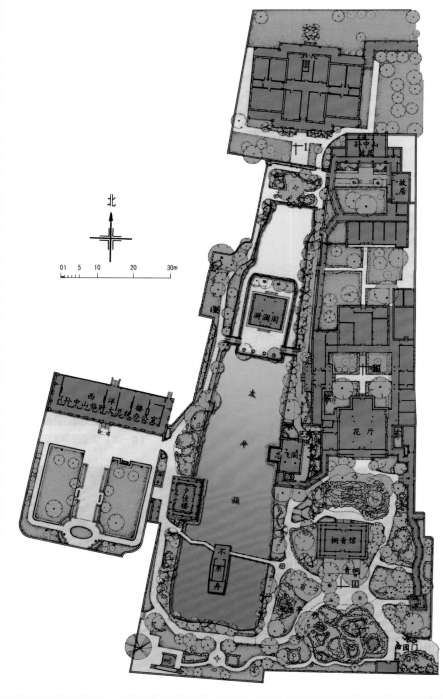

南京煦园平面图
Plan of Xu Yuan, Nanjing

宫，之后豁然开朗，可见一石碑，上有道光皇帝亲题"印心石屋"四个大字，周围刻有游龙、寿纹图案。出洞拾级而上，即至山岗平台，可见四周峰峦重叠，形态奇异，此处曾为赛诗胜地。山西有茅亭一座，亭中存有巨石，形似怪兽。

花园西南角有一方亭，上虚下实，下有印心石碑，故称印心石亭。此系太平天国时的建筑，是当年眺望深宫所在。登亭俯视，全园景色尽收眼底。

relationship, it is called "Yuanyang Ting" (Note: mandarin duck pavilion), showing its pairing relationship. There is a north rockery behind the building. There are caves in the rockery and they are all inter-connected with various turns and elevations, just like a three dimensional maze. After you are through the dark caves you suddenly see the light and it becomes widely open. There is a stone inscription with four large characters "Yin Xin Shi Wu" (Note: "seal center stone lodge"), which was hand written by Daoguang, a Qing dynasty emperor. Surrounding the inscriptions there are flowing dragon and longevity decorations etc. Out of the caves and walking up the stairs, one reaches a terrace on the rockery, where there is a surrounding view of mountains with various shapes and forms. This was once a place where poetry contests were held. West of the rockery there is a pavilion, in which a monster-like huge rock sits.

At the southwest corner of the garden there is a square shaped pavilion with empty upper part but solid lower part. Since it has a seal stele at the lower part it is called Yinxin stone pavilion. This is an architecture relic left in the Taiping Heavenly Kingdom times. Being at the pavilion and looking downward, one can have an overall view of the whole garden.

苏州环秀山庄 Huanxiu Shanzhuang in Suzhou

环秀山庄是苏州历史悠久的园林。始建于唐末，原为唐末吴越广陵王钱元璙的金谷园故址。宋代为文学家朱长文（朱伯原）药圃，后为景德寺，又改为学道书院，为兵巡道署。明代万历年间为大学士申时行住宅。清康熙初年，属吴门申勖庵药栏——"蘧园"，有来青阁闻名于苏城。清乾隆年间，为刑部蒋楫所有。蒋氏于厅室之东，盖"求自楼"五楹，以贮经籍。并于楼后叠石为小山，掘地三尺余，得古甃井，有清泉流出，合而为池，命名为"飞雪泉"，这就是环秀山庄山、池、泉、石的雏形。其后为尚书毕沅宅。不久又为相国孙士教所居。孙氏于清嘉庆十二年前后，请叠山大师、常州人戈裕良在书厅前叠假山一座，被称为"奇礓寿藤，奥如旷如"。清道光二十一年前后，孙宅入官，县令信笔批给汪氏。在汪小村、汪紫仙的倡议下，建汪氏宗祠，立"耕荫"义庄，重修东花园部分，改称"环秀山庄"，以为游憩之所。1860年太平天国庚申之役，有所毁伤。抗日战争前，沦为住宅，后来大部分园林建筑被拆卖，并在花园之西盖起小洋房，仅存山、池和"补秋山房"。"今海棠亭"为解放后从西百花巷程宅移建而来，体形别致，形式独特。现列入世界文化遗产。

环秀山庄占地约三亩左右。假山以南

Huanxiu Shanzhuang (Note: Mountain Villa with Secluded Beauty) has a long history among Suzhou gardens. It was built at the end of the Tang dynasty, originally as the site of Jingu Garden of Wuyue's Guangpo King Qian Yuanliao in the end of the Tang dynasty. It was also the medicinal garden of the Song dynasty writer Zhu Changwen (Zhu Boyuan) and was later changed into Jingde Temple, Xuedao College, and an army patrol station. In the Ming dynasty's Wanli period, it was scholar Shen Shihang's residence. In Kangxi's first year in the Qing dynasty, it belonged to Wumen's Shen Xuan as a medicinal garden called "Qu Yuan", and there was a Laiqing Ge made well-known in Suzhou. In Qianlong's time in the Qing dynasty, it belonged to a prison official Jiang Yi. East of the hall, Mr. Jiang built a five-unit "Qiu Zi Lou" to store ancient and classic books. He also piled up a rockery at its back and dug more than three Chis (Note: One Chi is about one foot long.) to have got the Guzhou well, from which spring water flowed out to converge to a pond, named "Feixue Quan" (Note: flying snow spring). This is the original form of Huanxiu Shanzhuang's mound, pond, spring and rock. After that it became Chancellor Bi Yuan's house, and not soon after that it was the residence of another official Sun Shijiao. At around Jiaqing 12th year in the Qing dynasty, Sun invited Ge Yuliang, a rockery master from Changzhou, to have built a rockery in front of the hall of the study and it was described as "rare rock intertwined by old vines with an air of mystery and wilderness". Around Daoguang 21st year in the Qing dynasty, Sun's residence was taken by the government and a county magistrate awarded the property to Mr. Wang. Under Wang Xiaocun and Wang Zixian's suggestions, Wang's ancestral hall was built, as well as "Geng Hu" cemetery. The eastern garden was also renovated and was renamed as "Huanxiu Shanzhuang", for recreational use. It was damaged partially in Taiping Heavenly Kingdom's Gengshen war

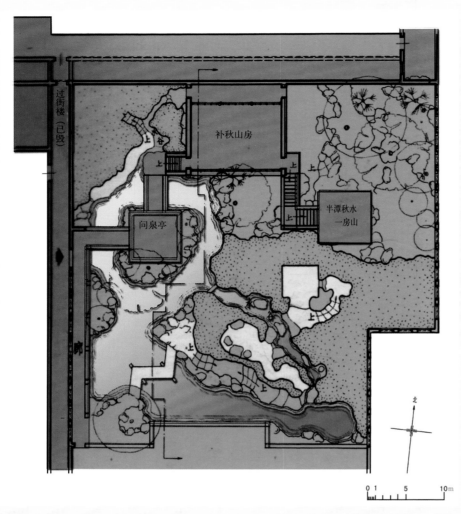

苏州环秀山庄平面图
Plan of Huanxiu Shanzhuang, Suzhou

为园林厅、堂、庭院。假山部分如真山，入画意。以一亩之地而岩峦耸翠，池水映天，气势磅礴，巧夺天工，可谓咫尺山水，小中见大，虚中有实，实中有虚，是乾隆以后叠山艺术趋于工巧的典型代表。

山池东面，砌高墙与外界隔断，如同绘画的边框，构成与周围环境互不相涉，得天独厚的空间。山池西部，为边楼和廊道，游人可以循廊欣赏山水，也可登楼俯视整个山池。西北部为精巧的石壁，北部有临水的"补秋山房"，东北部为"半潭秋水一房山亭"，是平岗小坂式的余脉。假山和房屋面积约占全园四分之三，水面约占四分之一，是一个以山为主，以水为辅的空间。山池的主山近其中，高出水面约七米，高出地面约六米，为南北向的山涧和东西向的山谷分为三部分，在山涧之上，用平板和石梁连接起来。《园冶》说："池上理山，园中第一胜也。"

从水池东南角看主山，峥嵘峭拔，气势宏大。水池西南有片山、短矶，作为游山起点，经三折板桥，沿主山岩脚旁起伏的磴道向前，有洞宛然，洞口在山涧旁。进洞道，经曲折转入正洞，构造颇为自然，

in 1860. Before the anti-Japanese war it was reduced to common residence. Thereafter, most of the garden architecture components were taken apart and sold and some western styled small dwellings were built in the west part of the garden, and only the rockery, the pond and "Buqiu Shanfang" (Note: Autumn Replenishing Mountain House) remained. After the Liberation, "Jin Haitang Ting" (Note: modern Begonia pavilion), was moved here from Cheng's house in West Baihua Lane. With unique form and characteristic style, it has been added into the World Cultural Heritage list.

Huanxiu Shanzhuang occupies about three Mus. Garden buildings and courtyard are located south of the rockery, which feels like a real mountain and very picturesque. With about only one Mu's area, it managed to create an environment that simulates real mountains with vegetation and water scenes with magnificent effect. This workmanship excels nature by using limited physical extent but offer unlimited natural sensation and by mixing the real with the surreal. Indeed, it represents typically the rockery art and workmanship in post Qianlong times.

East of the rock pond, a tall wall was built to separate the garden from the outside world, and it acts as a picture frame to define an uninterrupted and dedicated space. West of the rock pond are the side buildings and corridors for visitors to enjoy the garden landscape; they could also go up a story to overlook the whole rock pond. Northwest part of the garden is the exquisite rock cliff. There is a water front "Buqiu Shanfang" and in the northeast, "Bantan Qiushui Yifangshan", a pavilion at the extended foothill. The coverage of the rockeries and buildings occupy about three quarters of the whole garden, and the water body occupies about one quarter. So this is mainly a garden dominated by mountain and complemented by water scenes. The main body of the mount is located in the central area with an elevation of about seven meters above water and six meters above ground. It divides north-south oriented stream and east-west oriented valley into three parts. Over the stream stone slates and rock beams are linked together. According to "*Yuan Ye*", "The first attraction in a garden is to manage rockery over water."

Looking from the southeast corner of the pond towards the main mound, we could feel the towering and imposing air. Southwest of the pond there is a partial mound and a short jetty as a starting point to tour the mound. Along the zigzag-shaped slate bridge walking by the foothill's rocky path, you see a cave appearing by a mountain stream. Entering the cave path and meandering to the main cave,

无刀砍斧凿态。洞侧凹入地下，微见池水，设意奇特、空灵。出洞即幽涧，两边岩石壁立，与真山洞无少差。山涧中虽然无水，但是曲折幽深，涧中点缀散点峰石和汀步，有小洞凹入山内。由山涧透过石梁，看西北隅石壁，是一个很美的框景。越汀步进入山谷，谷左有石室可供稍事休息。循山谷向右，过石板山梁，即为主山顶峰。山道险峻，主山叠成岩峦状，动势朝向西南，自为环抱，成为整个山、池部分的构图中心。穿过峰顶峦头洞道，由湖石叠成的石梁架于山涧之上，左瞰池水，右临曲涧，一面开敞，一面闭合，意味全不相同。下主山，过石梁，一带子岗短阜，成为主山的龙脉。半潭秋水一房山亭建于岗阜之上，周围古木成荫。下亭便是补秋山房。山房面南临水，与池南书厅遥相对应。其西角隅叠石壁。此处假山占地甚微，却有洞、有壑、有涧谷、有悬崖，十分玲珑。构筑自成一体，而又和主山相呼应，有名的飞雪泉即在此处。石壁下即为"问泉亭"，由亭往西，登边楼可以俯瞰全景。

总观全局，环秀山庄的山、池布局委婉曲折，极尽变化。一开一合、一收一放、

you feel everything is natural without artificial trace. The side of the cave dips downward and lightly reveals the water. This is truly a unique and clever design. Stepping out of the cave one sees a quiet stream with steep cliff at its side and it is hard to tell the difference from a real one. Although there is no water in the stream, it is tortuous and deep, with scattered rock peaks, stepping stones, and small recessed cat holes. Looking at the rock wall in the northwest corner from the stream and through the stone girder one sees a beautiful picture frame. Leaping over the stepping stone and entering the valley, on the left there is a stone room for short break. Along the valley towards the right and over the slate girder, you could reach the peak of the main mound. The rocky path is steep and arduous. The main mound simulates a rocky mountain with a southwest moving momentum. It has a self-formed circle acting as the mound and water section's composition center. Passing the mound crest's cave path, you see a lake rock made stone girder spanning over the stream. It overlooks the pond on the left and faces the tortuous stream on the right. With one side open and the other closed, it creates a completely different feeling. Going down the main mound and passing the stone girder, one meets a stretch of hillocks and that forms the main mound's remaining range. Banfang Qiushui Yifangshan pavilion was built on top of a mound and it is surrounded by thick old shady trees. Out of the pavilion it is Buqiu Shanfang. The house faces water in its south and echoes the study hall that is located south of the pond. There is a rock wall or rockery at its northwest corner. Here the rockery's area coverage is rather small, yet there are caves, gullies, streams, and cliffs, forming a quite exquisite composition. The construction is self-contained yet at the same time works in concert with the main mound. The renowned Feixue Quan, or Flying Snow Spring, is located here. Under the rock wall it is "Wenquan Ting" (Note: Spring Examination/Interrogation Pavilion). Going west from the pavilion and ascending to the side building one can have an overlook of the whole view.

Looking at the general plan, we see the layout of the garden to be very tactfully complex and full of changes. It gets the rich rhythms of open vs. close, contraction vs. expansion, and emptiness vs. fullness. Although it has rich and diversified landscape scenes it is not complicated. Although it has a great number of scenes it does not have redundancies and duplications. This is in line with the principle to have a complete and unified layout that consists of a start, transition, change, and conclusion continuum. It got a well-knit composition that combines mounds, ponds, buildings and trees. Each is an integral part

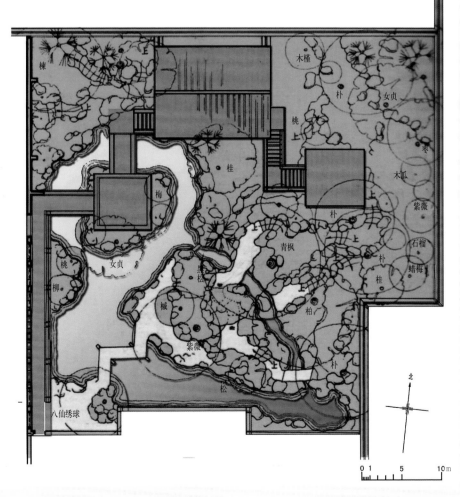

苏州环秀山庄屋顶平面图
Roof Plan of Huanxiu Shanzhuang, Suzhou

一虚一实，节奏性强。虽有多样变化的山水景象，却变而不繁，多而不复，符合起、承、变、结连续构图的原理，布局完整统一，结构极为谨严，把山、水池、建筑和树木组合成了不可分割的有机的整体，处处耐看；每个局部和每个组成部分都不散漫，经得住推敲。在苏州园林中独树一帜。

环秀山庄假山之所以能组成一个整体，是因为强调主景突出，主次分明的构图法则。叠山大师戈裕良灵活地运用了"宾主朝揖法"，使主山西北部石壁和东北部平岗短阜陪衬主山。又因山体脉络统一，自东北而来呈连绵之势，使主山不仅有峻拔高耸之感，且有奔腾跃动之势。"山得水则活"，山与水结合，是其获得成功的另一特点。主山直出水上，山高与宽比例恰当，增加了主山感染力；山根、岸脚，上突下空，叠成悬岸，使山、水互相渗透，紧密融合。并以东南部主山山涧和西北部石壁，集屋面雨水，形成两个人工瀑布，尤为匠心独运。

环秀山庄游览路线的处理，也有独特之处：时而跨水，时而登山，时而穿洞，时而越涧，有险有夷，有隐有现，通过导游路线的起、伏、转、折，满足了俯视、仰视、平视的要求，移步换景，在风景和画面上，形成了高远、深远、平远的三远之境。

of the inseparable whole and each is very engaging. You do not feel any slackness in any of these parts as each of them could stand up against close scrutiny. This is indeed a unique case among Suzhou gardens.

The reason why its rockery and mound are so well-integrated is that it stresses prioritization in composition by emphasizing the main scene dominating subordinate ones. The great rockery master Ge Yuliang applied with flexibility the approach of "guest and host", making the northwest rock wall and northeast hillock serve as foils to the main mound. It is also because the mountain range is consistent from the northeast continuation and this makes the mound not only appear elevated but also equipped with running and moving momentum. "A mountain is made alive with water." The integration of mountain and water is another trait of success. The main mound is right out of water, with its elevation and spread in proper proportion. This enhances its charm. It is protruding on top and receding at the bottom where the foothill and water meet. This overhung bank treatment makes the mountain and water blend well and integrated tightly. And using the southeast main mound's stream and northwest rock wall to collect rain water to form two artificial waterfalls is especially a brilliant idea.

The design of garden path is also unique. It sometimes steps across the water and other times climbs up the mound; sometimes passing through a cave and other times steps astride a stream. Sometimes it appears dangerous but other times safe; sometimes it is widely exposed but other times hidden. Through the garden paths' up and down orientation and zigzag turning, the need for view changes in looking up and down, or left and right, is satisfied. As one moves along, the scene changes accordingly. In terms of landscape scene and picture composition, it realizes the state of "three kinds of distances", that is, high and far, deep and far, and flat and far.

苏州怡园 Yiyuan of Suzhou

怡园位于苏州市中心人民路上，建造于清同治、光绪年间，距今一百余年，原为封建官僚顾文彬的私人花园。相传顾氏花二十万两白银，以七年时间才建成。园东部本为明代尚书吴宽的旧宅，西部为顾氏所扩建。园中布局，为顾氏之子——当时的名画家顾承规划，其好友任伯年等著名画家，亦曾参与设计。

在苏州古典园林中，怡园建造时间最晚，吸收了各园的长处，自成一格。园中复廊采取"沧浪亭"的手法，假山参照"环秀山庄"的章法，荷花池效法"网师园"，画舫斋则仿拙政园"香洲"。因此，园虽小而山池花木，亭榭廊舫，布局精巧，玲珑雅致。

怡园占地东西狭长，中间由南北向的复廊将全园划分东西两部分。东部以建筑庭院为主，配置峰石，点缀花木，景境幽

Yiyuan (Note: Joyful Garden), is located on People's Road in the center of Suzhou city. It was built in Tongzhi and Guangxu's reigns in the Qing dynasty over a hundred years ago. It used to be a feudal bureaucrat Gu Wenbin's private garden. Legend said that Gu spent two hundred thousand Liangs (Note: One Liang equals to 50 gram.) of silver and seven years to build it. The eastern part of the garden is the old residence of Wu Kuan, a chancellor of the Ming dynasty, and the western part is what Gu expanded. The layout of the garden is done by Gu's son, a well-known artist and painter Gu Cheng and his good friend and well-known artist Ren Bonian also participated in the design.

Of all Suzhou classic gardens, Yiyuan was built last; therefore it absorbed the strengths of all other gardens and formed its own style. The double corridor adopted that from "Cang Lang Ting" and its rockery referenced that in "Huanxiu Shanzhuang". Its lotus pond copied that of "Wangshi Yuan" and its boat house mimics the "Xiang Zhou" of Zhuozheng Yuan. As a result, although the size is small, all its mound, water, buildings, and plants are planned and designed exquisitely.

Yiyuan is elongated east-west wise and at its center, a double corridor is used to divide the whole garden

苏州怡园平面图
Plan of Yiyuan, Suzhou

雅。"玉延亭"、"四时潇洒亭",院内翠竹玉立潇洒;"坡仙琴馆"与"拜石轩"互为对景,异峰怪石巧置,松竹花木添景。

西部是全园重点,以山池为主景。环池点缀峰石、花木,配置建筑,参差自然,景色秀美。"藕香榭"临水北眺,山林葱茏,奇峰异洞,"小沧浪"高踞山岭,"螺髻亭"巧立洞上,"抱绿湾"幽深曲折,"画舫斋"建筑精雅,山池掩映,颇有山野风趣;"碧梧栖凤"青桐竹筠,环境清幽。

怡园湖石多,立峰、横峰、花坛驳岸,均系湖石佳品;花木多,白皮松、梅花、牡丹,皆是名卉嘉木,书法石刻多,廊壁上嵌满"怡园法帖",为历代名家的书法艺术珍品。

into east and west parts. The eastern part is mostly consisted of buildings and courtyards, with rocks and garden plants scattered within to create a very quiet and elegant atmosphere. "Yuyan Ting" and "Sishi Xiaosa Ting" are planted with green bamboos. "Poxian Qinguan" and "Baishi Xuan", or Rock Worshipping House, pose as each other's opposing scene, with rare and exotic rockeries and pine, bamboo and other garden plants.

The focus of the whole garden is in the west part, and the mound and water make the main scene. Around the pond there are scattered rocks and plants, coupled with buildings it looks natural with varying beautiful scenes. "Ouxiang Xie" (Note: Lotus Fragrance Building) is near water. Looking towards north, verdant mountains with fine peaks and caves meet the eye. "Xiao Canglang" perches high on the ridge, while "Luoji Ting" cleverly sits on top of a cave. "Baolù Wan" (Note: Verdant Embracing Bay) is deep and tortuous. "Huafang Zhai" (Note: Painting Boat House) has elegant architecture tightly integrated mound and water and it feels natural and wild. "Biwu Qifeng" enjoys a quiet atmosphere among Indus and bamboo trees.

There are a large number of lake rocks in Yi Yuan. Vertical, horizontal, and flowerbed bordered banks are all made up of fine lake rocks. There are much garden plants. bungeana, plum flower, and peonies are all precious species. There are also quite some calligraphy stone inscriptions. "Yi Yuan Fatie", or the calligraphy art treasures of past times, are all embedded along the corridor walls.

苏州怡园园景鸟瞰图
Bird's Eye View of Yiyuan, Suzhou

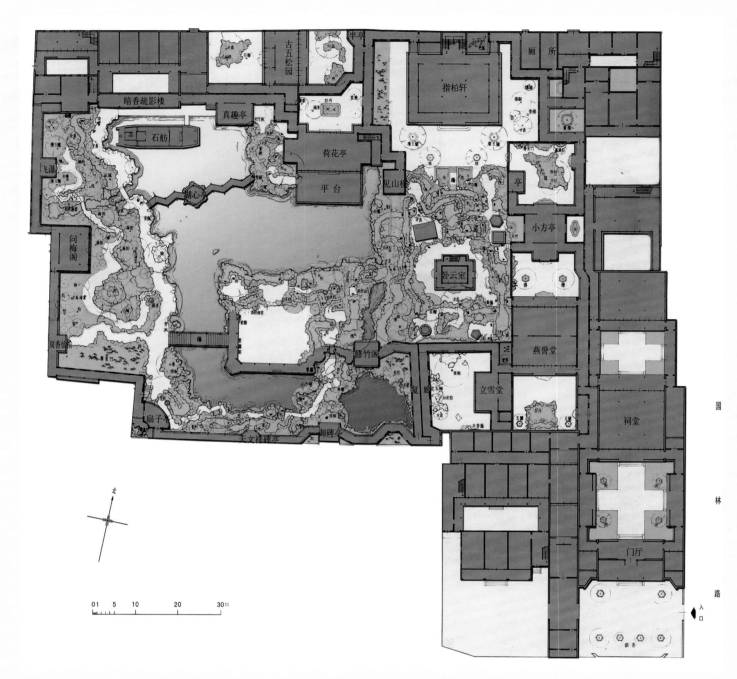

苏州狮子林平面图
Plan of Shizilin, Suzhou

苏州狮子林 Shizilin in Suzhou

狮子林原为寺庙园林，《吴县志》记载："狮子林在城东潘儒巷，元至正间，僧天如惟则延朱德润、赵善长、倪元镇、徐幼文共商叠成，而倪元镇为之图，取佛书狮子座名之，近人误以为倪云林所筑，非也。"

此园实为元代至元二年（公元1336年）僧天如禅师为其师中峰和尚而建，名为菩提正宗寺，亦称狮子林，园在寺的北侧，后改名为狮林寺。

此园当初范围较小，因佛陀说法称狮子吼，其座谓狮子座。故卧云室四周及其西部临池叠石，均模拟狮形，而卧云室旁狮形巨石，形态飞动，以示为佛陀说法的

Shizilin (Note: Lion's Grove) used to be a temple garden. According to Wu County Annals, "Shizilin is located at east city's Panru Lane. Monk Tianru Weize invited Zhu Derun, Zhao Yingchang, Ni Yuanzhen, Xu Youwen to have it co-designed and built. Ni Yuanzhen made the drawings and picked Leo from the Buddhist classics as its name. Nowadays people think that it was built by Ni Yunlin. Not true."

In fact this garden was built by the Zen master Tianru for his master monk Zhongfeng in the second year of Zhiyuan in the Yuan dynasty (1336). It was named as Authentic Bodhi Temple, also referred to as Lion's Grove. The garden is at the north side of the temple, and it was renamed later on as Shilin Si (Note: Lion Grove Temple).

At the beginning, this garden was pretty small, since according to Buddha it is a lion's roar and its base is supposed to be a lion's seat. Due to this reason, the area surrounding Woyun Room and in the west where the rocks close to the pond are all shaped after lions, and the lion-shaped boulder by Woyun Room show a flying gesture and momentum, representing the Buddha's statement;

场所，故有狮子林之称。虽有"含晖"、"吐月"、"玄玉"、"昂霄"等名峰，而以"狮子峰"为诸峰之首。其峰石之形象，均似狮子起舞之状，极有宗教艺术的价值。而座下的石洞，实为中国的叠石迷宫。所以此区叠石，并非假山，而系寓有宗教意识的狮子石林。称为叠石则可，称为假山则非。

清康熙、乾隆曾多次来游，并在长春园和避暑山庄内先后仿建。此园其后又有兴废，到1918～1926年，改为私园，东部建造宗祠。园内建筑，基本重建，有立雪堂、卧云室、指柏轩、修竹阁、荷花厅、真趣亭、暗香疏影楼、问梅阁等园林建筑。水池西面，掘池积土堆山，并叠石为瀑布涧谷，上设飞瀑亭，置人工瀑布，泻注湖中。

现在所见狮子林，其中石舫及某些建筑部分，掺糅不少西洋手法；园林空间，四周围墙及建筑封闭性太强，颇有局促之感；水池四周叠石，琐碎零乱。沈复曾在《浮生六记》中指为煤渣堆砌，不足取法。现今寺庙部分，久已废去，不复存在；而园林部分的造景，其后来增建部分，为文人园林的"暗香疏影"、"飞瀑流泉"具自然山水之美，与原先寺庙园林的含晖、吐月、昂霄等狮子吼动态和佛陀说法的宗教意识，共处于一个艺术构图之中，实难调和。以上种种，均为此园缺点。

另外，狮子林的漏窗，其纹样多为花鸟图案，颇具特色。西部及南部回廊壁间，嵌置《听雨楼帖》等书法石刻共六十七块，并有文天祥碑亭，颇得书法爱好者的赞赏。1997年狮子林作为苏州古典园林的杰出代表被列入世界文化遗产名录。

therefore the site is called Lion's Grove. Although there are several well-known rocks such as "Hanhui", "Tuyue", "Xuanyu", and "Angxiao", "Lion Peak" dominates them all. The images and shapes of the rocks are all mimicking dancing lions, and have high religious artistic value. The stone cave below is actually a Chinese versioned maze. Therefore the rock work in this area is actually not rockery in the traditional sense but a lion-shaped rock grove that has a religious connotation. We can refer to it as piled rock work but not a typical garden rockery.

In the Qing dynasty, emperors Kangxi and Qianlong had visited here numerous times and they later emulated this garden in Changchun Garden and the Summer Palace in Chengde. It had experienced ups and downs later on and was rebuilt as a private garden between 1918 and 1926. The eastern part was renovated as an ancestral hall. The garden buildings were all essentially rebuilt and those include Lixue Tang, Woyun Shi, Zhibai Xuan, Xiuzhu Ge, Hehua Ting, Zhenqu Ting, Anxiang Shuying Lou, Wenmei Ge etc. West of the pond, dirt was excavated to make mound and rock works were done to make waterfalls and streams. Above it Feipu Ting (Note: Flying Waterfall Pavilion) was built, and there is an artificial waterfall that spills into the pond below.

The Shizilin that we see today, especially the stone boat house part and some others, has quite some western architecture influence. One feels too closed and narrowness in its garden space, the surrounding walls and buildings. The rock work around the pond also appears to be in disorder. Shen Fu once wrote in his *Fusheng Liuji* that it was built with cinder and thus not worth modeling after. The temple part has now long gone and abandoned. The garden scenes and later expanded parts are literati garden's "Anxiang Shuying" and "Feipu Liuquan", both having beauty of natural landscape. However when they co-exist in the same artistic composition with the temple garden's Hanhui, Tuyue, Anxiao and other roaring lion features and other Bodhi context and religious sense, it is difficult to achieve harmony or compatibility. These are the weaknesses of the garden.

Moreover, the lattice windows use mostly flower and bird motifs and patterns, which is very unique. There are sixty seven embedded calligraphy stone inscription tablets showing works such as *Tingyu Loutie* and others in the double corridor in the west and south parts and there is a Wen Tianxiang stele pavilion; these are all favored by calligraphy enthusiasts. They are listed in the World Cultural Heritage Archive in 1997.

苏州鹤园 He Yuan in Suzhou

鹤园是清末民初文人雅集的园林，苏州市级文物保护单位。园近古城中心，在人民路西侧韩家巷4号，和曲园、听枫园南北为邻。宅园面积4.66亩，其中花园2.8亩。

清光绪三十三年（1907年），华阳洪鹭汀观察，以"宦囊"所得，筑宅韩家巷，在宅西建园。园原无额，取门厅屏门原缀七言篆联，上下联首字分别为"鹤"与"园"，而名为"鹤园"。园中池水清澈，修廊曲

He Yuan, or Crane Garden, was a place where literati met in the end of the Qing dynasty and the beginning of the Republic of China. It is a Suzhou city-level cultural relics protection site. The garden is close to the old city center and is located at No. 4 Hanjia Xiang at the west side of Renmin Road, adjacent to Qu Yuan and Tingfeng Yuan to the south and north respectively. The total area is 4.66 Mus and the garden part occupies 2.8 Mus.

In Guangxu 33rd year in the Qing dynasty (1907), Hong Luting from Huayang visited the site, and acquired the property as his government job benefit. The residence was built at Hanjia Lane and the garden was built

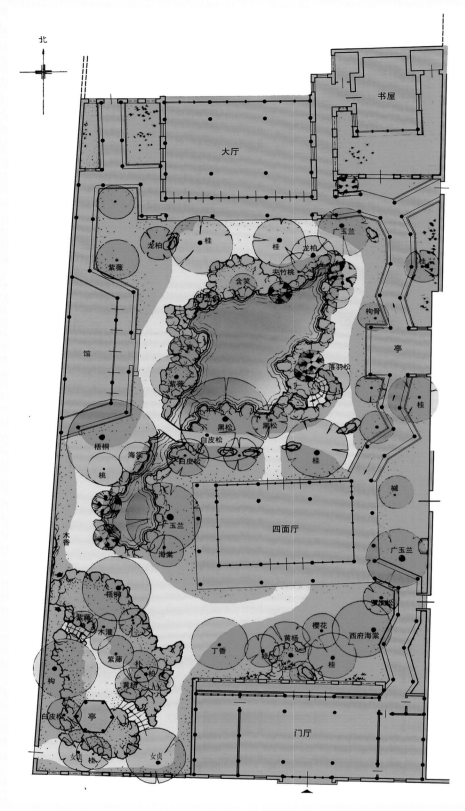

苏州鹤园平面图　Plan of Heyuan, Suzhou

折，风亭月馆，红栏小桥，地虽不足三亩，而具有林壑之美。但园宅工未竣，就一度归农务局。后归吴江庞屈庐，传其孙庞蘅裳（号鹤缘）。

民国12年（1923年）再度修建，有住宅五进。花园主厅为"携鹤草堂"，堂额为俞樾所题，雅朴劲道，后又名"栖鹤堂"。小斋"听秋山馆"在东廊尽处小庭中，植梧桐二棵，高可三寻，山馆故名"听秋"。掌云峰为全园最高立石，所镌"掌云"，为汪旭初题。四面厅后即为桂花厅。

by small buildings and pavilions helps enhance a feeling of natural mountain forest landscape on a small piece of land occupying only three Mus. However before the completion of the construction it became the property of the Farm Bureau. Later on, it was transferred to Pan Qulu of Wujiang, and then to his grandson Pan Hengshang (He Yuan).

In the 12th year of Republic of China (1923), it was built with five courts of residence. The main garden hall is entitled as "Xiehe Caotang", or Crane Carrying Thatched Hall, whose calligraphy on horizontal inscription tablet was written by Yu Yue. The characters are elegant, natural, yet powerful. It was later renamed as "Xihe Tang", or Crane Perching Hall. The small house "Tingqiu Shanguan", or Autumn Listening Mountain Hall, is at the end of the east corridor in a small yard, in which there are two Indus trees that can reach as high as three Xuns. So the mountain building was named as "Tingqiu", or Autumn Listening. Zhangyun Peak is the tallest rock erected in the whole garden. The inscribed characters "Zhang Yun", or Palm Cloud literally, was written by Wang Xu. Behind the Simian Ting, or Rectangular Hall, is the Osmanthus Hall. In the end of the Qing dynasty and beginning of Republic of China time, the famous literati Zhu Zumou (Ou Yin, or named as Gu Wei) once lived here. In Simian Ting's courtyard, he hand planted a lilac tree that was from Xuannan. The eight characters under the flower bed "Ou Yi Ci Ren Shou Zhi Ding Xiang" was written by Deng Bangshu on seal script style. The main garden house has two runs in depth. Its original inscription board "Ju Si Tang" was written by Lu Runyang after the Qing dynasty's Wang Shizhen's "Juyi Siming", meaning self-cultivation. Liao Nancai, President of Suzhou Agriculture College, rented the third unit Nǚ Ting, or Lady Hall. It was

west of it. Originally there is no title, or horizontally inscribed board, for the garden. Later on the first character of each part of the original couplets was taken and the two words are "He", meaning crane, and "Yuan", meaning garden. Therefore the garden's name became "He Yuan". In the garden, the pond is clear and clean and the long corridor winds zigzags through. Red railings with miniature bridges dotted

清末民初一代词宗朱祖谋（沤尹，又号古微）曾居园中，在四面厅庭中，手植引自宣南紫丁香一本，由邓邦述为篆题"沤尹词人手植丁香"八字于花坛下方。园宅二进为大厅，原额"居俟堂"为陆润庠所题，取意清·王士禛"居易俟命"。苏州农校校长廖楠才赁居住宅第三进，即女厅。后租予苏州银行公会。鹤园虽为家园，居住者都择雅士，第四进假三层为庞氏自用。因园近城区中心，以其小而雅静，故苏邑人士常假此为宴饮雅集之所，一时有"少长咸集，群贤毕至"之盛。每逢集会，随以唱吟。邓邦达、汪旭初、王佩诤、吴瞿安、张紫东及园主人庞蘅裳各显所长，声闻林池，有如兰亭雅集。叶恭绰、张善子、张大千、梅兰芳等各行家曾访鹤园。民国26年抗战开始，庞氏举家迁沪。新中国成立后，鹤园献予人民政府，始为政协办公之所。"文化大革命"中，政协机关为"三八战斗司令部"强占，园中匾额楹联皆遭毁坏。其后又为印刷厂、物资局、汽车配件等单位占用。1978年鹤园复归市政协。1980年，国家拨款十万元，由苏州古典园林建筑公司承担施工，全面修复鹤园，始复旧观。1982年10月，苏州市人民政府再度公布鹤园为市级文物保护单位。

鹤园布局简雅，园景以平坦、开朗为特色。园门南向，门厅五间，以粉墙花窗为屏障，以免一目了然。园北是主厅栖鹤草堂，堂前廊东西两侧，有砖额"岩扉"与"松径"，为园主自题，典出唐·孟浩然《夜归鹿门歌》，隐含栖隐之意。廊前以白瓷片作鹤形铺地。小庭梧桐生机盎然，听秋山馆修建如旧。水池居中央，西南置小石桥，环池叠湖石，立峰"掌云"题书清晰可读。配植迎春、含笑、海棠、木樨诸花木。池南四面厅，庭中紫丁香花香沁心，其后则为桂花厅。池东长廊曲折有致，联系三厅。东通住宅，廊间有静寄亭，廊与院墙构成几处小院，都植以花木、修竹、芭蕉，雅静多姿。池西小扇厅，和静寄厅相对，厅内摆设原都为扇形图案，故名。全园惟此无额无联。北廊连主厅，壁嵌金天翮所撰《鹤园记》，完好无缺。鹤园纵横平直，造景布局似有一览无余之嫌，但不失清幽秀丽。

later rented to Suzhou Bank Guild. Although He Yuan is a home garden, its residents always chose literati with good taste. The fourth unit false storey was used by Mr. Pang himself. Since the garden is close to the city center and it is small and quiet, local people often used this place as party and social venue. The get-togethers often attracted old and young and those with good virtues. Whenever there was a party, there were poem readings. Deng Bangda, Wang Xuchu, Wang Peizheng, Wu Quan, Zhang Zidong and the garden's owner Pang Hengshang each presented what they were good at and became well-known. These events can almost match the social literati gatherings at the ancient Orchard Pavilion. Ye Gongchuo, Zhang Shanzi, Zhang Daqian, Mei Lanfang and other celebrities once visited He Yuan. In the 26th year of Republic of China, the anti-Japanese war started, Pang and his family all moved to Shanghai. After the founding of the People's Republic of China, the garden was donated to the people's government, and it became an office of Chinese People's Political Consultative Committee (CPPCC). During the "Great Cultural Revolution", this CPPCC office became a "March 8 Fighting Headquarter", and the garden inscriptions and others were all destroyed. The place was later occupied by the printing house, the Commodities Bureau, and the auto parts factory. In 1978, the garden was returned to CPPCC. In 1980, the national government allocated a hundred thousand Yuans for its full restoration. The construction work was carried out by Suzhou Classic Garden Construction Company and the garden started to resume its old look. In October, 1982, Suzhou People's Government once again declared its city level cultural relics protection site.

Hu Yuan's layout is simple but elegant. The garden scene is mostly flat and open. The garden gate opens to the south, with five units. It uses painted wall and lattice window as a screen, to prevent from over exposure of garden scenes. The main hall in the northern part of the garden is Qihe Caotang, or Crane Perching Thatched Hall. In front of the building at the east and west sides, there are horizontal brick tablets "Yanfei" and "Songjing" - meaning rock door and pine path respectively. The source is from the Tang dynasty's Meng Haoran's "Yegui Lumen Ge", suggesting secluded living life style. At the front of the corridor white porcelain tiles in a shape of crane are used for pavement. The Indus trees in the courtyard are thriving and vibrant. Tingqiu Shanguan was restored just as the old one. The pond is in the center of the garden and there is a small stone bridge in the southwest. Lake rockeries are piled around the pond, and the inscription "Zhang Yun" on the upright rock is very clear and legible. Jasmine, Banana shrub, Begonia, and Osmanthus flowers are planted. South of the pond is Simian Ting, whose front yard has fragrant Lilacs flower and back yard Osmanthus. The long corridor east of the pond zigzags its way to connect three halls. To the east, it connects to residence buildings and in the middle, it is dotted by Jijing Pavilion. A few small yards are formed between the corridor and the garden wall, and garden plants such as bamboos and banana trees are planted, creating quiet and multi-various scenes. The small Fan Hall west of the pond faces Jijing Pavilion. Since its furniture are all fan-shaped, hence its name. This is the only building in the garden that has no inscriptions, horizontal or vertical. The north corridor connects to the main hall and it has an embedded wall inscription of Jin Tianyu's writing "He Yuan Annal" and it remains as intact. He Yuan is flat throughout and so its design seems to suffer from over exposure, although it is quiet and beautiful.

扬州小盘谷 Xiaopangu in Yangzhou

小盘谷亦为扬州名园。清光绪年间两江总督周馥购自徐氏旧园，重修而成。因园内假山峰危路险，苍岩探水，溪谷幽深，石径盘旋而得名。

园在住宅东部。自大厅东侧可入园门，门上嵌有清书画家陈鸿寿（曼生）书"小盘谷"石额。园分东西两部分。园内有花厅、假山、鱼池、水阁、游廊等建筑。此园虽无复道崇楼，占地不多，但园内布局紧凑，石径迂回，观赏游览线蕴藉含蓄，是一个登山可以畅怀，探洞可以寻幽的佳园。

小盘谷布局严密紧凑，因地制宜，随形造景。园中山、水、建筑和岸道安排，无不别具匠心。岩壁险峻，谷洞深曲，廊、厅随地形曲直，小径隐现通幽。既以强烈的对比手法，使园景变幻多端，景玄境妙，富有诗情画意；又做到对比协调，使小筑与崇山的大小，假山与鱼池的高低，静与动，幽深与开朗等等，对比谐和。虽为人工所筑，却宛如天然图画。

园内北部临池依墙的湖石假山，叠艺

Xiaopangu is also called Ming Yuan. In Guangxu's reign in the Qing dynasty, Zhou Fu, the governor of the Two Rivers Region, purchased the property from Mr. Xu's old garden and had it renovated. It became known for its dangerous rocks and paths, water dipping rocks, deep and quiet streams and gullies and winding rock paths.

The garden is located east of the residence. One enters the garden gate from east of the main hall, on which a stone inscription "Xiao Pan Gu" was written by calligrapher and art painter Chen Hongshou (Mansheng) of the Qing dynasty. The garden is divided into east and west two parts. Inside the garden there are flower hall, rockeries, fish pond, water hall, and promenade corridors and other garden buildings. Although this garden does not have complex path and tall buildings and it does not occupy large piece of land, it has a compact layout and enjoys roundabout rock paths. Its visiting routes are reserved and veiled and visitors can either climb up to a higher place to view a widely opened view or go down deeper to seek beautiful and serene scenes.

Xiaopangu's composition is tight and compact and its scenes are created according to the physical context. The arrangements of its mound, water, buildings and bank paths were all unique and carefully designed. The rock wall is steep and the gully cave is deep and tortuous. The corridors and halls' shapes vary according to the local terrain and small paths loom and lead to unseen and quiet places. Its modus operandi is sharp visual contrast through ever-changing garden scenes that are abstruse and picturesque; at the same

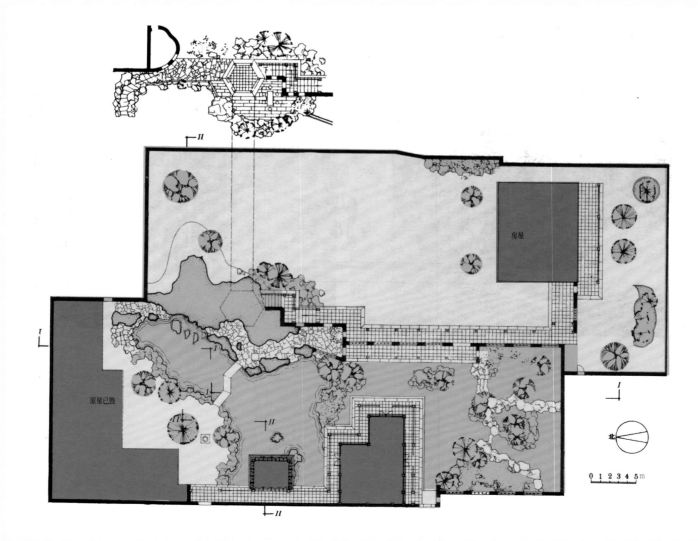

扬州小盘谷平面图　Plan of Xiaopangu, Yangzhou

高超，朴实自然，小中见大，气魄雄浑。其空间虽小而山峰峻峭，曲洞危崖，濒水有桥，远近成景。视其气度和布局，能叠成如此以假胜真的数丈高山，堪称假山中的精品之作。

小盘谷东、西园间以围墙相隔，有桃形门相通。园门上有石额"丛翠"，可惜今东园只残存回廊一道，余者皆废。

扬州小盘谷亦称小盘巢，旧有三处。一为南河下棣园的前身；一在旧城，堂子巷秦氏意园中；再者就是坐落在大树巷中的本园了。

time, it seeks to achieve a harmonious balance via contrast by placing a small building against sizable mountain, juxtaposing rockeries' and fish pond's different elevations, still vs. moving momentum, deep close vs. wide open etc. Although it is artificially built, it looks as if it is created by nature.

The lake rockeries against the wall in the north part of the garden show great level of artistry. It is natural, forceful and one can imagine big via its small extent. Although the spatial extent is limited, it has shrill peaks, tortuous caves, and steep cliffs. With a waterfront bridge as a nearby scene along with those in the distance, a beautiful scene is formed. From its magnanimity and overall composition with all these man-made high peaks that can beat the real ones, we can judge that it is among the elite of all the rockery design.

A wall is used to separate Xiaopangu's east and west gardens and they are connected through a peach-shaped door. On the door there is a stone tablet "Cong Cui" (Note: thicket green). Unfortunately, nowadays there is only a meandering corridor remaining in the east garden, with all the rest ruined.

Xiaopangu in Yangzhou is also referred to as "Xiaopanchao" and there were three locations. One is the precursor of Xiadi Yuan in Nanhe. One is located in the old city in Mr. Qin's Yi Yuan in Tangzi Lane. The last one is this garden located in Dashu (Note: Big Tree) Lane.

常州近园 Jin Yuan in Changzhou

近园，在今常州市长生巷内，为清进士杨青岩的家园。造园朴素洗练，得体合宜，"存自然之理，得自然之趣"，为江南园林代表作之一。

顾近园之名，已觉雅趣自生。杨青岩《近园记》云："买废地六、七亩，经营相度五年，于兹近似乎园，故题名近园。"其平正中出奇巧，无雕琢斧凿之痕，亦即在近似乎园。

近园山水花木，台榭亭廊，玲珑雅致，疏朗宜人，处处体现出"淡语皆有味，浅语皆有致"的匠意，清莹幽邃，能于"方寸之中，而瞻千里之遥"。笪重光曾称："其亭槛台阁，石峦花径，布置深曲，一一出诸指挥，可谓胸中具有丘壑矣！"园主杨青岩居官福建期间，曾遍游江、浙、闽一带名山大川，兴造时又得画家王石谷、恽南田，以及诗人笪重光等协助，遂将隙穴之地经营成城市山林。

近园的造景以结构紧凑，格局不凡，朴素大方，意境深远见胜。虽亦属山水园的布局，但安置妥帖，大胆地在中部的"鉴湖一曲"耸立黄石假山一座，按平岗小坂堆叠，高四点二米，借助古木荫翳，插入云霄，顿觉岩壑陡峭，气势磅礴。且有此

Jin Yuan (Note: Quasi Garden) is located in Changsheng Lane in today's Changzhou city, and it was the home of Yang Qingyan, a scholar in the Qing dynasty. The design is natural and succinct with appropriate design arrangement. "It keeps nature's law and thus has nature's fun." It is one of the representative garden design in Southern China.

Therefore by simply hearing its name Jin Yuan, one already feels its elegant taste. In his "Jin Yuan Annal", Yang Qingyan wrote, "I purchased a waste land of about six or seven Mus, managing and maintaining it for five years, and almost made it a garden. So I named it Jin Yuan." Its extra-ordinary out of ordinary approach and a garden art with little trace of artificial interference embody its meaning.

All landscape features, mound and water, flowers and trees, buildings and corridors, are all exquisite and elegant, clear and appropriate, and reflect the philosophy that "Even plain and simple words can carry deep and profound thoughts." Its clearness and depth enable one to "see thousands of miles from within inches of confinement". Dan Chongguang once said, "Its garden buildings, rockeries and flower paths were designed very sophisticatedly and tortuously, and it looked like they were all at his disposal at will. He must have a well-thought-out plan in his mind!" When the garden owner Yang Qingyan served his official post in Fujian, he toured all the famous mountains and rivers in Jiangsu, Zhejiang and Fujian. He also got assistance from art painters Wang Shigu, Yun Nantian, and poet Dan Chongguang, and these helped him to have managed to transform his small lot into an urban forest garden.

The structure of Jin Yuan is tight and extraordinary. It is simple yet munificent, and excels with its deep artistic conception. It is a layout of a

一隔，一泓池水则有了大小、方长、聚散的变化，呈现出深潭、溪流的气氛，迂回曲折的情趣，恰如文震亨《长物志》所谓"一勺则江湖万里"。虚舟一叶坐东面山，又给水系平添了无限生气。

近园假山自成丘壑，概括自然山川，用远近疏密、似断似续的手法，使"山体百里之回"，悬崖峭壁临水，颇具突兀之势。峰下筑垂纶洞，增山势之深邃。峰侧构山磴，叠蹬阶，垂藤萝，植芳草，显峰色自

mound-water garden with appropriate arrangement. It boldly placed a yellow stone rockery in the middle section "Jianhu Yiqu". It was built up as a small hillock. With a height of 4.2 meters and with the aid of heavy shadow from ancient trees, it feels tall, steep, and with a commanding air, and with it as a separation, the water pond then has a change in its size, shape, and degrees of openness. It has an air of deep pond and flowing stream with winding stream path, just as what Wen Zhenheng said in his "Chang Wu Zhi", "a spoonful of water can represent thousands of miles of rivers and lakes". A faux boat is at the east mound, adding some lively atmosphere to the water scene.

The rockery in the garden forms its own mountain system that generalizes

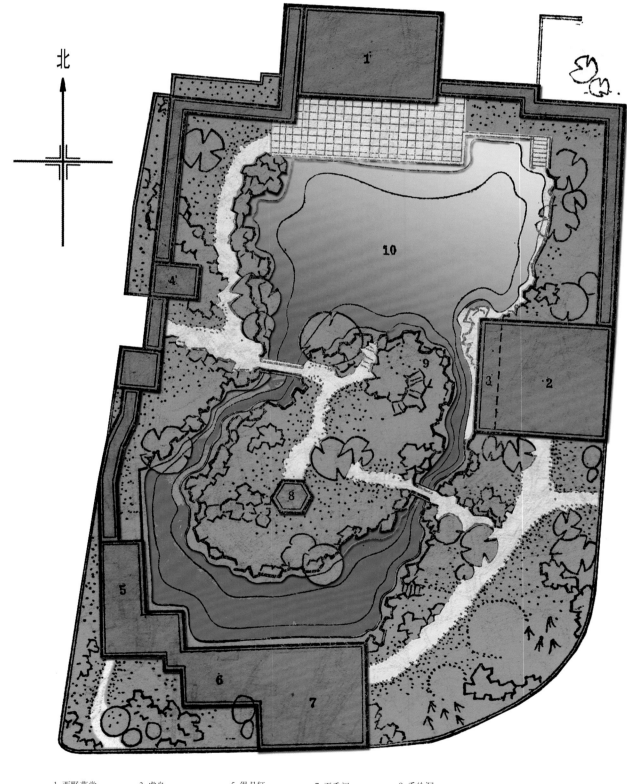

| 1 西野草堂 | 3 虚舟 | 5 得月轩 | 7 天香阁 | 9 垂纶洞 |
| 2 容膝居 | 4 秋水亭 | 6 安乐窝 | 8 见一亭 | 10 鉴湖一曲 |

常州近园平面图　Plan of Jin Yuan, Changzhou

然。南部循主山脉络理石，使山体有连绵不绝的气韵，从各点透视，都有群山环抱，林壑幽美之感。"见一亭"居水崖之上，踞势幽胜，可享"近水远山皆有情"的意趣，如画家把千里山河浓缩在数尺见方的纸上，耐人寻味。

近园的亭台堂阁多环山绕水，特别注意居山临水之势，参差错落，比例适度。北面傍水为"西野草堂"，系宴客之地。西南有"天香阁"、"安乐窝"，临池"得月轩"，似取苏轼的"惟江上清风，山间明月，耳得之而为声，目遇之而成色"之意命名，波光潋滟，锦鳞游泳，为吟诵垂钓佳处。向北是"秋水亭"，向东过"虚舟"入"容膝居"。为体现山情，以起伏的回廊（又名爬山廊）匝绕，游者步履盘桓，似入山庄。综观七亩小园，却有如盆景，小中见大，境妙而意深，表达了园主人自喻良材，寄情山水，借景述怀的思想感情。

natural landscapes. It applies different degrees of density with various distances, some mountain ranges are connected and others not. The cliff overhang dangles over water, appearing to be abrupt and pressing. Beneath the peak there is a Chuilun Cave enhancing the depths of the mountain. At the peak side a gully was built with climbing steps. With hanging vines and planted vegetation the peak appears to be natural. In the south along the main mound range, the rock work lets you feel that it is continuous. Looking from any view point, you can have the feeling of being embraced by the surrounding mountains with verdant quietness. "Jianyi Pavilion" is on top of a waterfront cliff. From this vantage point, one could enjoy the fun of "feeling the emotions of nearby water as well as faraway mountains", just as the intriguing art of landscape painters to condense and distill thousands of miles of landscape onto a few feet of paper.

At Jin Yuan, the landscape buildings are mostly close to water. With special attention to the adjacency of mountain and water, it has got proper profiles with appropriate scale. The waterfront "Xiye Caotang" in the north is a banquet venue. In the southwest, there are "Tianxiang Ge", "Anle Wo", waterfront "Deyue Xuan", simulating Su Shi's writing: "When the clear breeze over the river meets my ear it becomes pleasant sound, and when bright moon in the mountain meets my eyes it morphs into an enjoyable view." With the shimmering water and joyful koi fish, this is a perfect place to chant and go fishing. To the north is "Qiushui Ting" (Note: Autumn Water Pavilion) and to the east passing "Xu Zhou" (Note: Virtual Boat) is "Rongxi Ju" (Note: Tiny Lodge). To enhance the momentum of the mountain, corridors in wavy up and down motion were used – they are also referred to as mountain climbing corridor – and so when visitors spiral in, it feels like a mountain villa. Looking at this seven Mu small garden, you feel it is a bonsai-like landscape and could imagine large scenes out of small. A subtle and indigenous scene can have deep meanings. This reflects the garden owner's high evaluation of himself and his way to express his thoughts and feelings via landscape.

苏州艺圃 Yipu in Suzhou

艺圃始建于明代，园位于苏州城西北阊门内文衙弄5号，今入园处为十间廊屋10号。宅园面积5.7亩，其中花园占4亩，已列为世界文化遗产。

明嘉靖年间，初为袁祖庚之醉颖堂。崇祯间为文震孟（文徵明之曾孙）所有，宅有世伦堂。园名药圃，有青瑶屿，垂桥修竹，方亭崇阜。明亡，园一度废为马厩。清初莱阳姜毅侨寓吴门，辟废园为别业，改名敬亭山房。从此抱膝读书于山房中，不与世事30年。长子姜安节筑室思嗜轩，仲子姜实节辟为艺圃。弟兄皆读书好士，时园最为兴盛，有广池弥漫，奇花珍卉，幽泉怪石，轩馆亭廊。园有念祖堂、红鹅馆、

Yipu (Note: Art Garden) was first built in the Ming dynasty. It is located at 5 Wenya Lane inside Beichangmen in the northwest of the Suzhou city. At present the entrance is 10 Shijian Langwu. The residence size is 5.7 Mus, with the garden occupying 4 Mus. It has been listed in the World Cultural Heritage archive.

It was Zuiying Tang that was owned by Yuan Zugeng in Jiajing's reign in the Ming dynasty. In Chongzhen's reign it belonged to Wen Zhenmeng (great grandson of Wen Zhengming) and had Shilun Tang in its residence. The garden name was Yaopu (Note: Medicine Garden) and it had Qingyao Islet, bamboos, bridges, pavilions and a high mound. After the ending of the Ming dynasty, it was once reduced to a stable. At the beginning of the Qing dynasty, Jiang Zhenyi from Laiyang once lived here and renovated this derelict garden into a villa and renamed it as Jingting Shanfang. He had since started his reclusive scholar's life and severed ties

博饪斋、爱莲窝、香草居、绣佛阁、谏草楼、响月廊、乳鱼亭、南村、鹤柴、垂云峰、度香桥诸胜。四方骚人墨客络绎来园，为艺圃胜景题咏绘图。康熙年间，王翚（石谷）绘《艺圃图》，池北为平台。西为敬亭山房（今已不存）。道光间，艺圃归绸业，名七襄公所，重加修葺，规模遂小。筑博雅堂、听幽室、双声室。亭台池榭，参差有致，别饶幽趣。池植名荷湘莲，粉红重台，花色娇艳。之后公所房屋，租为住宅，园中亭阁，失修圮塌。

新中国成立后，艺圃先后为苏昆剧团、桃花坞木刻年画社及民间工艺厂等单位使用。住宅散为民居，住居民十余户。1963年，艺圃公布为市级文物保护单位。"文化大革命"期间，艺圃迭遭破坏，损失惨重。假山毁坏，湖石峰竟为烧制石灰。水池填塞，榭阁倒塌，全园荒芜。1979年苏州市编制城市总体规划，艺圃列为古典园林修复规划项目。

1982年，市人民政府决定修复艺圃，投资50万元。园内山池布局与明末清初旧况大致相同，遵循文物修缮"修旧如旧"之原则，使之复原。为此园林设计室设计人员，广泛搜集资料，精心测绘设计。按图制作修复模型（1/100），经专家会审，报市文管会审定。9月古建公司负责修复工程，能工巧匠悉心施工。主厅博雅堂落地翻修，换除部分腐蚀之柱、梁、桁、椽，保持原梁架结构、青石阶沿、覆盆木鼓磴之明代建筑特色。水榭已坍塌二间，采用补换石柱，加筑水下地梁、砖细台口。五间水榭全部落架修复。乳鱼亭于原位竖架整修，以保护明式彩绘。清挖被填水池，池底增挖水井一口，以利池水洁净，恢复湖石池岸。从池中挖得原石梁，修复乳鱼亭石拱桥，并修复三折之度香桥。按原貌堆叠湖石假山、石壁山径（所用湖石采购于安徽巢湖），重建对照厅、响月廊、思嗜轩，新建朝阳亭及管理用房与花房。全园补植大量花木，恢复全部家具陈设、匾额对联与书画布置。历时两年，艺圃花园修复工程竣工验收，于1984年国庆节开放游览。

艺圃水面集中，池岸低平，布局简雅开朗，风格自然质朴，存有明代园林规制，具有一定历史与艺术价值。水池居园中心，集聚坦荡。池南延伸水湾，蜿蜒曲折，上架小桥，断续分隔成五六小池。缘池南岸，堆土叠山，湖石嶙峋，石壁峭拔，洞壑宛转，

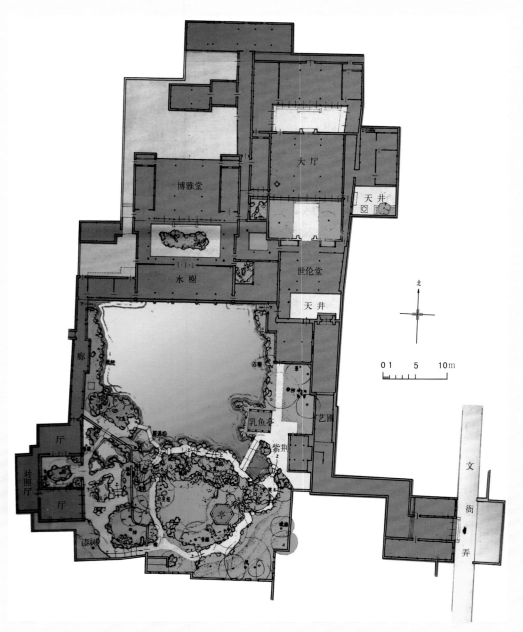

苏州艺圃平面图　Plan of Yipu, Suzhou

from the outside world for thirty years. His oldest son Jiang Anjie built Sishi Xuan and his second son Jiang Shijie created Yipu. Both brothers enjoy reading and treasure talents and this was the heyday of the garden. There were ponds, rare and beautiful flowers, exotic rocks and tranquil springs, and garden buildings of all kinds. There were numerous attractions in the garden including Nianzu Tang, Honge Guan, Ren Zhai, Ailian Wo, Xiangcao Ju, Xiufo Ge, Jiancao Lou, Xiangyue Lang, Ruyu Ting, Nan Cun, He Chai, Chuiyun Feng, Duxiang Qiao etc. Literati from everywhere came to visit and contributed their poems and paintings to the garden. During Kangxi's reign, Wang Hui (Shigu) painted "Yipu Tu" (Note: picture of Yipu), in which there was a terrace north of the pond and Jingting Shanfang in the west (non-existent as of today). During Daoguang's reign, Yipu belonged to the silk industry and was renamed as Qixiang Gongsuo after renovation and reduction in size. There were Zhubo Yatang, Tingyou Shi, and Shuangsheng Shi. Interests arose because of various garden buildings with different designs and positions. Lotus was planted in the pond and the pink flowers served as great background of the terrace. After

山径盘旋，垂云峰屹立池畔。山巅树木葱茏，白皮松虬干苍翠。山林中旧有朝爽台，今为朝阳亭，居高俯览园景。池北水榭五间，挑临水面，为延光阁，砖细台口，和合明窗，一览山池全景，为品茗赏景佳处。西厢小屋思敬居，小院有翠竹花坛，植有枇杷。院北置接待室，其境清静幽雅。东厢为旸谷书堂，天井小院，巧置峰石。水阁之北为主厅博雅堂，面阔五间，三明两暗。明式构制。堂内白缮墙门，陈设雅致。堂前庭院，牡丹湖石花坛，周绕回廊。池之东岸花木扶疏，乳鱼亭三面临水，梁架有明式彩绘，古朴雅致，为全园唯一之明代建筑，苏州园林之珍品。亭之东沿墙有思嗜轩。旧时园主曾植枣数株，以思其亲。稍北为书室爱莲窝。池西相对者为响月廊，长廊贴墙沿池，中置半亭，留有隙地。廊前水波涟漪，亭后翠竹映窗。廊接西南之园中园，高墙相围，洞门相对。园中浴鸥小池，曲桥花石，古木参天，幽然成趣。香草居与南斋组成对照小厅，中有山石花坛小庭，精雅小巧。现暂辟为入园处，艺圃世伦堂三进住宅位于园之东北。门厅三间，东临文衙弄，照壁相对。门厅与住宅间有曲折小弄相通，旧时由住宅入园，今为居民所用。

that, the public buildings were rented as residence and the garden buildings went into disrepair and collapsed.

After the founding of the People's Republic of China, Yipu first was used as the base of Suzhou Kunqu Opera Troupe and then one for Peach Woodcut Folk Art Association and local arts and crafts factory etc. The residence buildings were used by about ten families. In 1963, Yipu was declared as a city-level cultural relic protection site. During the "Great Cultural Revolution", Yipu suffered heavy loss as it was damaged several times. The rockery was destroyed and the lake rock peak was even burnt to make quicklime. The water pond was stuffed, buildings collapsed, and all garden went derelict. In 1979, the city of Suzhou worked on its urban master plan, making Yipu as one of the classic gardens restoration project.

（三）村庄地

古之乐田园者，居于畎亩之中；今耽丘壑者，选村庄之胜，团团篱落，处处桑麻；凿水为濠，挑堤种柳；门楼知稼，廊庑连芸。约十亩之基，须开池者三，曲折有情，疏源正可；余七分之地，为垒土者四，高卑无论，栽竹相宜。堂虚绿野犹开，花隐重门若掩。掇石莫知山假，到桥若谓津通。桃李成蹊，楼台入画。围墙编棘，窦留山犬迎人；曲径绕篱，苔破家童扫叶。秋老蜂房未割，西成鹤廪先支。安闲莫管稻粱谋，沽酒不辞风雪路。归林得意，老圃有余。

(3) Village Land

In ancient times, those who love rural scenery actually lived in the wild country; however today those who love nature choose to live in a village with natural beauty. Use tree branches to make fences and plant mulberries and hemps everywhere. Dig ditches to channel water and hill up to make embankment to plant willow and poplar trees. Standing on a gatehouse to enjoy the crops and know about the season, and the buildings are all connected with my vegetable gardens. If the total area is ten Mus then the pond should occupy three tenth. It is convenient for dredging and at the same time satisfies the needs for naturally formed shapes. The rest seven tenth land areas can be used for hilling up to make a mound. It does not matter how tall the mound is and it is proper to plant bamboos. The hall should be open and face the broad green pasture, and flowers and plants can be employed to act as a shield or separation. To make a mound, it is important to make it like a real one. Use a bridge for crossing and at the end of a road there should be a ferry dock. Paths can form under peach and plum trees and all garden buildings can be made picturesque. One can cut a hole in a bramble fence to let his dog out to greet people. Winding paths going around the fence; family children mess up with the moss while sweeping falling leaves. Although bee honey has not been harvested in deep autumn, mature crops and saved ones can be used first. We are at ease and do not need to worry about getting enough to live comfortably. I buy liquor to entertain myself and do not fear cold wind and snow. My secluded life in the country is satisfied and my garden life is more than self-sufficient.

扬州白塔晴云 Baita Qingyun of Yangzhou

扬州剪纸艺术馆系爱国侨胞陈伸先生及其夫人和子女捐款，于1983年设计并建造完成。获1986年江苏省优秀设计一等奖。

扬州剪纸艺术馆系在瘦西湖风景名胜区"白塔晴云"景点基址上建造的，依据风景名胜区的建设宗旨，首先要恢复白塔晴云景点的历史风貌，再现史籍上记载的主要精华。同时亦考虑投资者的要求：陈列剪纸的艺术展览馆。

作为风景名胜区内的一个景点，设计成败的关键在于"因借"。稽考史料、参照古图将景点设计为面湖而筑。选择性的恢复白塔晴云景点内"积翠轩"、"林香榭"、"花南水北之堂"三个主要建筑，并按照别墅园林的格局来经营，组成两条面湖的轴线。林香榭与花南水北之堂为全园的主景，榭、堂之间围以回廊贯通，中为天井，堂北为花木庭院，林香榭前伸以平台，围以石栏临水，由于位置安排得当，在此可将瘦西湖有代表性的建筑五亭桥、白塔借景入围，拓展了景点的景深。林香榭与花南水北之堂组成长方形的格局，为使其不呆板，除建筑本身造型变化外，在堂南设小憩平台，点缀石凳，榭后丛置山石爬以附藤，产生空间的变化。特别在围廊东将一间廊改为歇山顶的门亭，既强调入口，又得到不均衡对称的构图。由于主体建筑结体的方整，故在相邻空间积翠轩为主的庭院尽量活泼，空间组合力求丰富。积翠轩北墙向东与内围墙相连，将整个空间分隔为前庭后院，前庭因其广而点景疏朗，后院因其窄而布置愈加稠密，产生疏密的强烈对比。与相邻空间的门亭呼应的是揽尖四方半亭，半亭和积翠轩之间以曲廊相连，在相邻围墙前留出长带状绿地，修竹石笋拔地而起，这样不仅使前庭空间中又孕有小空间，修篁的摇曳又隐蔽了主体空间的建筑立面，加深了景深。两个相邻空间入口处理的门亭与半亭除建筑系不同结构形式外，还在半亭"半青"南侧围墙下开一水门，引入瘦西湖水一泓进入亭底，亭南配以相应的美女靠临水，利用亭侧点石形成"坐雨观瀑"的景观，顿使庭院景色丰富，寓意深远。

白塔晴云作为剪纸艺术馆展览馆亦是

Yangzhou's paper-cut art museum's construction fund was donated by patriotic compatriots Chen Shen, his wife and his children. It was completed in 1983 and had won first place in Outstanding Design in Jiangsu Province in 1986.

The museum was built at the site of "Baita Qingyun", an attraction in Shouxi Lake scenery zone. According to the mission of the scenic area, the historical style of Baita Qingyun ought to be restored first to reproduce the essence recorded in history. In the meantime, the donor's request for the paper-cut art museum should also be considered.

As an attraction in the scenery zone, the key success factor mainly lies in "leveraging". After much research on historical records and drawings, the attraction was designed as facing the lake. Three main buildings, "Jicui Xuan", "Linxiang Xie", and "Huanan Shuibei Zhitang", were selectively restored. Also, based on the villa garden's general layout, two lake-facing axes were formed. Linxiang Xie and Huanan Shuibei Zhitang are main scenes of the whole garden. Between them is a terraced corridor for connection. In the middle there is a patio. North of the hall there is a landscaped yard. Linxiang Xie protrudes into a terrace and is surrounded by waterfront stone railings. Because of its proper positioning, from here, the representative scenes in Shouxi Lake's Five Pavilion Bridge and the white pagoda can be viewed inside the garden, extending the vista of the attraction. Linxiang Xie and Huanan Shuibei Zhitang form a rectangular shape. In order to break its rigidness, besides the changing appearance of the building proper, a rest area terrace was built south of the hall. Dotting stone benches and clustered rocks with vines create variations in spatial feeling. Especially at the east of bordering corridor, a one unit corridor's renovation using a Xieshan roof on the one hand stresses the entrance and on the other helps achieved asymmetrical balance in composition. Because of the main building's square and tight shape, the adjacent space Jicui Xuan's yard is made more vivacious, to achieve rich spatial complementary effect. The north wall of Jicui Xuan veers east and connects to inner border wall, splitting the whole garden space into front court and backyard. The broadness of the front court makes its scenes sparse while the narrowness of the backyard dense, creating a sparse vs. dense sharp contrast. Echoing the entrance pavilion of the adjacent space is a Zuanjian four-sided semi pavilion, connected by a curved corridor to Jicui Xuan. A long stripe of lawn is allotted between adjacent walls and slender bamboos and rocks shoot up from the ground. This way a sub-space is spawn in the front court area and waving bamboos also partially screen the main façade in the main space, adding depths of field. In addition to the different treatments of architecture style with the gate pavilion and half pavilion at the entrance area at the two adjacent spaces, a water gate was added at the south wall at the half pavilion "Banqing". It channels in water from Shouxi Lake to the bottom of the pavilion. At the south of the pavilion some water front curved chairs were employed, and through pavilion side scattered rocks formed a scene "Zuoyu Guanpu" (Note: sitting in the rain watching a waterfall), adding the richness of the court scene with deep meaning.

It is very appropriate to use Baita Qingyun as the Paper-cut Art

十分适宜，三个不同体量造型的建筑可分别为序馆、主展馆、配展馆，回廊与曲廊的连接使参观路线明确，得到投资者高度评价。

著名古建筑专家潘谷西教授实地考察白塔晴云景点亦倍加赞赏，认为总体上是符合瘦西湖风景名胜区的特点，在具体处理上认为："可见设计者对古建筑的设计已到得心应手的地步"。

Museum. Three different sized and styled buildings can be used as the introduction hall, the main exhibit hall and the subsidiary hall. The connection between the two corridors makes the visiting route clear and apparent, and had won high acclaim from the project donors.

When the renowned ancient Chinese architecture expert Professor Pan Guxi visited Baita Qingyun he complimented it immensely, praising the planning fits the characteristics of Shouxi Lake in general and for the concrete design work he extolled, "From this design I can see that the designer is very proficient in ancient Chinese architecture."

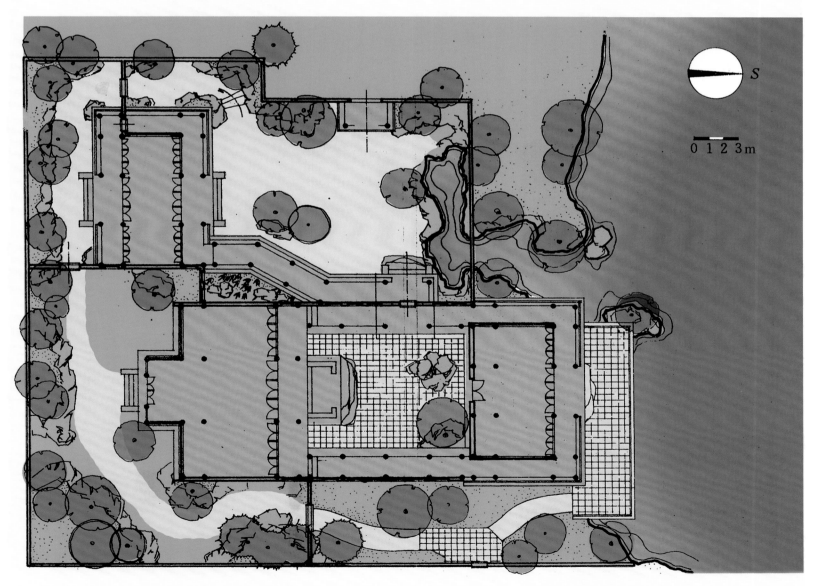

扬州白塔晴云平面图　　Plan of Baita Qingyun, Yangzhou

（四）郊野地

郊野择地，依乎平冈曲坞，叠陇乔林，水浚通源，桥横跨水，去城不数里，而往来可以任意，若为快也。谅地势之崎岖，得基局之大小；围知版筑，构拟习池。开荒欲引长流，摘景全留杂树。搜根带水，理顽石而堪支；引蔓通津，缘飞梁而可度。风生寒峭，溪湾柳间栽桃；月隐清微，屋绕梅余种竹。似多幽趣，更入深情。两三间曲尽春藏，一二处堪为暑避。隔林鸠唤雨，断岸马嘶风。花落呼童，竹深留客。任看主人何必问，还要姓字不须题。须陈风月清音，休犯山林罪过。韵人安亵，俗笔偏涂。

(4) Suburban Land

To build a garden in suburban area, one must follow the natural terrain, such as gentle sloped hill, winding col, undulating hill, and thick forest. A water channel must be un-obstructive and has a source, and over a river there must be a bridge. With only a few Lis (Note: One Li is a Chinese measurement that is half a kilometer in length.) from a city, what a joy to have to be able to travel back and forth whenever one wants? For planning, one must observe how flat a site is to get an idea on its size. A good way to make walls is by using rammed earth between two planks and to build pond, learn from Xijia Pond. To prepare an unused land one must first decide how to diverge water flow; to create sceneries one must preserve assorted trees. The base of a rockery is prone to erosion so it must sit on solid rocks for support. Over a water channel leading to a pond a bridge needs to be built across it. In early spring when it is still cold one may plant some peach trees among willow groves at a stream turn. When it is at night and the sky is clear one may plant a few bamboos among plum trees. Red peach among green willow creates quiet and elegant interest; scattered shadow cast on garden features makes it more picturesque. A couple studios and ateliers bring warmth when it's cold while a few pavilions by water can help dispel summer heat. One can enjoy birds' chirping through forest and hear horses neighing at the cliff. All that falling leaves call for servants cleaning; deep and quiet bamboo forest keeps guests wandering. There is no need to ask for permission for guests to enjoy the scene and no need to report their name to the owner either. To build a garden in suburban land, its pristine natural flavor must be preserved. Never commit a crime to dilapidate natural landscape. People of good taste would never profane natural landscape; only vulgar ones would impose their graffiti in the wild.

苏州拙政园 Zhuozheng Yuan of Suzhou

苏州拙政园，始建于明代正德八年（1513年）前后。御史王献臣，因与权贵不合，弃官还乡，退居林下，将愤世嫉俗之心寄托于"娄齐门之间"的一处"隙地"。由于当时这里"有积水亘其中"，故"稍加浚治，环以林木"，并寓晋代潘岳《闲居赋》"灌园寓鬻蔬，是亦拙者之为政也"之意，取名拙政园。其时园以沧浪池为主，园中建筑有"若墅堂"、"梦隐楼"等，"凡诸亭槛台榭，皆因水为面势"。植物配置，则"夹岸皆佳木"，多成片栽植，形成好几个景区。从文徵明所著《王氏拙政园记》可以看出：明代拙政园从"逍遥自得，享闲居之乐"出发，淡泊自然是其特色。

以后，几经变迁，换了好几个园主，明崇祯四年（1631年），园东部荒废，为侍郎王心一购得，于崇祯八年建"归田园"。清乾隆三年（1738年）前后，又归太守蒋诵先，在修葺中"因阜叠山，因洼疏池"，使"山增而高，水浚而深，峰岫互回，云天倒映"，全园更显"丰

Suzhou Zhuozheng Yuan (Note: Clumsy Administrator's Garden) was first built around Zhengde 8th year (1513) of the Ming dynasty. Wang Xianchen, a royal censor, had conflicts with some dignitaries, so he gave up his government official position and returned to his home town. After retrieving to his "forest", he vented his discontent and found his sustenance in an "interstice" between "Louqimen". Since there "was stagnant water in there", so he "did some dredging and planted some trees around". He referenced the Jin dynasty's Pan Yue's "Xian Ju Fu" (Note: Ode on Leisure Living), "I choose a hermitage pastoral life style and grow food and vegetables self-sufficiently. After all, this may be accepted as a humble and incompetent person's way of life." That is why he named the garden the Clumsy Administrator's Garden. At that time, the garden is focused around Canglang Pond, with buildings such as "Ruoshu Tang", "Mengyin Lou" etc. "All the garden buildings are designed around the pond." On plant design, "they are all planted with fine species", and they are planted in tracts, forming quite a few scenic areas. According to "Mr. Wang's Zhuozheng Yuan" by Wen Zhengming, the Ming dynasty's Zhuozheng Yuan is based on "free and unfettered life with leisurely complacency and pleasure", characterized by being natural and nonchalant.

Later, after several changes in the garden and its ownership, in the Ming dynasty's Chongzhen 4th year (1631), the eastern part of the garden fell into disrepair and it was bought by Assistant Minister Wang Xinyi, who built "Guitian Yuan" in Chongzhen's 8th year. Around the Qing dynasty's third year (1783), it belonged to Prefecture Jiang Yongxian. During renovation, "mounds were made at higher ground while ponds were dug

拙政园中部园景鸟瞰之一
Bird's Eye View I of Zhuozheng Yuan, Suzhou

拙政园中部园景鸟瞰之二
Bird's Eye View II of Zhuozheng Yuan, Suzhou

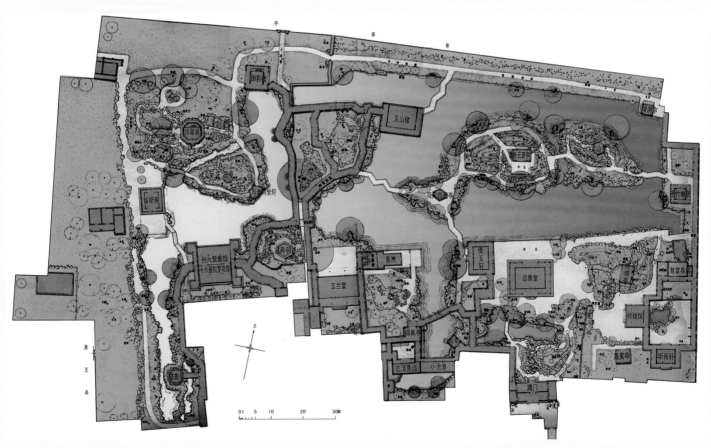

苏州拙政园平面图　Plan of Zhuozheng Yuan, Suzhou

而不侈，约而不陋"，至此"旧观仍复"取名为"复园"。同治十年（1871年），被巡抚张之万改为八旗奉直会馆，恢复拙政园旧称。其西部却为张履谦割为"张氏补园"，至此，园分为三，直至新中国成立后方始合璧，获得新生。并列为首批全国重点文物保护单位之一。1997年列入世界文化遗产。

because of its depression", making "the mound appearing higher and the pond deeper; the peaks responding among themselves and cloudy sky being reflected in the pond." The whole garden appeared to be "abundant but not extravagant, simple but not crude". Now, the "old appearance was restored" and so it was

named as "Fu Yuan" (Note: Restored Garden). In Tongzhi 10th year (1871), Governor Zhang Zhiwan changed it into Eight Banners Fengzhi Hall, and its name was changed back to Zhuozheng Yuan. Its western part was separated out by Zhang Lǚqian and named as "Mr. Zhang's Bu Yuan". So up to now,

苏州拙政园鸟瞰图　Bird's Eye View of Zhuozheng Yuan, Suzhou

今日拙政园，特别是园的中部，虽历经变迁，但风骨犹存，明代遗风依稀可辨。

一进腰门，迎面假山壁立，翠嶂为屏。穿曲径就到"远香堂"，四面空灵，为全园之冠。其西为"小沧浪"、"小飞虹"所成水院，匠心独运。其东边粉垣一带，隔院别馆，为"枇杷园"，二者虚实相生，情趣各异。出远香堂，豁然开朗。"雪香云蔚亭"露而据巅，"待霜亭"藏而稍隐。西北"柳荫路曲"，画楼掩映。过"倚玉轩"南行，浮廊可渡，水槛凌波。折至"西半亭"，池水分流，深远不尽。登"见山楼"，视野开阔，园景尽收眼底，绕道到"倚虹亭"，隔院楼台，北寺塔移置眼前。进"枇杷园"，园中之园，明净典雅。自成一体。透过月洞门，雪香云蔚亭巧得怀中。

综观拙政园艺术布局，自明至清，"久在樊笼里，复得返自然"的思想感情，一直渗遗其中。并因地制宜，创造了"山花野鸟之间"的自然风光，成为苏州园林的代表之作。

the whole garden is divided into three parts. They were all merged and started a new life after the Liberation. It was first listed as a key National Cultural Relics Protection site and was later listed as a World Cultural Heritage site in 1997.

Today's Zhuozheng Yuan, especially the mid-section, even with much changes, still maintains its strength of character, and the influence of the Ming dynasty can still be felt.

Once entering the inner entrance, you are greeted by a verdant wall-like rockery that serves as a screen. Passing a winding path you arrive at "Yuanxiang Tang" (Note: Distant Fragrance Hall). The vacant feeling on all its four sides tops the whole garden. West of the hall is a water court formed by "Xiao Canglang" (Note: Mini Azure Water) and "Xiao Feihong" (Note: Mini Flying Rainbow), which shows unique design ideas and implementations. At its east in the painted wall area separately there is "Pipa Yuan" (Note: Loquat Garden). These two, one with a void feeling with water and the other solid with buildings, show different interests. It is wide open out of Yuanxiang Tang. "Xuexiang Yunwei Ting" (Note: Snow and Cloud Pavilion) is exposed and perched up high while "Daishuang Ting" (Note: Frost Anticipation Pavilion) is somewhat hidden. The northwest "Liuyin Luqu" (Note: Willow Shadow and Curved Path) is covered by decorated building. Passing "Yiyu Xuan" (Note: Jade Leaning Building) to the south, one could cross water on a bridge corridor. Turning to "Xiban Ting" (Note: West Half Pavilion), you see the water diverges and tributaries seem to extend to infinity. Climbing up "Jianshan Lou" (Note: Mountain View Building), you have a broad panoramic view to see the whole garden. Detouring to "Yihong Ting" (Note: Rainbow Leaning Pavilion), over the yards and buildings in the middle, you see Beisi Ta

园内水景，整体简洁，脉络清晰，主次分明，主体且近端方。其理水又有着丰富的细部，聚中有散。通过地形、建筑，以及植物的分隔与掩映，形成诸如湖、池、河、涧等不同景区，故能于端方中见曲折之趣。

理水离不开掇山。园中之山多"平岗小坂"，以池上理山为主。池山形状简单，外廓单纯，土山带石，没有巉岩怪石，亦无奇峰异洞。山林情趣的获得，主要借助于林木的烘托。然而，简洁中又有丰姿。池山以一水"破"之，实中有虚，为山林增添不尽幽趣。不仅以可见的形象丰富了人们的感受，还巧妙地渲染了"蝉噪林愈静，鸟鸣山更幽"的深邃境界。

拙政园的山水，通过以少胜多、以简略繁的手法，再现了"南山低小而水多，江湖景秀而华盛"的神韵。其空间处理同样如此。园中主要空间（池岛形成的山水空间），通过水湾、港汊，渗透到园林的各个角隅，成为彼此流通，互相串联的完整空间。在高度统一中有丰富的层次，园中有园，景中有景，如枇杷园分隔为"玲珑馆"前后及"海棠春坞"等几个不同局部，构成了多层次的空间结构。

closing by. "Pipa Yuan", a garden inside a garden, is bright clear, elegant, and self-contained. Through Yuedong Men (Note: Moon Gate), Xuexiang Yunwei Pavilion is right in the picture frame.

Looking at the design and layout of Zhuozheng Yuan, since the Ming dynasty to the Qing dynasty, the main theme of "returning to nature after being caged and shackled for long" has been permeated throughout the garden. Its re-creation of natural landscape to enjoy "mountain flowers and wild birds" based on site-specific physical and cultural context is representative of Suzhou gardens.

The water system is generally simple and shows a clear priority order. The main water body is mostly square shaped and it has much rich detailed treatments. There are divergences among the convergence. By using terrains, buildings, and plants as means of separation and screening, different kinds of scenic types such as lakes, ponds, rivers, and streams are created. This is how it created diverse and sophisticated interests, breaking the initial rigidity from a square shaped lot.

Water scene creation is not complete without mountain building. There were mostly gentle sloped hillocks in the garden and handling was mostly mound building above the pond. A mound by a pond has a simple figure; it is mostly dirt with some rocks and there is no exotic rocks around. Neither are fantastic peaks and caves. The wild mountain scene creation is mostly dependent upon trees and forests. However, there are rich luxuriant scenes among the general simplicity. A mound is "broken" by a pond, adding a void entity into a solid one and creating infinite interests in the mountain landscape. This is realized not only by visible physical features that entertain the eyes but also rendering a deep zen-like state of mind, as exemplified in the poem, "The forest appears even quieter with cicada yelping; the mountain seems more tranquil with birds chirping."

The landscape design of Zhuozheng Yuan adopts a less-is-more and simple-over-complex methodology, realizing the charm of a scene with "more water area over the south flatter and smaller mound and the river and lake scenes are pretty yet sumptuous". Same applies to its spatial design. The main garden space – those formed by the ponds and islets – via inlets and branches, reaches each corner in the garden. They are all inter-connected and form an interlocking integral space. Under a high degree of unity, there are levels of depths. There is a garden inside a garden and a scene inside another scene. For instance, Pipa Yuan is divided by the front and back parts of "Linglong Guan" (Note: Exquisite Building) and "Haitang Chunwu" (Note: Crabapple Spring Cove) and a few other parts, making up a multi-layered spatial structure.

苏州耦园 Ou Yuan of Suzhou

耦园在苏州小新桥巷。正宅居中，有东、西两个花园，故称耦园；两人耕种称为耦，这里指夫妇皆隐居归田的意思。

东花园原为清初保宁太守陆锦所筑，名为涉园，后为祝氏别墅。清末沈秉成扩建，始成现状。新中国成立前为刘国钧所有，新中国成立后为振亚丝织厂一部分。1962年，由苏州市园林管理处维

Ouyuan (Note: Dual Garden) is in Xiaoxinqiao Lane in Suzhou. The main residence is in the middle and there are east and west two gardens; hence the name Ouyuan. A way of cultivation by two people is referred to as Ou. Here it refers to a husband and wife leading a secluded rural agrarian life.

The east garden was built by Lu Jin, a prefecture in Baoning in the early Qing dynasty. Its name is She Yuan, and later on became Mr. Zhu's villa. It was expanded by Shen Bingcheng at the end of the Qing dynasty and became what it looks as of today. Before the Liberation it belonged to Liu Guojun and after the

修开放。2000年列入世界文化遗产。

西花园在正宅西部。由园门进入，一带湖石假山挡住视线，假山轮廓完整，山上有云墙相隔，山下有山洞相通。假山北面，用湖石围成花台，很是自然。花台北，为面阔三间的书斋，名为"织帘老屋"，书斋前，有宽敞的月台。假山、花台成为书斋前很好的画面。书斋后有一院落，点缀峰石、花木，其后为一不对称的凹形书楼。

东花园在正宅东部，南、北、东三面为河滨包围，再东是苏州城墙遗址。入园门是一个大院落，院东一带花墙，院北为一月洞门。进门后起伏曲折的走廊向西、北行进，即为"城曲草堂"，堂二层，为园主宴客处，体形庞大，企口杉木板壁，装修很精美。堂前有平台，堂北有小院三处。堂的两边是"樨廊"和"筠廊"。再南是用黄石叠成的假山，气势雄伟，以"邃谷"一分为二，东部较大，有石径可通山上平台和石室，峰顶为留云岫，平台东，有峭壁下临水池，转东南顺磴道而下，即为"受月池"。山上树木葱郁，苍翠欲滴，环境异常安静。受月池为一南北狭长的水池，西面临山，东面靠廊。过折桥"宛如虹"，为"吾爱亭"和"枕波双隐"，这里有曲廊北接"双照楼"（书楼），南有四面厅跨

Liberation it became part of a silk mill. In 1962, it was renovated by Suzhou Garden Management Office and opened to public. In 2000, it was listed in the World Cultural Heritage.

The west garden is at the west side of the main residence. Entering the garden gate, you see a line of lake rockery blocking your sight. The silhouette of the rockery is complete with cloud-shaped wall on top for separation and cave or grotto at bottom for connection. North of the rockery is a flower bed bordered by lake rocks, which is very natural. North of the flower bed is a study of three units across, named "Zhilian Laowu" (Note: Curtain Weaving Old House). In front of the study there is a wide open platform. nice view from the study. A courtyard is behind the study and it is dotted with rock peaks and flowers. Behind that there is an asymmetric concave study building.

The east garden is to the east of the main residence. Its south, north, and east sides are surrounded by a river. Further east is the relics of the Suzhou city wall. After entering the garden entrance one sees a large courtyard. East of the yard is a line of decorated lattice wall. North of the yard is a moon gate. Entering the gate and walking along the zigzagged corridor towards west and north, you see "Chengqu Caotang", which is a two-storied building and was where the owner hosted banquets for his guests. The building is sizeable. The wood panel wall is made of Chinese fir with exquisite decorations. There is a platform in front of the hall and in the north, there are three smaller yards. At each side of the hall there is "Xi Lang" and "Yun Lang". Further south is a yellow stone rockery with imposing grandeur. "Sui Gu" divides the rockery into two parts. The eastern part is bigger and has a stone path leading to the terrace and stone room on top, which is climaxed by Liuyun Peak (Note: Could Retaining Peak). East of the terrace there is a steep cliff facing the pond. Toward the southeast stepping down the rock steps you see "Shouyue Chi" (Note: Moon Receiving Pond). The trees on top of the mound are verdant and exuberant, creating an extraordinary

苏州耦园平面图　Plan of Ouyuan, Suzhou

水上，称"山水间"。山水间东南，用黄石叠成平岗花台，台中植牡丹。园的东南角为"听橹楼"，有阁道和走廊与西北部的月洞门相通。这里吾爱亭表达了这一对双隐归田的夫妇，在山水之间枕波、顾影、听橹、织帘、读书的高尚情操和理想生活。

耦园居城弯一隅，特别是双照楼和听橹楼，楼上借城墙和内、外城河景色，十分突出。园内黄石假山和湖石假山，也有独特之处。

quiet environment. "Shouyue Chi" is an elongated one that is north-south oriented. Its west is against a rockery mound and its east by a corridor. Passing the zigzag bridge "Wan Ru Gang" (Note: As If Bridge), you meet with "Wuai Ting" (Note: My Love Pavilion) and "Zhenbo Shuangyin". Here there is a curved corridor connecting to the north "Shuangzhao Lou" (a study building), and to the south the four-sided hall over the pond, referred to as "Shanshui Jian" (Note: Between Mountain and Water). Southeast of Shanshui Jian, there is a flower bed made by yellow stones, in which peonies are planted. At the southeast corner of the garden it is "Tinglu Lou" (Note: Scull Listening Building), and it has indoor passage and corridor connecting to the northwest's moon gate. Here Wuai Ting expressed the high ideal life style of this couple who, between mountains and waters, lived in secluded and self-appreciated live, and listened to scull rowing, weaved curtains, and read books.

Ouyuan is situated at the city corner. Especially, the upper parts of Shuangzhao Lou and Tinglu Lou can leverage the city wall with inner and outer city river scenes. This is unique and extraordinary. The yellow stone and lake rock rockeries inside the garden are also unique indeed.

苏州沧浪亭　　Canglang Ting of Suzhou

"清风明月本无价，近水远山皆有情"。这是沧浪亭内清代诗人俞曲园手书的楹联，写尽了沧浪亭情景交融的艺术境界。

沧浪亭是苏州历史最为悠久的名园。原为唐末吴越广陵王钱元璙亲戚的花园。宋庆历甲申年（一○四四年），诗人苏子美（舜钦）于废地临水建沧浪亭。后几易其主，转至章氏，扩大基地，增掇西山，添置建筑。南宋时为韩世忠住宅，又称"韩园"，于东西山间架"飞虹"，连为整体。后为庵。由元至明又变为僧房。清同治改为"五百名贤祠"。太平天国后缩小规模，改修而始呈现状。现为世界文化遗产。

沧浪亭的造园艺术不落凡响，敢于破格，一反高墙深园的常规，将园内、园外融为一体。未进园门，已是绿水回环，垂柳迎风，平栏依岸，石坊点染，园林意趣油然而生。这是起景，就是沧浪亭的一个序幕。再看，临水山石嶙峋，其后山林隐现，仿佛后山余脉延伸水边，又有复廊介于山山水水之间。在这里，复廊漏窗起着沟通园林内外空间的作用。

沧浪亭之破格，还在于以山为主，不落造园多以水为主的陈法。造园以水为主，这是由于水体是虚，宜于扩大空间，造成深远空灵之趣。沧浪亭由于园外已有水景，

"There is no price in clear wind and bright moon; however, affection exists in waters here and mountains there." This is a poem inscribed on a couplet and written by Yu Quyuan, a poet in the Qing dynasty. It expresses fully Canglang Ting's artistic state where people's feelings and landscape scenes commingle.

Canglang Ting is the oldest renowned garden in Suzhou's history. It was originally a garden owned by a relative of Qian Yuanliao, a regional governor in Wuyue in the end of the Tang dynasty. In Qingli Jiashen year (1044) of the Song dynasty, poet Su Zimei (Shunqin) built Canglang Ting on a piece of waterfront badlands. After that, it changed hands several times. When Zhang took it over, he expanded the base area, added the west mound and buildings. In the South Song dynasty, it was the residence of Han Shizhong, so it was called "Han Yuan". A bridge "Fei Hong" (Note: Flying Rainbow) was built to make it a whole. Afterwards it became a nunnery. From the Yuan to the Ming dynasties it became a monk house. In Tongzhi's reign in the Qing dynasty it was renamed as "Temple of Five Hundred Virtuous People". After the Taiping Heavenly Kingdom it was reduced in size and renovated into the current state. Now it is included in the World Cultural Heritage list.

The garden art of Canlang Ting is outstanding. It dares to break the rules and it is against the convention of using tall wall with deep garden yard. It integrates inside and outside of the garden into one. When you have yet entered the garden gate, you already see meandering water with wavy willows. With railings on water bank and dotted stone houses, you already feel the garden atmosphere. This is the introductory scene, i.e. a preface of Canglang Ting. Upon further examination, there are rugged rocks by the water, and the mountains behind loom through as if the remaining mountain ranges reach the water. There are also double corridors located in the landscape. Here the double corridor and the lattice window act as a liaison agent between the inner and outer spaces of the garden.

Another extraordinary trait is that it uses predominantly mountain, as opposed to water, which is used in

若再加水，则陷重复，故借水而掇山，一虚一实，互呈对比，实为因地制宜之举。沧浪亭之东山，相传为宋时所遗，土山带石，有谷道飞梁，山上古木森森，箬竹遍地，加之藤萝掩映，山林野趣横生，景色苍润。其亭居山之西侧，采用方形，且以石为柱，古朴拙重，融溶于山林的一片宁静之中。与东山相比，西山较为平庸，格调不高，然其西一池如临深渊，既与园外之水呼应，又与之情趣各异。

沧浪亭敢于开门见山，也是其破格之处。从所处位置特点出发。进园之前，园外景色，已先一路铺陈渲染；进园以后，又渐入佳境，恰到好处。由于曾经是佛寺、公祠，故园内建筑严正，似嫌单调；但从整个园林所呈现的意趣来看，从建筑空间与山林空间的对比来看，这种格局还属相宜；在整体严正之中，建筑又有所变化，如复廊中穿插"面水轩"、"观鱼处"、"御碑亭"等。在建筑空间上，"瑶华境界"

most other conventional gardens, as the main structure of the garden. This is due to the fact that water represents void and is appropriate at expanding space to create a deep and vacant spatial feeling; however, there is already water outside Canglang Ting and if more water body is added it would be redundant. Therefore, by leveraging water outside and building mountains inside, a void with a solid contrasting each other, it accomplished the goal of achieving flexibility on context. The east mound of Canglang Ting is said to be from the Song dynasty. The mound is mostly dirt with occasional rocks. It has got ancient roads and bridges and its vegetation is thick with bamboos and vines, full of wild mountain natural interests. The square-shaped pavilion is at the west side of the mound. It has stone columns, appearing ancient and humble, and blending well into the tranquil mountain atmosphere. Contrasting to the east mound, the west mound is pretty plain without high taste; however a pond in its west side makes one feel standing on the edge of an abyss, which corresponds well to the outside water bodies, and complements well to the east with different taste.

The fact that Canglang Ting dares to be straight to the point is its other exceptional trait. From the location point of view, before entering the garden, the external scenes already make some preparations. After entering the garden, you gradually and appropriately proceed to get better and better scenes. Due to the fact that it was once a Buddhist temple and an ancestral hall, the buildings in the garden tend to be rigorous and a bit monotonous. However, from the point

苏州沧浪亭平面图
Plan of Canglang Ting, Suzhou

之南翳以修竹，环境清幽，与"明道堂"的豁达适呈对比。明道堂北的乱石铺地也与其南的庭院铺地互异其趣，且与山林取得呼应。此外，在明道堂、"清香馆"这两组建筑之间，又有"看山楼"与"翠玲珑"作为穿插。看山楼昔日可于封闭的庭院中将视线引向其南，远眺南园的田园风光及天平诸山的晴峦耸秀，从而使空间顿扩至无限。翠玲珑取于"日光穿竹翠玲珑"的诗意，四面环翠，亦于严正肃穆中呈现曲折幽邃之趣。园内多碑刻，更得古朴清新之意。

of view of general taste of the garden and that of spatial contrast between buildings and mountain landscapes, the layout is appropriate. Among the rigidity and sternness of the general layout, there are variations in the architecture, such as "Mianshui Xuan" (Note: Water Facing Building), "Guanyu Chu" (Note: Fish Watching Spot), "Yubei Ting" (Note: Royal Stele Pavilion) and so on that are interspersed in the double corridor. In architectural space handling, bamboo grove south of "Yaohua Jingjie" creates quiet and serenity, properly in contrast to the broadness and openness of "Mingdao Tang". The assorted stone paver north of Mingdao Tang also differs in taste from those in the south court, and corresponds well to the mountain and forest. Moreover, between Mingdao Tang and "Qingxiang Guan" two buildings, there are "Kanshan Lou" (Note: Mountain Watching Building) and "Cui Linglong" (Note: Verdant Exquisite) as inserts. In the past, from Kanshan Lou, one could direct his view to the south, where the south garden's pastoral scene and Mount Tianping's green landscape can be enjoyed, increasing the spatial extent to infinity. The name Cui Linglong is taken from the poem "sun light passing through green and exquisite bamboos". With all the surrounding exuberant verdant bamboo scenes, the deep serene feeling counters well to the rigidity and austerity of the general genre. There are many stone inscriptions in the garden, adding ancient yet fresh feeling.

扬州卷石洞天 Juanshi Dongtian of Yangzhou

卷石洞天在扬州蜀冈－瘦西湖风景名胜区内，由市区绿杨村西行过新北门桥即达"卷石洞天"。据《扬州画舫录》载："卷石洞天在'城闉清梵'之后即古郧园地。"郧园以怪石老木取胜。其内假山与楼阁被誉为"郊外假山，是为第一"，"楼之佳者，以夕阳红半楼……为最"，是清代瘦西湖二十四景中的佳品。

随着时代的变迁，园内假山、古木、建筑，焚于咸丰年间兵火。1949 年后，园地一直被占用。1987 年划归园林部门，次年开始规划。1989 年市政府根据规划设计投资建设，至 1990 年 5 月建成，以其独具风韵的艺术魅力再现昔日风采。

卷石洞天的特色是"卷石"与"洞天"。造园者巧妙地利用了瘦西湖北岸的土阜平岗，因地制宜，在土丘上堆叠湖石山子，其章法以洞壑幽深取胜，洞内仿自然石灰岩溶蚀景观，玲珑剔透，根据湖石天然景态因势利导，诸如岩壑、水岫、溶洞、裂隙，应形而生。卷石洞天假山贵在山水相互依存，相得益彰。潜流、叠泉分别从溶隙、岩窟曲曲而下，呈现了忽断忽续、忽隐忽现、忽急忽缓、忽聚忽散的不同水景，产生不同音响。水音与岩壑共鸣，不啻八

Juanshi Dongtian is located in Shugang-Shouxi Lake scene area in Yangzhou, and it can be reached from the city from Lùyang Village west through New Northgate Bridge. According to *Yangzhou Huafang Lu*, "Juanshi Dongtian became the old Yun garden after 'Chengyin Qingfan'". Yun garden is well-known for its exotic rocks and old trees. Its rockery and buildings won the compliment as "number one in suburban rockery", and "best architecture, Xiyang Hongban Lou". It is one excellent work among the twenty four scenes in Shouxi Lake in the Qing dynasty.

With the changing times, the rockery, old trees and buildings were destroyed in wars during Xianfeng's reign. After 1949, the garden site had been occupied. In 1987, it was transferred to the garden design and management department and the planning work started the next year. In 1989, based on the planning and design work, the city government invested funds for construction, which was completed in May, 1990, and its past unique characteristics and artistic charm were again shown.

The distinguishing features of Juanshi Dongtian are "Juanshi" and "Dongtian". The garden designer masterly took advantage of Shouxi Lake north bank's flat hillock and created on top of dirt mound a rockery made of lake rocks. It scores successfully in mimicking a deep and quiet feeling. Inside the cave, it simulates a natural limestone Karst landscape and based on natural lake rock formation, exquisitely carving out rock gullies, water cavern, and rock crevices. The value of Juanshi Dongtian lies in its nurturing of dependency between mountains and waters and they in turn complement each other perfectly. Undercurrents and stacked spring water flow down from crevices and caves, with on and off, visible and hidden, fast and slow, converging and diverging varying water scenes, contributing to differing

音齐奏，似有左思"非必丝与竹，山水有清音"的诗意。当游人盘旋攀跻于此，可领略到"洞内有洞，洞中有天；水中有洞，水中有天"的洞天福地胜景。扬州八怪之一郑板桥有"移花得蝶，买石饶云"的集句，卷石洞天之"卷石"实为"卷云"。园内假山十分注意以山外形圆浑勾勒出连绵的云态，遵循"山欲动而势长"的画理叠山，体现"卷云"奔趋磅礴的动态。"夏云多奇峰"，每组山峰仿佛朵朵飘浮跳跃的卷云，拱揖追随"云层"。整个山体经营成上实下虚、上明下暗、上散下整，力求体现"卷云"飘逸轻盈与浑厚深远的层次。卷石洞天的湖石山子与西面平岗小阪的土山有机组合，形成"山外有山"的结体，用"混假于真"的手段取得"真假难辨"的造景效果，湖石山宛如东北走向土丘的余脉蜿蜒延伸至此。湖石山呈现石灰岩构造特点似水流的溶蚀结果。此山人工开辟的泉流与瀑布的作用，可使人们联想到仿佛湖石裸露与形态是由于流水冲刷泥土并溶蚀的结果，故取得"做假成真"的效果，"得自然之理，而呈自然之趣"。

卷石洞天由东部的水庭、中部的山庭与东北部的平庭三部分组成。入园，古木

acoustics. The sound of water resonates with those from rock gullies, forming interesting natural octave sonata, somewhat like what Zuo Si wrote, "Beautiful music does not have to come from musical instruments since it can be created from the natural mountains and waters". When visitors manage to get here, they could savor this blessed cavern landscape of "cave inside cave, sky inside cave, cave in water, and sky in water". Zheng Banqiao, one of the eight Yangzhou eccentrics, said, "transplanting flowers to get butterflies, and purchasing rocks to add clouds". The "Juanshi" in "Juanshi Dongtian" actually means "Juanyun" (Note: curled clouds). The garden designers paid great attention to the shapes of rockeries in the garden, as they wanted them to appear rounded to simulate clouds. According to the Chinese landscape painting theory that "mountains should possess the rhythm of music", "Juanyun", the rockery, should exhibit dynamic and majestic moving trend. "There are more exotic peaks in summer clouds." Each peak mimics floating and jumping curly clouds, forming Chinese greeting gesture. The whole rockery is made solid upper and abyss lower, bright upper and dark lower, dispersed upper and whole lower, striving to realize "curly clouds" elegantly light and deep and far-reaching layers. Juanshi Dongtian's lake rockery is combined naturally with the western section's dirt hillock to have formed a complexity of "mountain beyond mountain" and, using a method of "mixing fake among real", to have achieved a landscape effect that "cannot tell the real from the fake". The lake rockery feels like an extension of the northeastward dirt mountain foot, and its limestone construct looks like it has been eroded by water. The artificially created springs and waterfalls make people think of exposed lake rocks eroded constantly by flowing water. This accomplished the job of "using fake to make real" and exemplifies "presenting natural wonders after understanding how nature works".

Juanshi Dongtian is consisted of a water court in the east, mountain court in the middle, and a flat yard in the northeast three sections. Entering the garden, a visitor is greeted with an old tree with twisting trunk and a huge monolith. This is a typical approach in Chinese garden design and it is here to serve a "scene blocking" purpose. With this view obstruction, the sceneries inside the garden cannot be seen at a glance. This would arouse visitors curiosity for exploration and it is in line with the "conceal before reveal" design rule. Turning to walk north, entering Qunyu Shanfang with Baosha roof, one sees a wall-mounted stacked spring with three laps and a half filled pond. With slate paver, rush dotted moss and a picture frame with shadowy flower in the east window, emerges a classic landscape "as if a vignette". The bamboo couplets "A peak can suggest long range of mountains; A spoonful of water can represent miles of rivers

A-A 剖面　A-A Section View

B-B 剖面　B-B Section View

C-C 剖面　C-C Section View

扬州卷石洞天剖面图　Section View of Juanshi Dongtian of Yangzhou

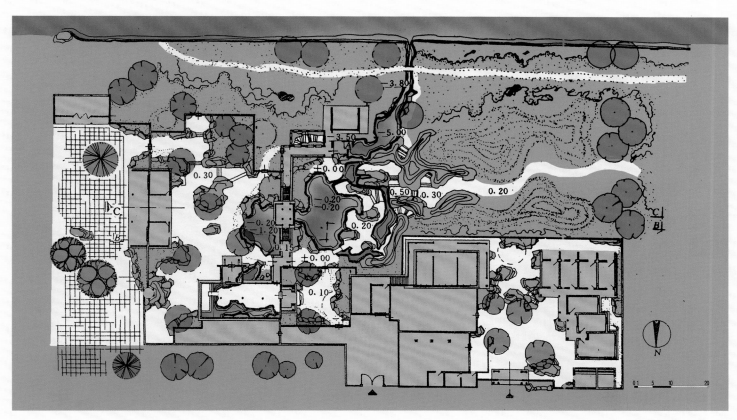

扬州卷石洞天平面图
Plan of Juanshi Dongtian of Yangzhou

龙蛇起舞，巨石兀立为屏，为中国园林特有的"障景"手法。有此一障，使园内景物不得一目了然，激发游人寻幽访胜，一探究竟之欲望，符合"欲扬先抑"的艺术布局法则。折而北行，入群玉山房抱厦，可见壁泉三叠，潭水半泓；片石铺装，蒲草点苔，东部画幅借窗来浮动花影，"仿佛片图小李"；竹联"一峰则泰华千寻，一勺则江湖万里"，点出景意。渡石梁进大厅，迎面是一幅"粉墙为纸，竹石为绘"的立体山水画卷。右侧丛置的山岩滚水泻玉，藤萝扶疏，推远了粉墙上贴壁山的空间，加强粉墙的云雾之感。西侧耸立的叠石，毛竹丛中引出一股清流，形成"盂泉"，增添吊古之情。人们看到中间粉墙下部陈列的云片状的苍松翠柏盆景，不难悟出"扬派"盆景的艺术特征，"云片"原来系生长在层峦叠嶂的植物，经历了百余年暴风骤雨的洗礼，为了生存而被自然剪裁成最小的受风面——云片。

"山房"东边限于空间，点缀画舫半只，名曰"知音"，作为接待与品茗之用，满墙镜面使画舫的艺术形象得以完整。知音舫与"山房"西面角隅窗外的象形山石"听琴峰"遥相呼应，取意"俞伯牙砸琴谢知音"的民间佳话。循长廊至"水庭"，面对着蓄聚停泓的碧波、潆洄缭绕的曲溪，以及

and lakes" demonstrate this central theme. Across a stone bridge entering a large hall, a visitor is greeted by a three-dimensional landscape painting – bamboos and rocks as the scene objects against white painted wall background. The clustered rocks on the right, intertwined with vines, appear to be jade stone rushing down with boiling water, pushing further apart the space of wall-nestling mound above the painted wall and enhancing the foggy feeling of the painted wall. A clear stream is introduced from among the erected stacked stones on the west and bamboo grove and forms "Meng Spring", adding a nostalgia feeling. When people see cloud layer shaped pine bonsai displayed at the bottom of the painted wall in the middle, it is not difficult to comprehend the artistic feature of the "Yang School" bonsai. "Yunpian" originally grows out of layered rocks. After experiencing hundreds of years of violent storms, to survive, it is pruned by nature to form the thinnest layer – Yunpian (Note: cloud flake, literally).

East of "Shan Fang" (Note: Mountain Dwelling), due to space constraint, only half a boat house was built. It was named "Zhi Yin" (Note: Bosom Friend) and serves as a place for reception and tea serving. The mirrors mounted on all walls complete its artistic image. Zhiyin Fang echoes the pictographic rock "Tingqin Feng" (Note: Music Listening Peak) outside the southwest corner window of "Shanfang" (Note: Mountain House), hinting the folk legend "Yu Boya smashed his music instrument in memory of his friend who really understood and appreciated his music". Walking along the long corridor and arriving at the "Water Hall", one faces a full pond and winding brooks. Hearing clangorous spring from within "Shanfang", watching whirling falling water and partially fog (i.e. painted wall) covered mountain, one immediately feels the elegant taste of "high mountains with flowing water" (Note: a piece of ancient classic Chinese music).

The mound and water courtyards are separated by a long corridor, which is punctuated in the middle by a pavilion, whose horizontal tablet inscribed as

"山房"内铿锵的泉鸣、盘涡的泻瀑、被云雾（粉墙）遮隐的峻岭，"巍巍乎高山也，洋洋乎流水也"，高山流水的曲调呼之欲出。

山庭水庭以长廊为分隔，廊中部为廊亭，匾额点出"卷石洞天"景名。廊亭北为爬山廊，南为跌落式廊。起伏的爬山廊前为婀娜婆娑的青枫，折线顿挫的曲廊旁则为亭亭玉立的劲松，一高一矮，一刚一柔，顾盼生情。

山庭最南端为二层结构的薜萝水阁，下层观瀑，上层观山，阁面北的中间为画幅窗，将山庭尽收眼底，恰似一幅"风壑云泉"山水中堂。风：山顶苍松翠柏斜出石隙，为听松看涛之佳处；壑：假山以洞壑造型，得到曲折深邃的效果；云：假山乃堆云，其叠石依皴合撮，山体笔意为云头皴法；泉：水皆从石隙渗流而出，化为清溪曲调，蛇行斗折于崖根石角，游人至此，如鼓琴瑟，幽趣丛生。"云生润户衣裳润，风带潮声枕簟凉"联句描写得十分贴切。

山庭以西土阜蜿蜒，竹径通幽。修竹之北庭园逶迤，花木扶疏；花墙后红楼抱山，气极苍莽，尤以夕阳红半楼飞檐峻宇，斜出山体，联云"渔浦浪花摇素壁，玉峰晴色上朱栏"。楼东为委宛山房，隐于山后，花脊卷棚，四面各异。与委宛山房相对的丁字楼，呈曲尺形旋律构筑，别具韵味，三楼环抱的平庭仅略施笔墨，实中求虚。临街的门厅采用挑梁式，与宏伟庄重的夕阳红半楼产生矫与健的对比。门侧贴壁山石，虽千石嵌合，然天衣无缝，尽衣薜萝，不见石骨，加之苍古遒劲的摩崖石刻，气度非凡。

"Juanshi Dongtian" (Note: Curling Rocks with Sky Light) scene title. North of this pavilion is a mountain climbing corridor, whose southern part is a stage-falling one. In front of the curved climbing corridor is a graceful dancing young maple tree. By the zigzag shaped falling corridor is a straightened up bold and vigorous pine tree. One is tall and one short; one is rigid and one flexible. They develop a loving relationship while gazing into each other.

In the south most of the mountain court it is two-storied Biluo Shuige. The lower floor is used for watching the waterfall while the upper level for enjoying the mountain. The north facing middle window is a picture window that would let a viewer see the whole mountain court at a glance, just like a frame of "Feng He Yun Quan" landscape hall. Feng (Note: wind) - Where the old slant pine and cypress tree grown out of rock cracks is a great place to watch the forest and listen to wind blowing through them. He (Note: gully) - The rockeries are modeled after caves and had achieved deep and tortuous effect. Yun (Note: cloud) - Rockeries can be thought of as stacked clouds, and its textures simulate those in rocks in landscape paintings. Quan (Note: spring) - All water comes from rock crevices and becomes clear and melodic stream, meandering around cliff base rocky corners. When visitors are here, they hear the water sound like music, creating great interest. This poem describes this scene aptly, "Stream spawn clouds moisten my clothes; Wind with tide sound makes me feel chilly."

The dirt mound west of the mountain court meanders through and a path through bamboo grove leads to a quiet place. North of the bamboo grove is a yard with thriving plants. Behind the decorative wall a red-colored building surrounds a mound with vast extent. Especially when at sunset time, a flying cornice sticks out against the mountain silhouette forming a scene described by the poem, "White walls are shaken by ocean spindrift; red railings are painted by clear mountain color". Weiwan Shanfang at the east of the building is hidden behind the mound with Juanpeng style decorative ridge that shows different look at all its four sides. Dingzi Lou (Note: T-shaped Building), which faces Weiwan Shanfang, displays an L-shaped rhythmic construct with unique taste. There is little manipulation in the embracing flat court on the third floor, showing a case of "less among more". The street front hall uses an overhung girder style architecture, contrasting nicely to the power of grand and solemn style of the half building Xiyang Hong (Note: Sunset Glow). At the side of the gate, large amount of rocks were used; however they appear seamless. With all the climbing vines and other plants, the rock becomes non-visible. Engraved with forceful calligraphic carvings, it boasts a majestic air.

（五）傍宅地

宅傍与后有隙地可葺园，不第便于乐闲，斯谓护宅之佳境也。开池浚壑，理石挑山。设门有待来宾，留径可通尔室。竹修林茂，柳暗花明。五亩何拘，且效温公之独乐；四时不谢，宜偕小玉以同游。日竟花朝，宵分月夕。家庭侍酒，须开锦幛之藏；客集征诗，量罚金谷之数。多方题咏，薄有洞天。常余半榻琴书，不尽数竿烟雨。涧户若为止静，家山何必求深。宅遗谢朓之高风，岭划孙登之长啸。探梅虚寒，煮雪当姬。轻身尚寄玄黄，具眼胡分青白。固作千年事，宁知百岁人。足矣乐闲，悠然护宅。

(5) Residence Side Land

If there is available land by a residence or behind it, it can be used for building a garden. It is not only convenient for recreational use but it can also serve as a best environment to protect the residence. One can dig up and dredge a pond or gully, and pile up dirt and rocks to make a mound. A garden gate can be set up to receive visitors and a special path can be made for family member to go to inner rooms. There can be bamboo groves and other beautiful trees and flowers. Even if it is as small as five Mus it can still model after Sima Wen's "Du Le Yuan" (Note: Solo Pleasure Garden). The flowers are in blossom all year round and it is best to tour the garden with young maids. One can enjoy the flowers and plants all day long and can have feast and party all night long on the mid-Autumn festival. For such a family banquet, there is no need to set up tents and screens anymore. Guests are invited here to write poems and read them aloud and for those who cannot do their jobs, we should impose some penalties by making them drink. Poems and articles are written and intoned at many places and interesting scenes are created. There are books on bed and foggy drizzles outside. If constructing a building by water is for its serenity, why do we need to build a mound at home that is tall and deep? A residence house should have Xie Tiao's integrity and a mound created should have Sun Deng's wisdom from out of this world. There should be no need to travel far just to enjoy the fragrance of plum flower, and when making teas with boiled snow water we should have concubines around. Even out of government job we still live in the world; dealing with people we do not need to tell that clearly between our likes and dislikes. We would like to be well-established and well-accomplished; however our life span is short. We should learn to be content and leisurely enjoy our home gardens.

扬州个园 Ge yuan in Yangzhou

个园，是扬州市内以石斗奇的著名古典园林。

进园修竹临门，石笋参差，构成一幅以粉墙为纸、竹石为绘的天然图画，起笔就点出春景。竹旁点放的山石，仿佛是雨后破土而出的春笋，使人联想到一片生机勃勃的春天景象。月门两旁修篁弄影，门上横额有"个园"两字，"个"字为竹叶的形状，愈觉其切合竹石图的主题。

该园系清嘉庆年间，两淮商总黄至筠购买马元琯的小玲珑山馆修筑而成，他在园内广植修竹，颜其颤曰"个园"，乃取古人"宁可食无肉，不可居无竹；无肉令人瘦，无竹使人俗"的诗意，以此标榜清高。

进月门过春景，绕"桂花厅"，前面出现一座以湖石叠成的玲珑剔透的"夏山"。云峰临水，清流环绕，山顶秀木繁荫，有柏如盖；山下水声潺潺，涧谷幽邃；山

Geyuan, is a well-known classic garden in Yangzhou for its rare and unique rock work.

Entering the gate you're greeted with bamboos and rock peaks. Rocks and bamboos as the foreground objects against the painted wall background, it forms a natural spring picture scene that starts the tour. Rocks placed by bamboos make people feel like bamboo shoots sticking out of dirt after a spring rain and brings forth a sense of lively spring scene. Bamboos and their shadows cast on the moon gate, whose horizontal inscribed tablet displays "Ge Yuan" two characters. Since the word "Ge" is in the shape of a bamboo leave, it is particularly apt to the bamboo and rock scene.

The garden was built from Ma Yuanguan's small Linglong Shanguan, purchased by the Huai area merchant Huang Zhiyun, in Jiaqing's reign in the Qing dynasty. Huang planted large amount of bamboos, and Yan Qichan named it as "Geyuan", which is derived from the ancient poem "I'd rather have no meat to eat than no bamboo to live with; If I have no meat, it would at most make me thin, but if I have no bamboo, it would make me grow more vulgar", to show the owner's aloofness from politics and material pursuits.

Entering the moon gate and passing the spring scene and the "Osmanthus Hall", one faces a

腰蟠根垂萝，草木掩映，令人顿生眼底丘壑，咫尺山林的感觉。其特点就是运用"夏云多奇峰"的形象来叠石，通过灰调的石色，绿树披洒的浓荫，山洞的幽深，予人以苍翠如滴，夏山多态的感觉。山分东、西两峰，与"秋山"相连。夏、秋两山通过天桥（长楼廊）相接。由于夏山体现南方的秀丽，秋山突出北方的雄健，当人们过天桥到秋山，大有从南方飞渡到北国之感。

秋山亦即黄石山，气魄雄伟，峻峭依云，着笔气势非凡，用石泼辣，似有石涛画意。全山立体交通组织极妙，磴道多置洞中，山路崎岖，时洞时天，时壁时崖，时涧时谷，引人入胜，上下盘旋，造意极险。秋山环园半周，二十余丈，分三峰处理，每峰又有各尽其妙之洞，并都突出了北方雄健的风格，如中峰就分二洞，下洞如入深山石林，众峰环抱，复置石室，点缀山石几案，留有石门、石窗，四季皆十分干燥，至西北洞口并有风沙呼啸之感，纯属北派

lake rock made exquisitely carved "Summer Mound". The tall peak is surrounded by water and there are thick trees on top, such as a cover-like cypress tree. Below is gurgling water in deep and quiet gullies. At the middle of the rock there are deep-rooted and hanging vines with thick vegetation, giving a feeling of a miniature deep mountain. It is characterized by "more rare shaped rock-like clouds in summer" and the rockery was modeled after this using gray colored stone color. Shadows cast by green trees and the quietness of the caves make one feel lush green and summer mountain's variant appearances. The mound has east and west two peaks and it connects with "Autumn Mound". A flyover bridge (a long corridor) connects the Summer and Autumn Mounds. Since the

Summer Mound reflects the beauty of Southern China while the Autumn Mound highlights the vigor of the northern country, when people pass the flyover to the Autumn Mound, they somewhat have a feeling of migrating from the south to the north.

The Autumn Mound is a yellow stone mound with an imposing grandeur. It is tall and shrill and poses extraordinary momentum. It is bold and vigorous, somewhat with an impression of Shi Tao's landscape painting. The whole mound has complex and wonderful traffic network. The step stone paths are mostly inside the caves. They are rugged, sometime in the form of cave and other times open to the sky. Sometime it is a wall and other time a cliff. Sometime it is a gully and other time a valley. It is up and down, spiraling and twisting, forming a seemingly dangerous yet fascinating experience. The Autumn Mound is a semi-circle of about twenty Zhangs and is consisted of three peaks. Each peak has its own indigenous cave and they all manifest the northern vigorous style. For instance, the mid-peak is divided into two caves, and when you enter the lower one, it feels like being in a stone forest in the mountains. It is surrounded by all other peaks and there are a stone room with a few stone-made tea tables, doors and windows. It is dry throughout the four seasons and when you walk to the northwest hole you feel the whistling of sandy wind, characterized with the northern weather. The cave is referred to as fairy cave. With all its four sides hanging in the air it is extremely exquisite. Above it there is a pavilion and kiosk and a two-storied platform. When you are on top of it, you would feel way higher. Overlooking the distant Summer Mound, you would feel it as a sea of clouds. Looking downward, you see all the peaks

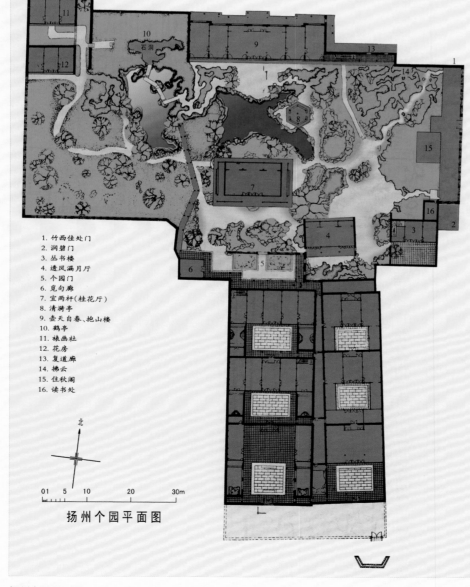

扬州个园平面图　Plan of Geyuan, Yangzhou

风味；中洞称仙人洞，四面凌空，极为雅致，上有飞阁凉亭，并有二层平台，登临其上，如入九霄云天，远眺夏山，似一片云海，俯视，则群峰皆罗足下；中峰最高，峰谷险峻，仰视悬崖削壁，突兀惊人。高不可攀。秋山又是全园的制高点。造园家采用了画家"不了"二字的巧妙手法，使最高的秋山有尽而不尽之感，给观赏者以更多想象回味的余地。

从黄石山东峰而下，招风漏月厅南围墙下，宜石（雪石）堆起的冬景，给人以积雪未化的感觉。部分山头则借助阳光照射，放出耀眼的光泽（宜石主要成分是石英），既突出峰头，也增加雪的质感，这是画家塑造雪山景色的匠心之作。当人们正在叹赏瑞雪之余忽而寒风呼啸，这是利用雪山的墙面上开了四排约尺许大的圆洞，犹如口琴音孔所造成的音响效果。造园家并将"琴孔"对着扬州常年主导的东北风方向，使人工制造的凛冽北风常年不断，真是技巧妙绝，既有高超的艺术形式，又甚合科学原理，是我国罕见的造园手法。

冬景结束，西墙开有两个圆形漏窗，远远招来春天的景色。修篁、石笋，重又映入眼帘。冬、春两景既截然分隔又巧妙地互相应借连接起来，表达了"冬天来了，春天还会远吗"的含意。由于游览是沿环形路线前进的，春、夏、秋、冬景色，周而复始，好似经历着四季气候的循环变化，可谓匠心独运。

据刘凤诰所撰《个园记》云：园内假山原为此园的前身寿芝园遗物，其原来叠石，相传出于大画家石涛之手。"扬州以名园胜，名园以叠石胜"。天造地设的四季假山是现存扬州园林的杰出代表，其中某些手笔，也有可能是石涛的山水精品。由于画师具备游遍名山大川的经历和高超的艺术技巧，故能在小块境地里布置出千山万壑的写意山水境域。并根据扬州假山石皆从外地运来，品种繁杂、体量较小的特点，大胆运用他的绘画理论："凡写四季之景，风味不同"，独出心裁地设计出分峰用石的四季假山。四季虽各具特色，但又有一气贯注之势，颇能表达出"春山淡冶而如笑，夏山苍翠而如滴，秋山明净而如妆，冬山惨淡而如睡"的诗情画意。

under your feet. The mid-peak is the tallest, steep and arduous. Looking at the steep cliff from down under, it is amazingly lofty and unattainable. The Autumn Mound also holds the commanding height of the whole garden. The garden designer adopted landscape painter's ingenious "unending" trick, making the tallest Autumn Mound appear to be unfinished and giving the viewer room for more imaginations.

Going down from the east peak of the yellow stone mound, one reaches the wall south of Zhaofeng Louyue Hall (Note: Wind Attracting and Moon Leaking). The winter scene made by Yi rocks (snow rocks) gives people a feeling of accumulated snow not yet melted. Some of the peaks under the sun shine brightly (Yi rocks are mostly composed of quartz). It stresses the peak and enhances the feeling of snow as well. This is the painter's ingenuity in this snow mountain scene's creation. Just as people enjoy the propitious snow, suddenly they hear the cold wind blowing. This is achieved by opening four rows of one-foot diameter holes on the snow mountain's wall. The acoustic effect is just like that of a harmonica. The garden designer oriented the "instrument's hole" and let them face Yangzhou's dominant northeasterly wind. This makes this man-made cold north wind always persist. This marvelous technique is superb in its art form as well as in accordance with scientific principles. It is a rarely seen garden construction method in China.

At the conclusion of the winter scene, the spring scene is introduced from the distance through two decorative tracery windows on the west wall. Long and slender bamboos and stalagmite-shaped rocks are visible again. The winter and spring scenes are both completely separated, cleverly connected, and leveraged mutually, suggesting the soon arrival of spring as in "if winter is here, can spring be far behind?" As visitors tour the garden along a looping path, spring, summer, autumn and winter scenes circle around, implying four season changing cycle with ingenuity. According to Liu Fenggao's "Geyuan Annals", the rockeries in the garden are the remains of its predecessor Shouzhi Garden, and its previous rock work is said to be the work of the great master landscape painter Shi Tao. "Yangzhou is well-known for its famous gardens and those famous gardens are well-known for its rock works." These natural looking four season rockeries are exemplar representatives of the existing Yangzhou gardens, with some of its works very possibly being those from the great master Shi Tao. Since the master had extensive experiences touring all famous mountains and rivers and great mastery of artistic landscape painting, he could re-create a world of natural landscapes within a very limited spatial extent. Because all the rocks were transported from other places and they are generally small in size and of various types, he could deploy them discriminately and apply his painting theory to show "different characteristics depending on different season". He ingeniously divided them into four different seasons. Although the four seasons are unique on their own, they do have a consistent air and express well the poetic illusion that "spring mound is like a light smile, summer mound dripping verdant, autumn mound is clear with adornment, while winter mound is dimly sleepy".

扬州片石山房 Pianshi Shanfang in Yangzhou

片石山房，在扬州花园巷，寄啸山庄东南部。《履园丛话》卷二十记载："扬州新城花园巷，又有片石山房者。二厅之后，湫以方池，池上有太湖石山子一座，高五六丈，甚奇峭，相传为石涛和尚手笔。"石涛在扬州造了万石园、片石山房两处园林，前者早已焚殁，遗迹难寻，后者的部分假山和楠木厅保留至今，其假山被誉为石涛叠石的"人间孤本"，1989年扬州市人民政府决定修复片石山房。由扬州市古典园林建设公司高级工程师吴肇钊主持修复工作，本着以史料为参考、遗迹为依据、画论为指导三结合的方法，经半载日夜耕耘，完成研究、设计与施工，"人间孤本"得以再现。

因为没有更多的遗迹和资料可供查考、复原，故此次片石山房的修复，不是文物的复原，也不是利用现存遗物就事论事地整修开放，更不是重新设计一个园林。修复设计是在对石涛的艺术思想与绘画进行较全面的研究基础上进行的，以其画论为全面规划的指导理论，汲取其山水布局的精髓；认真搜集、研究了《履园丛话》、清嘉庆和光绪《江都县续志》、续纂光绪《扬州府志》等史料及当代人——寄啸山庄主人何芷舠之孙何适齐在92岁（1984年）时所著《记扬州园林片石山房》等资料；认真、细致地进行片石山房遗迹的现场调查，开挖后慎重研究，以诗为意境、画为蓝本而作出的。

修复前，片石山房面积780平方米。现场调查、开挖可知楠木厅北部的水池痕迹基本可寻，楠木厅亦是史籍记载的，在厅的西北部水池旁，发现一段约80厘米高的石柱，并有石鼓两个，按石柱与石鼓的比例尺度应是亭子所用的，说明西北部原来很可能有个跨水的半亭。厅本身是四间，但西侧的一间木材、结构形式一看就是后来拼接的。厅南由于原为工厂仓库与车间，因此没有任何遗迹可寻。结合史料稽考，基本格局大致可知，北部三面均为太湖石山所环绕，山前为水池，与楠木厅相接，西北部有半亭伴水，厅南还有一厅。水池从现场开挖情况看最底层是弯曲的湖石，上面有方整形的三合土池壁，由此可证实清朝后期修整时，确实在曲池上重新改成方池的，而且面积是缩小的。

Pianshi Shanfang (Note: Slate Rock Mountain House) is located in Yangzhou's Garden Lane, southeast of Jixiao Shanzhuang. According to "Lǔyuan Conghua" volume twenty, "In the new town of Yangzhou there is a Garden Lane, and there is Pianshi Shanfang. Behind two halls, there is a square pond, in which there is a Tai Hu rockery of five or six Zhangs tall. It is precipitous and is said to be the work of the monk Shi Tao." In Yangzhou, Shi Tao built Wanshi Yuan (Note: Ten Thousand Rock Garden) and Pianshi Shanfang two gardens, but the former was burnt and its remains are hard to find. Part of the rockeries and Phoebe wood hall of the latter has been preserved. Its rockery is hailed as Shi Tao's "only masterpiece around". In 1989, Yangzhou People's Government decided to restore Pianshi Shanfang. The main designer for this restoration project is Wu Zhaozhao, senior engineer of Yangzhou Classic Garden Construction Company. Based on historical records, real site relics, and landscape panting theory, he worked on the project with his heart and soul and finally completed its research, design and construction and made this "only masterpiece around" re-appear.

Since there are no more historical relics and research literature to reference, the renovation of Pianshi Shanfang is actually not a restoration, neither is a repair based on publicly availabe records of historical relics, nor is a complete new design of a garden. The restorative design is based on a complete research of Shi Tao's artistic thought and his paintings. The guidance for planning is based on his landscape painting theory, especially the essence of his landscape layout. It is based on a meticulously collected and researched "Lǔyuan Conghua", "Jiangdu County Annals Continued" in Jiaqing and Guangxu's reigns in the Qing dynasty, and the continued work on *Yangzhou Prefecture Annals*, as well as the contemporary He Shiqi, grandson of Jixiao Shanzhuang's owner He Zhidao, when he wrote *About Yangzhou Garden Pianshi Shanfang* in 1984 when he was 92. The design was made after excavations with careful and meticulous researches on its relics at the site, taking poetic imagery and artistic painting as blueprints.

Before restoration, Pianshi Shanfang occupied 780 square meters. At the site after excavation, we traced the pond north of the Phoebe Hall, which was recorded in the annals. At the northwest side of the hall by the pond, an 80 cm tall stone column was found and it was with two stone drums. From their sizes and scales, we judged that they should be part of a pavilion, suggesting that there could be a half pavilion over water in northwest part. The hall itself has four units; however the west part one appeared to be an attached extension looking from its wood material and structure. South of the hall, because there used to be a factory warehouse and workshops, there was no relic to be found. Based on historical records, the general layout could be known. The three sides in the north were all constructed with Tai Hu lake rockeries. South of the rockeries is a pond that connects to the Phoebe Hall. There is a half pavilion in the northwest and there is another hall south of the Phoebe Hall. From the excavation findings, the bottom layer of the pond is curved lake rocks, and above it is tabia pond wall. This proves that the restoration work done in the late Qing dynasty made the change into a square-shaped pond with a smaller pond size. The field survey and cleanups provided some supporting materials for its restoration. In the old time, there was

现场的测绘与清理，为修复片石山房提供了部分素材。古代叠山没有施工图，仅按图稿施工，《扬州画舫录》卷二载石涛以绘画为主，兼工叠石。画家叠石实际上是将画的二度空间塑造成三度空间的实物，应该说两者是十分相似的。石涛山水画遗作流传至今是有限的，设计者在为数不多的画幅中，摹写五幅，与现存假山相对照，发现主峰的气势、体面、形状、虚实处理以及细部经营均惟妙惟肖，把画幅作为叠山的设计效果示意图，当获成功。现场调查表明，仅存的片石山房假山很能体现画家"一峰突起，连岗断堑，变幻顷刻，似续不续"的章法。其主峰高达9.5米的做法在江南园林中前无古人，正如他画论中申述："若以画图险峻，只在峭峰悬崖栈直崎岖之险耳。"只有"巉岏突兀"，才能产生出"岚气雾露，烟云毕至"的效果。石涛强调有灵气的画家，山的处理应注意到"内实外空"或"外实内空"，现存假山内理成石屋亦符合其画理。

根据石涛《苦瓜和尚画语录》画论中山水布局艺术"得乾坤之理者，山川之质也"、"山有拱揖，山水吞吐"、"对比衬托，相得益彰"等见解，以片石山房现存假山与石涛山水画相对照，从石涛《醉吟图轴》和《卓然庐图轴》悟出总格局的概貌。《醉吟图轴》临本与《履园丛话》中记载十分相似："二厅之后，湫以方池。池上有太湖石山子一座，高五六丈，甚奇峭，相传为石涛手笔。"此画石涛作于五十岁时，已侨居扬州九年之久了。另一幅《卓然庐图轴》是石涛五十九岁的作品，不仅画幅展现的景物和史料记载与现状极其吻合，题画诗："四旁水色茫无际，别有寻思不在鱼。莫谓此中天地小，卷舒收放卓然庐。"作为片石山房即兴诗亦是恰如其分的。遂以这两幅画和诗作为片石山房修复总体设计的蓝本。

为使修复设计更加正宗，设计汇集了石涛画中的建筑形式，收集画中各种式样的小桥，树木搭配的方式。在山水局部处理方面，诸如入山口、谷道、山峦石矶、坡岸及其点石等也都作了搜集、整理。至于如何将主峰延展，体现画论中所曰需有拱揖，山欲动而势长，贯山川之精英亦在画中寻思，作了初步构思，基本形成片石山房的雏形。

依照石涛绘画蓝本，"山房"南岸经no construction drawing for building a rockery, and the guide was only some literature and paintings. *Yangzhou Huafang Lu* volume two noted that Shi Tao mainly drew and painted, and doubled as a rockery builder. For a landscape painter, when he builds a rockery, it is mainly transforming a two-dimensional space into a three-dimensional one, and the spatial sense in each is very close. There is only a limited number of landscape painting work of Shi Tao that was spread up to now. Among this few, the designer copied five of them by hand and compared them with the current rockery work. Judging at its momentum, façade, shape, and the handling of void and solid, we see all details show high levels of resemblance with the paintings. Using landscape paintings as schematic diagrams proves to be successful. From the on-site survey, we found that the few remaining rockeries could really realize the painter's art of composition that "with one peak sticking up and followed by connected, abrupt, and varied mound ranges, it appears to be continuous although it is not." Its practice of using the main peak of 9.5 meters is unprecedented in Southern Chinese gardens. As he stated in his artistic painting theory, "a precipitous and dangerous feeling in a painting lies in close proximity of steep cliff." Only through abrupt rising rocks can one achieve "foggy, misty, and surrounding clouds" effect. Shi Tao stressed on painters with ingenuity and creativity, emphasizing that the handling of mound should be on making "solid inner and void outer" or "solid outer and void inner". The existing rockeries' inner treatment as a stone house is well in accordance with this principle.

According to Shi Tao's *Bitter Gourd Monk's Remarks on Painting*, the landscape composition art in painting reflects "the nature of mountains that reflects the whole natural environment". He argued that "mountains are like a person greeting others with his hands making a bow in front of him and a mountain would swallow water and push them out", and that they "set each other off and enhance each other". Comparing the existing rockery against Shi Tao's landscape paintings, we can realize the general layout of his *"Zuiyin Tuzhou"* and *"Zhuoranlu Tuzhou"*. A copy of *"Zuiyin Tuzhou"* matches very well with what is described in *"Lǔyuan Conghua"*, "Behind the two halls, there is a square pond. A Taihu Lake rockery of five or six Zhangs tall is in the pond. It is pretty rare and precipitous and is said to be a work of Shi Tao." When Shi Tao worked on this painting, he was 50 and had been living in Yangzhou for nine years. Another painting *Zhuranlu Tuzhou* is a work he did when he was 59. The scenes displayed in the painting match perfectly with what historical records show. The poem in the painting goes, "The extent of water is immense; however my mind is preoccupied with something other than fish. Don't complain that the space is limited as I can expand and collapse my scroll of painting at will here at Zhuoranlu." This improvised poem for Pianshan Shifang is very appropriate. Therefore, I used these two paintings and the poem and my blueprint for the overall layout for the restoration project.

In order to make this restoration work more authentic, the designer collected and compiled architecture styles in Shi Tao's paintings, as well as various styled small bridges and vegetation compositions. Design details like mound entrance, valley paths, hillocks, sloped banks and other spotted rocks, are also collected and categorized. As for how to extend the main peak to reflect his painting theory of making a bow gesture and taking advantage of mountain's momentum to enhance its power and realize the mountain essence in painting, the designer made conceptual design and

扬州片石山房平面图　Plan of Pianshi Shanfang, Yangzhou

营水榭三间，横跨水面，遥望主峰依云，形成山高水长的立体画轴。榭东伸出曲廊与主建筑楠木厅连接。为拓展空间层次，在楠木厅西山墙建面西的不系舟，在此可倚栏观鱼，又恰与西南半壁书屋呼应。浑厚端庄的楠木大厅与嵯峨雄健的贴壁山相对应，系铿锵结体。

片石山房的主景应是山体，而山体的精华则为"奇峰"。因石涛是在大自然中搜尽奇峰素材打画稿，修复画家的园林山水，就需要在他的作品中"搜"奇峰，使现存假山惟妙惟肖，尤其与"一峰突起，连岗断堑，变幻顷刻，似续不续"的神韵更为相同。

画山与叠山艺术相通，石涛画山注意旁边经营瀑布，产生虚与实、动与静的对比。片石山房现存假山将磴道紧贴在耸峭的主峰旁布置，这样处理就使山体产生出深邃曲折的虚处。磴山道又安排为层峦的汇水线，即雨水的明排水渠道，石阶陡缓不均的布置，"山洪"就形成了画论所述"决行激跃"的瀑布，可见其匠心独运。

遵画理循画幅，将瀑布之下设计为深潭。施工清理倒塌山石泥土时，果然是深潭，不谋而合，更证实石涛叠山是以"粉墙为纸，竹石为绘"的。奇峰之东，按石涛布局章法，叠成水岫洞壑，以虚亲实，以幽深烘托峻峭，得到相得益彰的效果。主山的西南与东南处理时，力求产生"未山先麓"、"山欲动而势长"的气韵，使"山的纵横体现出动感，山的潜伏表现静态"。"山房"占地较小，在占边布置假山的中间，留有较大水面使"水

formed Pianshi Shanfang's initial shape.

According to Shi Tao's painting blueprints, there is a three-unit water hall at the south bank of "Shanfang" (Note: Mountain House). It spans the water body and looks into the distance toward the main peak Yi Yun, forming a three dimensional landscape scene. East of the hall extends a zigzagged corridor connecting with the main architecture - the Phoebe Hall. In order to expand the spatial depths, at the west gable of the Phoebe Hall a boat-like building was built. People can enjoy the fish against the railings and it echoes well to the southwest's Banbi Shuwu (Note: Half Study). Deep and dignified Phoebe Hall corresponds nicely with the towering and strong walling mound, forming a solid-bond pair.

The main scene of Pianshi Shanfang should be the mound, whose essence is "Qi Feng" (Note: Rare Peak). Because Shi Tao searched exhaustively in the wild for rare peaks and made them in his sketches, to restore his work well, one needs to "search" among his works for those rare peaks and make existing mounds with a high level of resemblance, trying to revive the charm "with one peak sticking up and followed by connected, abrupt, and varied mound ranges, it appears to be continuous although it is not."

There is a connection between mountain painting and rockery art. When Shi Tao draws a mountain, he paid attention to manage waterfalls by the mountain side and created contrast between void and solid, and dynamic and static. At Pianshi Shanfang, the rocky mound path is placed very close to the precipitous main peak, creating a deep and tortuous undefined void spot. The elevated path is also coupled with a water collecting line, i.e. an open ditch. Due to the uneven design of the rocky slope, mountain flash flood can be created, described in the painting theory as "jumping water", to form waterfalls, showing its ingenuity.

Following the artistic painting principles, below the waterfall a deep pond was designed. During construction, dirt cleanup process revealed that it is indeed a deep pond, and this further proved that Shi Tao's rockery work used "painted walls as papers and bamboos and rocks as ink". East of Qi Feng, according to Shi Tao's design, water caves were constructed. This is to attach void to solid and to set deep and quiet off the precipitous, making them contrast and enhance each other. The handling of the southwest and southeast parts of the main mound attempts to achieve "foothills preceding hill" and "momentum carrying force" effect and charm to enable "dynamic movement of a mountain via layout and still image of a mountain via latency". "Shan Fang" occupies a small lot. While rockeries are placed along the perimeter, a larger area of water is in the center area. This makes "water follows mountains" and "mountains made alive because of water". The misty and ethereal water contrasts well against layers of mountains. If we use the door and window frames of hall in the south, facing the foggy mountains, a perfect picture is created for us to behold a high mountain with awe.

"Shan Fang" is where the garden owner lead his daily life and at the

随山转"、"山因水活"，水之缥缈烘托层崖叠嶂。若以南面厅榭门窗为框，面对岑岚，则画意盎然似寄高山仰止之思。

"山房"是园主的起居场所，同时又是主人寄情的环境，修复时大胆将水引入水榭，围栏成泉，搁琴台，置棋桌，设书房，雅趣顿生。修复时在水岫中利用光的折射原理，"明月"映入碧波，与摇曳芦花相映成趣，拟让游客去领略欧阳修诗句"清风明月本无价，近水远山皆有情"的意境。

修复后的片石山房与史料记载基本吻合，总格局合石涛的山水画创作理论，"得乾坤之理者，山川之质也"，贯山川之形神，体现出"莫谓此中天地小，卷舒收放卓然庐"的意趣和"一峰剥尽一峰环，折径崎岖绕碧湍"、"拟欲寻源最深处，流云飘渺隐仙坛"的诗情，展现在人们眼前的是一帧石涛的山水册页。横览能组成长卷，纵观亦不失为巨幅山水中堂，画意盎然。山、水、石、建筑、花木均不失画家的气度，并使之融为一体，不失为古典园林修复设计一例佳作。荣获1991年江苏省优秀设计一等奖。

附：著名学者同济大学陈从周教授为片石山房撰写碑记原稿。

重修片石山房记

世之叠石能手，皆工画。石涛高名，艺垂千秋，人所共鉴。欲求其构山之作，难矣。然余不信世间未有存者，曩岁客扬州成扬州园林一书，非敢步武画舫录，留真况耳。其时终于发现片石山房，考之乃出石涛之手，孤本也。小颜风范，丘壑犹存。近吴君肇钊就商于余，细心复笔，画本再全，功臣也。石涛有知，亦当含笑九泉。而扬人得永宝此园，洵清福无量矣。

一九九〇年庚午陈从周撰并书

same time, it was also the venue where he embodies his thoughts and feelings. During the restoration, water was daringly introduced into the water hall. The water is then surrounded by railings to form a spring source. A musical instrument table, a chess table, and a study were installed and that instantly created refined and elegant pleasures. During the reconstruction, in the water cave, light reflection and refraction were leveraged to let the "bright moon" into the green water. With swaying reed catkins, visitors can experience Ouyang Xiu's poem, "There is no price tag for clear wind and bright moon but both nearby water and faraway mountains have emotions".

The restored Pianshi Shanfang matches well with what has been recorded historically. Its general layout is in accordance with Shi Tao's landscape painting theory that "once you understand the way that the universe runs, you would know the essence of the natural landscape." To have the unity of form and spirit of the mountains, and reflect the interest and charm of "expressing a full landscape scene even within a limited spatial extent" and the poetic feelings in "peaks after peaks and winding paths surrounding green current" and "seeking deep for the immortal source amidst misty environ". What is presented here is a frame of Shi Tao's landscape painting. Viewing horizontally, we see a scroll; viewing vertically, it is a vertical one. Either way it is very picturesque. Mounds, water, rocks, buildings, and flowers and trees are all with art painters' disposition and they are well-integrated. This is indeed a great example of classic garden restoration work, as it was awarded the first prize of Jiangsu Province Excellent Design Award in 1991.

Note: The following is the original stone tablet inscription written by Chen Congzhou, a famous professor of Tongji University.

Annals on the Restoration Project for Painshi Shanfang

Great masters of rockery art are all good at artistic paintings. The famous Shi Tao's artistic accomplishments are well known to many through his paintings, but it is very difficult to find his rockery work. However, I do not believe there is none left. I once visited Yanzhou and wrote a book on Yangzhou gardens but did not dare to touch and follow what is described in Huafang Lu, just to keep its genuine and authentic status. During that time, Pianshi Shanfang was finally found to be Shi Tao's work after research and it is the only one of his real work left. Although it went into desolate and disrepair, its strength of character still exists via its remaining mounds and gullies. Recently Mr. Wu Zhaozhao consulted with me and through his careful and attentive restoration work the original look re-appeared. He is the man! If Shi Tao knew, he would smile in heaven. The Yangzhou people are fortunate to keep this treasure and would enjoy it immensely.

Written and inscribed by Chen Congzhou in Gengwu year, 1990.

扬州寄啸山庄 Jixiao Shanzhuang of Yangzhou

寄啸山庄位于扬州城东南徐凝门街，系清光绪年间何芷舠购得吴家龙别业片石山房，经扩大、增建而为宅园，故俗称何园。"寄啸"寓"倚南窗以寄傲"之意。

寄啸山庄虽在平地构筑，却有其特色。通过嶙峋的山石，磅礴而又连绵的贴壁山，盘山的蹬阶，置建筑群于山麓陂泽；因地势高低而点缀的厅楼、山亭，蜿蜒错落，逶迤不绝，山、水、建筑浑然一体。步履其间，假山环绕，奇峰异石，幽险之致，有城市山林之誉，这也就是山庄的涵义。"寄啸"二字则与苏州网师园的"网师"，无锡寄畅园的"寄畅"一样，借此表示园主人对当时统治者的不满。陶渊明的"久在樊笼里，复得返自然"，文徵明的"山居为上"，都是想隐居避世，寄情于自然山水的感情流露。我国古典园林的精粹就在于景中有情。寄啸山庄的假山花木等自然造景及亭台楼阁和山水花木的艺术布局，是该园的生境（自然美境界）和画境（艺术美境界），而"寄啸"的寓意，应视为意境（理想美境界）了。

寄啸山庄的总布局分三部分。东部以船厅为主景。南向明间廊柱上，悬有木刻联句"月作主人梅作客，花为四壁船为家"。厅事四周，以瓦石铺装，纹

Jixaio Shanzhuang (Note: Jixiao Mountain Villa) is located at Xuningmen Street, southeast of Yangzhou city. It was a residence garden by expansion after He Zhidao bought Wu Jialong's villa Pianshi Shanfang in Guangxu's reign in the Qing dynasty. So it is commonly referred to as Heyuan. "Jixiao" means "finding sustenance by leaning against the south window to express one's haughty feeling".

Although it was built on a flat ground, Jixiao Shanzhuang has its own characteristics. Through rugged rocks, majestic and continuous Tiebi rockery, winding and climbing step stones, buildings are placed by mound foothills and ponds. Buildings and mound pavilions dot at different elevations with variations and they integrate well as a whole with mounds and waters. Walking through them among all the rockeries and rare and exotic rocks, you could feel tranquil and danger at the same time. This is known as urban mountain and forest and that is what it means by mountain villa. The two words "Ji Xiao" are equivalent to "Wang Shi" of Suzhou's Wangshi Yuan and "Ji Chang" of Wuxi's Ji Chang Yuan, and they all express discontent with their contemporary rulers. Tao Yuanming's "staying in a cage for too long, one feels like returning to nature" and Wen Zhengming's "mountain living is the most favorable" both express the same idea of withdrawing from society and living in solitude, and seek sustenance in natural landscapes. The essence of Chinese classic gardens is to embed emotions in landscape scenes. The rockeries and plants and other natural scenes and all the building and terrain compositions are with the garden's physical realm (natural beauty) and artistic realm (artistic beauty), while the meaning of "Ji Xiao" should belong to the metaphysical realm (idealistic beauty).

The general composition of Jixiao Shanzhuang is made of three parts. The boat hall is the main scene in the east. To the south on the open unit columns hung a woodcut couplets showing "The moon is the host and the plum is the guest; flowers are our walls and the boat is our home". Surrounding the hall are tiled stone pavers with ripple patterns. North of the hall to the upper level of Chuan Lou, there is a half-moon platform, which

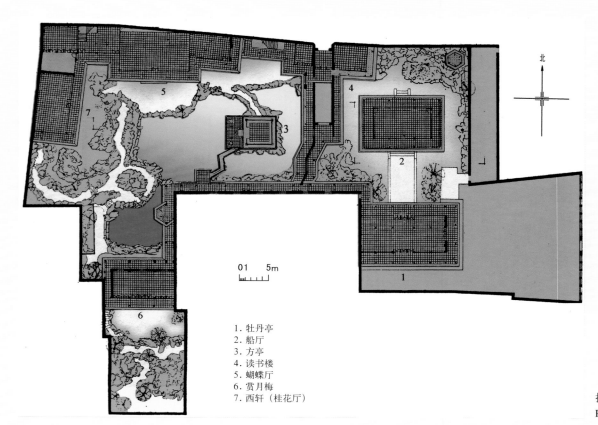

1. 牡丹亭
2. 船厅
3. 方亭
4. 读书楼
5. 蝴蝶厅
6. 赏月梅
7. 西轩（桂花厅）

扬州寄啸山庄平面图
Plan of Jixiao Shanzhuang, Yangzhou

似水波粼粼，厅北串楼上有半月台，与中部串楼尽头旧有半月台互相呼应。一是东观上月色，一为西观落月景。可能是由于唐人徐凝"天下三分明月夜，二分无赖是扬州"的诗句，使园景与明月的关系竟如此密切。

中部凿有鱼池，楼厅廊房环池而建，形成园林空间。池北楼宽七楹，屋顶高低有致。主楼三间稍突，支出两翼，形若蝴蝶，故俗称蝴蝶厅。楼旁连复道廊，与假山贯串分隔，上下脉络相承，形成立体交通、多层欣赏的园庭。廊壁间以漏窗，可互见两面景色，更觉空灵深远。东、中两部分虽隔楼廊，但"景物锁难小牖通"，两区空间通过漏窗得到互相渗透的效果。池东架有石梁，与水心亭贯通，亭南有曲桥一座，与平台相连，最是游人逗留处，亦为纳凉拍曲之地。此亭枕流环楼，兼做戏台，水与建筑的回音，增加了音响效果，四周的回楼廊，又可做观戏的看台，此亦为利用立体空间，而达到小中见大的目的，成为造园佳例。

园的西南，运用"水随山转、山因水活"的画理，池中突起湖石假山一座，拔地峥嵘，向北延展。随着山势层叠起伏，布置了牡丹、芍药台，得体而无做作之态。山南则峰峦陡壁，苍岩隔流，一泓池水，宛转多姿；更有厅事一叠，独占深山大泽一隅，幽粹多态，恰与中部开敞的空间形成对比，使全园透迤衡直，闾爽深密都恰到好处。

寄啸山庄的植物配置，亦经精心设计。半月台傍的梅花、桂花、白皮松、北山麓的牡丹、芍药，南山坡的红枫，庭前的梧影槐荫，建筑角隅补白的芭蕉，山石点苔的书带草，水畔披洒的迎春、黄馨皆情趣妙生，既体现一年的季相布局，又顾及一日之中的晨昏变化。并能在种植上顾及高低参差，枝叶扶疏，使之野趣横生，增添画意。

echoes to the existing half-moon platform at the end of Chuan Lou in the middle. One could be used for enjoying the moon rising in the east, while the other moon setting in the west. This close association between garden scenes and bright moon view is probably derived from the poem "Of three parts of bright moon night, two of which lie in Yangzhou", written by Xu Ning of the Tang dynasty.

A fish pond was dug in the middle section, which is surrounded by buildings and corridors to form garden spaces. The building north of the pond has a width of seven units with various building heights. The main building's three units stick out a bit and its two wings are spread out, shaped like a butterfly, so it is commonly called Butterfly Hall. A double corridor is attached to the side of the building, connected to and separated by the rockery. With the connections between the upper and lower sections, a three dimensional, and multi-layered routing and viewing networks are established for the garden court. Decorative lattice windows are used between the two corridors and as such, scenes at both sides can be enjoyed from either, adding airiness and spacious feeling. The middle and east sections are, although separated by corridors, interconnected through the latticed windows, just like the poem "a seemingly locked scenes could be opened up by small windows". A stone bridge spans east of the pond and connects to the pavilion in the center of the pond. A zigzagged bridge is at the south of the pavilion, connected with the platform and is a place where visitors hang about, enjoy the shade and practice traditional opera singing. This pavilion is built above water and is surrounded by buildings, and it doubles as a stage. The echoes from the water and buildings enhance the acoustic effect. The surrounding corridors can be used as opera watching stands. This is an excellent garden example of effective space utilization and seeing big from a small setting.

Southwest of the garden, in accordance with the drawing principle of "water following mountains and mountains made alive via water", a lake rock made rockery was erected from ground, loft and steep, extending to the north. Peony flower beds are placed on top of cascading mounds without affectation. South of the mound is steep cliff surrounded by water with changing appearance. There is a set of halls occupying a corner of deep mound and water, tranquil with various forms, posing interesting contrast against the mid-section's broad openness, and making an appropriate balance in the whole garden between broad and narrow, open and hidden, shallow and deep.

The plant design of Jixiao Shanzhuang is also well-conceived. Plum flowers, osmanthus, white bark pine by the half-moon platform, north hill's peonies, south slope's red maple, phoenix and locust trees in front of buildings, banana trees that fill the void around building corners, long stem grass by the rocks, spreading jasmine flower by the water bank, reflect the annual seasonal changes and at the same time, show the morning and evening changes throughout the day. Different plant heights are also factored in the plant design and because of this, luxuriant and well-spaced plants help greatly to create wild natural interest that contribute to the picturesque quality of the garden.

Drawing a mountain and building rockeries share some common artistic traits. When Shi Tao drew a mountain, he paid much attention at handling mountain side waterfalls, to create contrast between void and solid, dynamic and static. The existing rockeries of Pianshi Shanfang have its rock climbing path lean against the precipitous main peak to have created deep and tortuous and with an undefined ending void effect. Water catchment line was also lined up along the climbing rock path and it serves as a rain water draining open gully. With various slopes along the path, "mountain torrents" would be created to form what the drawing theory refers to as "jumping over" waterfall, showing its design ingenuity.

苏州壶园 Huyuan of Suzhou

庙堂巷七号壶园位于住宅西侧。门作圆洞形，入门即为走廊，北通一厅，南接一轩，走廊中部有六角半亭一座。园以水池为中心，北、东两面厅廊临水而设，池岸低平。北面厅前平台挑临水池之上，六角亭凌水而建，增加了水面的开阔感。园内不叠假山，仅在池周散置石峰若干，间植海棠、白皮松、蜡梅、南天竹和竹丛等，掩映于水石亭廊之间。池上架桥两座，以沟通水池两岸，小桥低矮简朴，能与水池相称，惟铁制栏杆与全园风格不相协调。园西界墙高兀平板，故在上部开漏窗数方，再蔓以薜荔之类的藤萝，沿墙布置花台、石峰和竹丛树木，形成较为活泼的画面。西北角厅前湖石花台与水池、小桥的结合也较别致。

此园面积仅约 300 平方米，但池水曲折多致，池上小桥及两岸树木湖石错落布置，白皮松斜出池面，空间富有层次变化，无论从南望北或从北望南，都有竹树翳邃的风景构图。小园用水池为主景者以此为佳例。

Huyuan (Note: Kettle Garden) is at Miaotang Lane #7 and it is located west of the residence. The entrance is round shaped and right after entering the gate, it is a corridor. To the north it connects to a hall and to the south, another one. In the middle of the corridor there is a hexagonal half pavilion. The garden is centered around a pond, with east and north side's corridor hall facing the pond with low bank. The platform in front of the north hall hangs over the water, and so does the hexagonal pavilion. This added the broad spread feeling of the water body. There is no rockery in the garden but just a few scattered rocks along the bank, which is interspersed with cherry-apple trees, white bark pines, wintersweet, and bamboo trees. These plants are blended with water, rocks, pavilion and corridors. Two bridges were built to connect the opposing banks. The small bridges are low and simple, very compatible with the pond, except for the wrought iron railings that show inconsistent style against the whole. The west fence wall is tall and flat, so on its upper portion some decorative lattice windows are installed. Vines climb on top of the wall, and flower beds, rock peaks and bamboo groves create a lively and vivacious picture. The combination of lake rock flower bed at hall front in the northwest corner and pond with bridge also forms a unique scene.

This garden only occupies about 300 square meters; however, with changing and tortuous pond, small bridge and lake rocks scattered along the banks with slanting white bark pines sticking out, the space is rich in layered changes. No matter from what direction you place your view, from north to south or otherwise, the scene is always rich with bamboos and other plants. This is a great example of small garden using a pond as the main scene.

壶园园景鸟瞰图　Bird's Eye View of Huyuan

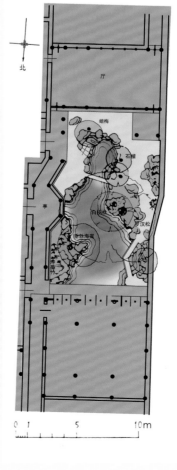

壶园平面图　Plan of Huyuan

苏州畅园 Chang Yuan of Suzhou

庙堂巷二十二号畅园位于住宅东侧。园以水池为中心，周围绕以厅堂、船厅、亭、廊，采用封闭式布局及环形路线，面积虽小（约一亩余），园景却丰富而多层次，是苏州有代表性的小园之一。

园门设于东南角，经门厅及小院至桐华书屋。过此展望全园池水亭廊，视界豁然开朗。水池居园内中心，南北狭长，大部以湖石为岸，疏植花木，近南端以曲桥分水面为二。池东傍水建长廊，曲折迤逦，高低起伏。廊间设小亭两座，南名延辉成趣，平面六角形，北名憩间，为方形半亭，皆一面临水。曲廊与院墙间留有小院，内置湖石，植竹丛、芭蕉，并于廊墙上开洞门和漏窗，构成小品图画。再北有较大的方亭，折西即至全园主厅留云山房，厅南设平台，宽敞平坦，面临水池。经曲廊至池西船厅"涤我尘襟"，此厅平面南北狭长，东向临池，惜其基座僵直，出水过高，权衡欠妥。由此往南过方亭，沿廊升至西南的待月亭。此亭建于假山上，是园内最高处，由此俯瞰，全园在目。由亭顺石级可下石洞，或沿斜廊往桐华书屋，循此便环园一周。

园内建筑物较多，局部处理手法细腻，比例尺度大体能和周围环境相配合，山石花木的布置也做到少而精，给人以精致玲珑的印象。

Located at #22 in Miaotang Lane, Chang Yuan is at the east side of the residence house. The garden is centered around water and at its perimeter, there are halls, boat houses, pavilions, corridors. The closed layout adopts a circular path. Although it is small (about one Mu), the garden scenes are rich with layers of depth, representing a typical small garden in Suzhou.

The garden entrance is at the southeast corner and after entering the entrance and through a small courtyard, you are at the Tonghua Shuwu. Passing this point, you suddenly see a broad view of the whole garden with pavilions, corridors and the pond. The pond is at the center of the garden, whose elongated shape runs from north to south. The bank is mostly dotted with lake rocks with plants. Toward the south part of the pond, a zigzagged bridge divides the water body into two sections. East of the pond a long corridor was built by the water with tortuous turns and as it rises up and down. Two pavilions punctuate the corridor. The southern one named Yanhui Chengqu is hexagonal shaped. The northern one named Qijian (Note: Rest Unit) is a square shaped half pavilion. Both face water at one side. Between the corridor and the garden wall there are small courtyard, in which there are lake rocks, bamboos, and banana trees. Doors and decorative lattice windows are on the corridor wall and they help to frame small vignette scenes. Further north there is a larger square pavilion. Turning west you see the main hall of the garden Liuyun Shanfang (Note: Cloud Retaining Mountain House). South of the hall there is a platform, which is flat and open, facing the pond. Passing the corridor to the west of the pond is the boat house "Diwo Chenjin" (Note: Wash My Dirty Clothes), whose narrow plan runs from north to south. It faces the pond at its east side. Unfortunately, its base is too rigid and too high above the water line, lacking balance. From here walking to the south and passing the square pavilion, one can walk along the corridor and climb up to Daiyue Ting (Note: Moon Waiting Pavilion) in the southwest. This pavilion was built on top of the rockery and is the highest point in the garden. Looking down from here, you can have a panorama view of the whole garden. From here you can walk down the rock steps to the stone cave below, or along the slant corridor to Tonghua Shuwu, and this concludes the circular route of the garden.

There are quite a few buildings in the garden and they have very refined details. Their size and scale generally match the surrounding installations and the rock and plant uses are scant but of high quality, appearing delicate and exquisite.

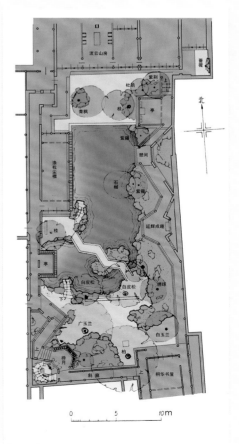

畅园平面图　Plan of Chang Yuan

苏州王洗马巷七号某宅书房庭院 A Study Courtyard at #7 Wangxima Lane in Suzhou

这所住宅建于清光绪年间，书房庭院一区位于住宅的东南一隅，处境僻静。书房四周装置透空的槅扇和槛窗，使室内空间不致有局促感。书房东面正对庭院，南北两侧接以亭廊，西侧设小院，内点湖石，植丹桂，使书房四向均有景可观。院中亭

This residence was built in Guangxu's reign of the Qing dynasty. The first zone of the study courtyard is located at the residence's southeast corner, which is very quiet. Surrounding the study there are hollowed partition boards and windows, avoiding the stuffy feeling in the indoor space. The study faces the courtyard directly to the east and is flanked by corridors at both the south and north sides. In the west there is a small yard with lake rocks dotted in it. In addition, orange osmanthus were planted, and these equipped the study with views at all four sides. The pavilion and corridor buildings' sizes are relatively small. The plants used include osmanthus, crape myrtle, cherry-apple trees, and elecampane and other ornamental seasonal species. Against the east wall, dirt is piled up to make a mound and lake rocks were employed to make rock caves, flower beds and tree pools. There is a winding path that you can use to get to the pavilion, climb up the mound and then go through the cave to get down to the ground. This courtyard uses a small piece of land

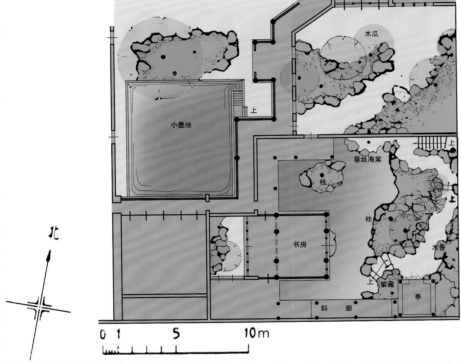

王洗马巷七号某宅书房庭园平面图
Plan of a Study Courtyard at #7 Wangxima Lane

王洗马巷七号某宅书房庭园剖视图
Section View of a Study Courtyard at #7 Wangxima Lane

廊建筑比例尺度较小，周围配植桂、紫薇、海棠、木香等四季观赏花木。靠东墙堆土为阜，用湖石叠成石洞、花台与树池，有曲径可由亭登山再穿洞而下。此院用地很少（约300平方米），但纡曲而有层次，建筑、花木与湖石的布置和空间尺度也较相衬，是当地旧住宅书房庭院有代表性的一例。

(about 300 square meters); however, since it is indirect and has depths of scenes with buildings, plants and rocks in proper composition and spatial scale, it becomes a representative example of local old residence study courtyard garden.

苏州残粒园 Canli Yuan of Suzhou

装驾桥巷三十四号残粒园建于清末，原为扬州某盐商住宅的一部分。住宅有中、东、西三路，园在住宅东路花厅东侧。全园面积很小，约140平方米，仅相当于拙政园远香堂的面积，但能利用空间，将亭子、假山、水池、花木组成曲折高下而有层次的景面。

由住宅后部经圆洞门"锦窠"入园，迎面有湖石峰作屏障。园内布局以水池为中心，花台树丛沿周布置。池岸用湖石叠砌，以石矶挑于池面，东南墙角和池岸边各立石峰与入门处石峰相呼应，错落布置桂、南天竹、蜡梅等花木，增加了风景层次。墙上开漏窗数方，蔓以薜荔、爬墙虎等，使呆板的墙面有所变化。池西紧靠界墙叠

Located at #34 Zhuangjiaqiao Lane, Canli Yuan was built in the end of the Qing dynasty. It was originally part of a residence owned by a salt merchant in Yangzhou. There are middle, east, and west three paths with the residence and the garden is located in the east side of the flower hall at the east road. The whole garden is rather small, occupying for about 140 square meters, equivalent in size of Yuanxiang Tang of Zhuozheng Yuan. However, through its effective use of space, it organizes pavilions, rockeries, pond, and plants into a complex and rich in depths landscape garden.

At the back of the residence, entering the garden through the round-shaped entrance "Jin Chao", you see a lake rock peak as a screen. The garden is centered around the pond and flower beds and trees are placed around the perimeter. The pond bank was built with lake rocks, with the rocky bank overhung over water. Rock peaks are erected at the southeast corner as well as at the pond bank, echoing nicely with the one at the entrance. Landscape plants like osmanthus, bamboos, wintersweet were planted, adding depths of scene. On the wall, there are a few hollowed lattice windows, climbed with vines and ivies,

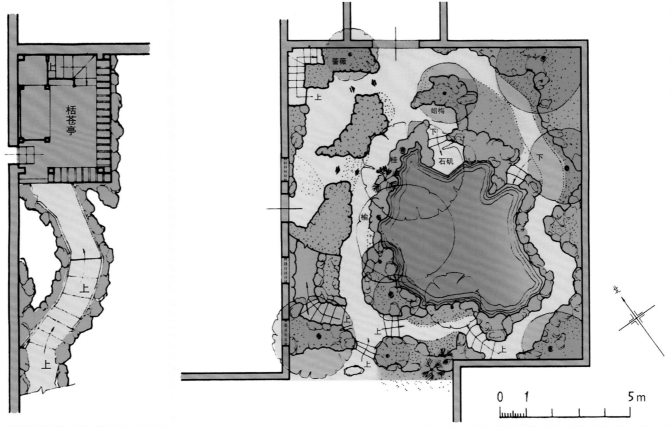

残粒园平面图　Plan of Canli Yuan

湖石假山一座，山中有石洞，入洞循石级上达一半亭名梧苍亭，亭侧有门可通花厅。此亭居全园最高点，是主要观赏处，也是园景的构图中心之一。

此园运用传统小空间处理手法较为成功，半亭、石洞、水池、花台的位置高下相称，尺度适当，组合紧凑。但池岸嫌高，且少起伏，四周墙面也缺少变化，是不足处。

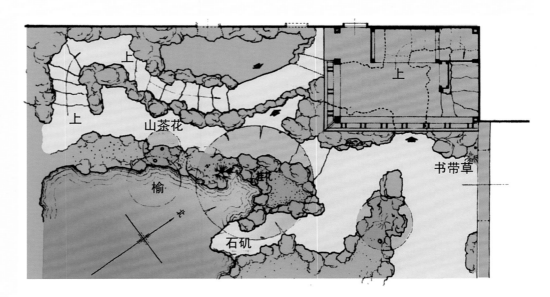

局部平面图　Plan

外观　Appearance

bringing some variations to the otherwise dull and rigid wall. In the west close to the border wall, there is a lake rock made rockery, in which there is a cave. Entering the cave and climbing up the rock stairs one can reach a half pavilion named Guacang Pavilion. Its side door connects to Tonghua Hall. This pavilion is at the highest elevation of the whole garden and thus is the main viewing point and also the garden's composition center.

This garden's success is with its use of traditionally small spaces. Also, its half pavilion, rock cave, pond and flower beds' positions and elevations are properly designed with appropriate scales and compact composition. However, the pond bank is a bit high and lack of elevation changes. The surrounding walls are also in need of variations. These are what make it less than perfect.

（六）江湖地

江干湖畔，深柳疏芦之际，略成小筑，足征大观也。悠悠烟水，澹澹云山；泛泛鱼舟，闲闲鸥鸟。漏层阴而藏阁，迎先月以登台。拍起云流，觞飞霞仔。何如缑岭，堪谐子晋吹箫；欲拟瑶池，若待穆王侍宴。寻闲是福，知享即仙。

(6) River and Lake Land

At the river bank or lake shore, where deep willow groves meet with desolate reeds, with a little construction, one can achieve spectacular effect. Large expanse of water shrouded in fog with misty cloudy mountains. Scattered fishing boats floating on water among flocks of water fowls leisurely flying over. From landscape buildings draped with spotted shadows cast by forest leaves to moon light lit mansions in early evening. Knock on the beat to make flowing cloud dance and drink to one's heart's content to retain some sunglow. Our mountain is just like Mound Gou, and we seem to play flute with Wang Zijin. Our lake is just like Yao Chi, and we seem to dine with Zhou Mu Wang there. If we could get some leisure time then it is happiness. As long as we could seek some pleasure then we are almost like supernatural being.

嘉兴南湖烟雨楼 Yanyu Lou in South Lake of Jiaxing

嘉兴南湖小瀛洲岛上的烟雨楼原是五代吴越国中吴节度使广陵王钱元璙所筑，当时的位置在湖滨。南宋建炎年间（1127—1130年）楼毁，到嘉定年间（1208—1224年）才恢复，以后就成了一方胜景。元末又毁于兵火。明嘉靖二十七年（1548年），知府赵瀛疏浚城河，将余土运入湖中填成小岛，于岛上建楼，并沿用"烟雨楼"旧名。万历十年（1582年），增筑亭榭，并拓楼的南面平台，称为"钓鳌矶"。此后，这里又称"小瀛洲"，和杭州西湖三潭印月所在之岛屿名称相同。其后又在岛之南面筑堤，既可对岛起到保护作用，又可在堤内池中养鱼，还增加了风景层次。

楼作二层，面南。前面除平台外，别无其他建筑遮挡，视野开阔，南湖景色，尽收眼底。最能动人的景色是细雨迷蒙之时，湖上远舟近树如笼于轻纱之中，故以"烟雨"命名此楼，最能起到点睛作用。楼中有楹联恰到好处地说出了楼的佳境所在：

如坐天上，有客皆仙，
烟雨比南朝，多少楼台归画里；
宛在水中，方舟最乐，
湖波胜西子，天边风月落樽前。

楼的后面是一个不规则的开阔庭院，三面环以屋宇及游廊，院中堆假山、植花木，内容较丰富，形成与前面相对比的"奥如"空间。在楼的南、西、北面，还加筑墙一周组成庭园空间，形成曲折多变且宁静的

Yanyu Lou (Note: Misty Building), on the Xiao Yingzhou isle in South Lake of Jiaxing, originally was built by Qian Yuanliao, a secretariat and governor in Guangling of the state of Wuyue in the Five Dynasties time. The location was at the lake shore at that time. It was destroyed in Jianyan's reign in the Song dynasty (1127-1130) and was restored in Jiading's reign (1208-1224). It has since been a local attraction. At the end of the Yuan dynasty it was destroyed again in war. In Jiajing 27th year (1548) of the Ming dynasty, Prefecture Magistrate Zhao Ying dredged the city moat and dumped the surplus dirt into the lake and formed the isle. A building was built on the isle and the old name "Yanyu Lou" was followed. In Wanli 10th year (1582), some other pavilions and halls were added, and in the south of the building, a platform was extended and named as "Diao Ao Ji". Thereafter it was called "Xiao Yingzhou", which is the same as the isle's name on which Santan Yinyue was built in the West Lake in Hangzhou. After that, at the south of the isle, an embankment was built. It serves as a protection to the isle and at the same time, it formed a pond in which fish could be raised. Not to mention it could add depths to landscape scenes.

The building is two storied and faces south. Besides the platform in the south, there is no other building to block the view. With broad field of view, the scenes of South Lake can be enjoyed all at a glance. The most enchanting moment would be when it is in fine and mild drizzles, where all objects, such as distant boats and nearby trees, are shrouded in fog as if seen through a light veil. Hence the building name "Yan Yu" (fog and drizzle), a very apt title. In the building there is an inscribed couplet that best illustrates this wonderful scene:

As if in heaven, all my guests are immortals; the misty rain, like that in south China, makes picturesque architecture.

Just like in water, the boat house is most enjoyable; the lake scenes, better than those of the West Lake, draw wonderful landscape to your wine goblet.

Behind the building is an irregular shaped broad and open courtyard surrounded at three sides by buildings and corridors. In the courtyard, there

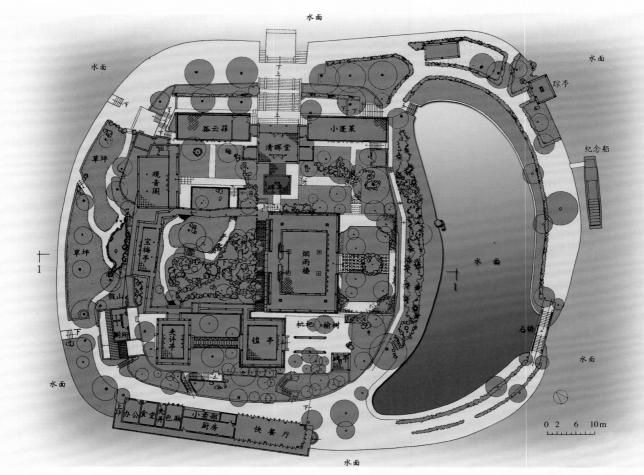

嘉兴烟雨楼平面图　Plan of Yanyu Lou, Jiaxing

嘉兴烟雨楼鸟瞰图　Bird's Eye view of Yanyu Lou, Jiaxing

意趣。楼前两株古银杏树，树干都已在二人合抱以上，枝叶茂密，直冲云霄，为岛的创建年代作了有力的注解。其余树木以朴树最有特色，数量多，树龄老。至若桂、白玉兰、牡丹、樱桃、蜡梅、香樟、槭、盘槐、榆、枇杷等则分布在楼的四周及庭院之中，使烟雨楼处在郁郁葱葱的林木环境之中。而楼周的清晖堂、菱香水榭、孤云簃、宝梅亭、凝碧阁、鉴亭、碑亭等一批辅翼建筑，又烘托着烟雨楼。所以无论从湖滨远望还是从长堤近观，此楼都是前拥后护，左辅右弼，主导地位突出而又不陷于孤立无依，建筑布局是十分成功的，假山花木的陪衬与大水体取得相得益彰的艺术效果。

are rockeries and plants and other rich contents, forming an "Ao Ru", or enjoying the scenes when you get deeper inside, a contrast against previous spatial sequence. At the south, west, and north sides of the building, a fence wall is used to from a yard space, adding interest in variety and tranquility. The two ancient gingko trees, so thick that requires two people to embrace to enclose, are tall and luxuriant, offering a powerful side note for the age of the garden. Other kinds of trees, such as characteristic hackberry tree in particular, are numerous and aged. There are cinnamon, magnolia, peony, cherry, wintersweet, camphora, maple, locust tree, elm and loquat trees and they are located around the building and inside the courtyard, making Yanyu Lou stand in a wild profusion of vegetation. The surrounding Qinghui Tang, Lingxiang Shuixie, Guyun Yi, Baomei Ting, Ningbi Ge, Jian Ting, and Bei Ting and so on act as auxiliary foils to Yanyu Lou. Therefore, no matter if one looks from afar at the lake shore or from the nearby long embankment, the building always looks like that it has large retinue with support from left and right. It has a dominant position but not in isolation. Its architecture layout is very successful and its foils of rockeries and plants complement well with the large expanse of water.

杭州湖园郭庄　Huyuan Guozhuang in Hangzhou

西湖上的私家园林，从唐朝起就开始有人营造了，经过五代吴越国和宋、元、明、清的兴衰交替发展变迁，形成了中国江南园林中风格鲜明、内涵丰富的流派之一——西湖园林，简称《湖园》。其特色，是在杭州天造地设的湖光山色大风景、大环境中，通过独到的选址、巧妙的构思和精心的建造，外师造化，内拓意境。尤其是借景真山真水，妙用名家名篇，汲取乡土风情，从而营造出园中有园、景外有景、情趣盎然的境界和格调。地处西湖杨公堤卧龙桥畔的郭庄，就堪称湖园的经典之作。

郭庄是清朝后期由富商宋端甫改建原先位于庄址所在的一处宋氏族祠堂而出现的。20世纪早期，这座私家园林转让到福建人郭士林名下。郭氏追随当时西湖园林别墅多采用中西合璧手法建造的时尚进行了改建，并据唐朝中兴名将郭子仪的封号《汾阳郡王》而雅称庄园为《汾阳别墅》，杭州地方上则按惯例称之为郭庄。

一九八九年，承上海同济大学陈从周教授的深情指点，曾长期被冷落的郭庄，按照湖园传统的风貌格调和特有的形神品

The private gardens on the West Lake were built as early as the Tang dynasty. Through Wuyue State of the Wudai dynasty all the way to the Song, Yuan, Ming and Qing dynasties, they went through rise and fall with much changes and has formed among Southern Chinese gardens West Lake School, abbreviated as "Hu Yuan", with rich and characteristic styles. Its distinguished features include grand lake and mountain views and through them ingenious siting, designs, and constructions were implemented. They learned externally from nature and built internally artistic conceptions. Especially notables are that real mountains and water were leveraged; famous literatures were referenced, and vernacular styles were adopted. From these, an extremely interesting style of garden within a garden and scene within a scene is created. Guozhuang, which is located by the Wolong Bridge at the Yanggong Dyke of the West Lake, is thought to be such a classic example.

Guozhuang was renovated by a wealthy merchant at the late Qing dynasty and it was originally an ancestral hall of the Song family. Earlier last century, this private garden's title was transferred to Guo Shilin, a gentleman from Fujian. Guo rebuilt it based on the then fashionable hybrid style combining Chinese and western styles and renamed it as "Fenyang Villa" after the title of "Fenyang Junwang" of the famous Tang dynasty general from Zhongxing. It was still locally referred to as Guozhuang customarily.

In 1989, with consultations with deep love from Chen Congzhou, a professor from Tongji University, this long time unfrequented garden was restored, added and dropped, adjusted, and re-created, based on its traditional style and character, from the general garden layout, architectural ornaments to plants and furniture design.

After a comprehensive renovation, a large area of water body that connects

性，从园林布局、建筑装饰到花木布置、室内陈设，一一加以修复、增删、调整、创新。

经过全面修整，在郭庄的北部，辟出与西湖水相通的大片水面。水面上一派天光云影，架设贴水平板曲桥以通往来。

在郭庄的南、东南、西南三部，建构了香雪分春堂、衍芬精舍、静必居、景苏阁、浣藻亭、翠迷廊。堂、舍、居、阁、亭、廊，全都面朝一泓清池，池水与北部水面引自西湖的清波脉络相通。

重建郭庄的北水和南池之间，以两宜轩为屏隔。从两宜轩向南，可见池岸的假山叠石错落参差，池中游鱼追逐。水池周围，远近高低地分布着楼台亭阁。从两宜轩向北，但见水面开旷，林荫披历，与轩南的情景和境界，形成空间上明显的虚实对比与协调的疏密互补。而当人们的目光和心思向郭庄之外的东、北、南方向延伸之时，园外见湖，湖外见堤，堤外见山，山上见塔。诚所谓波光柳态，云影峰姿，旦夕与共，风花雪月，诗梦画境，春去秋来。其艺术魅力可谓——不到郭庄，难识西湖园林。

（此文摘自《郭庄》一书）

with the West Lake was made in the north part of Guozhuang. A broad sky and cloud scene was re-created and a low flat bridge with turns was added for connections.

In its south, southeast, and southwest parts, Xiangxue Fenchun Tang, Yanfen Jingshe, Jingbi Ju, Jingsu Ge, Huanzao Ting, and Cuimi Lang were built. Tang, She, Ju, Ge, Ting, and Lang all face water, which is connected to the water body in the north that again connects to the West Lake.

Between its north water body and the south pool, Lianyi Xuan serves as a separation. From here to the south, one can see various rockery work at the bank and fish in the pond frolicking. Around the pond, buildings were placed with varying distances and elevations. To the north, the water surface is open and broad and with trees at varying heights. This contrasts well with the southern section with changes in void and solid and complements well with the balance in density. When people's views are directed toward outside to the east, north and south directions, scenes with lake beyond the garden, dykes beyond the lake, mountains beyond the dyke, and pagoda on top of mountain can be enjoyed. This is indeed a fantastic scene shifting from water and willows, to clouds and mountains, to mornings and evenings, to flowers in the wind and snow at moon-lit night, to poetic and idyllic dreams, and to the passing of spring and coming of autumn, or all of them together. Its artistic charisma is so great that there a saying that goes, "If one has not visited Guozhuang, he does not understand the gardens of the West Lake."

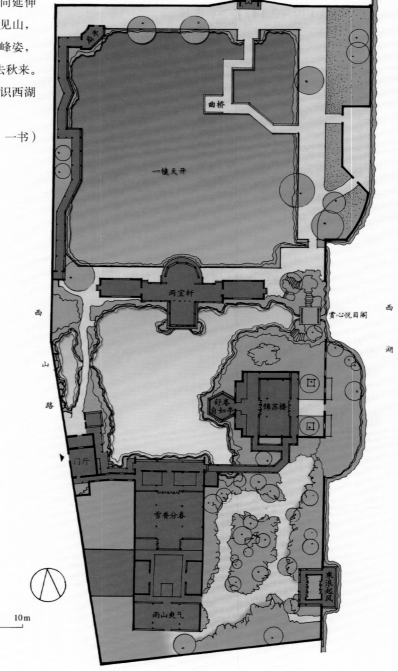

杭州郭庄平面图　Plan of Guozhuang, Hangzhou

杭州三潭印月 Santan Yinyue in Hangzhou

三潭印月是西湖十景之一，在杭州西湖小瀛洲岛。此岛始建于明万历三十五年（1607年），由浚治西湖时的淤泥堆积而成。万历三十九年（1611年），又周以环形围堤而成岛中有湖的格局。以后又在东西方向连以土堤，南北方向连以曲桥，使整个岛的平面呈"田"字形，遂成为别具一格的湖中有岛、岛中有湖的独特布局。早在宋代，西湖十景中已有"三潭印月"之名，可见小瀛洲岛屿堆成之前，此处已成名胜，但只是水上游览点。自从此岛堆成，三潭印月不但可舟游，而且可以陆游，还扩大了游览内容，使这一名胜更加充实丰满起来。全岛范围约300米×300米，水面约占65%，岛上、堤上遍植柳树，夹以石楠、槭、水杉、桂、重阳木、香樟、枫杨、白玉兰等花木。以南北向桥、岛为轴线，由北向南依次布置先贤祠、三角

Santan Yinyue is one of the ten scenes of the West Lake. It is located at the Xiaoyingzhou Isle. This Isle was first built in Wanli 35th year of the Ming dynasty (1607) and it was formed using the dredged out silt from the lake. In Wanli 39th year (1611), it was further surrounded by a circular dyke forming a structure of lake inside an isle. Later on a dirt dyke was built as a connection from east to west and bridges with turns were added to connect the north from the south, forming a pattern that looks like the Chinese character "Tian". This created a unique design of isle in the lake as well as lake in the isle. As early as the Song dynasty, "Santan Yinyue" was already among the ten scenes of the West Lake. This proved that even before Xiaoyingzhou became a scene, this place was already a scenic spot, but it was only one attraction on water. Since the completion of the isle, Santan Yinyue can be used not only for boat tour but also land tour; its range of tourism expanded, making this scenic spot enriched much more. The whole isle is about 300 meters by 300 meters, with water covering 65%. On the isle and the dyke willows are everywhere, interspersed with heather, maple, metasequoia, osmanthus, Bishop wood, camphor, and magnolia trees. Using the north-south oriented bridge and isle as an axis, from the north to the south there are Xianxian Ci, Triangle Pavilion (Kaiwang Ting), Square Pavilion (Tingtingting), Yingcui Xuan, Bird and Flower Hall, Hexagon Pavilion, Woxin Xiangyin Pavilion, Santan Yinyue and other buildings, rockeries, and bridges for sightseeing and resting. East of Xianxian Ci there is a three side walled and one side opened small yard, in which there is a four-sided hall styled building, named "Xianfang Tai". This is a place with a closure feeling on the isle among a predominantly wide open space and it makes people feel a change in spatial handling and adds a sense stability.

"Woxin Xiangyin Pavilion" at the southern tip of the isle is the place one could enjoy Santan Yinyue to the fullest. Santan Yinyue was first built in the Song dynasty when Su Dongpo stationed here in Hangzhou and dredged the lake. He erected three pagodas in the lake to prevent people from growing

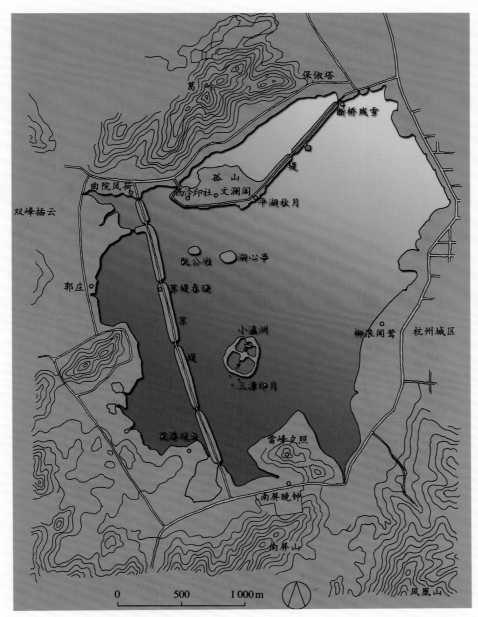

杭州西湖平面图　　Plan of the West Lake, Hangzhou

亭（开网亭）、方亭（亭亭亭）、迎翠轩、花鸟馆、六角亭、我心相印亭、三潭印月等建筑物及湖石峰与曲桥，以供游览休息之用。在先贤祠东侧则有一座三面围墙一面敞开的小院落，院中设四面厅式建筑一座，称为"闲放台"，这是在整个岛区极为开阔空旷的环境下设置一处封闭幽静的境域，使人感到一种空间气氛的变化，并从中获得心情上的安定感。

岛南端的"我心相印亭"是观赏三潭印月的最佳去处。三潭印月始建于宋代苏东坡守杭疏浚西湖时，于湖上立三塔为标志，禁止在此范围内种植菱芡，以免造成湖床淤浅，后此处成为湖中一景。明天启元年（1621年）重立三塔，即现存三石塔，塔高2米左右，作喇嘛塔式，塔身成球形，中空，开五孔圆窗，每当明月当空，塔中燃灯，窗孔明亮，映于水中，与天上明月倒影相伴成趣。从"我心相印亭"南望，三塔对称地布置在全岛的轴线上，使岛与塔连成一个整体。

water chestnuts causing silt deposit. It later became a scenic spot. In Tianqi's first year of the Ming dynasty (1621) the three pagodas were re-constructed and became the current ones we see today. The pagodas are 2 meters tall and they are in the form of Lama tower. The shaft of the pagodas are spherical and with hollow inside it has five round openings like windows. When there is a bright moon and lights are on from inside the pagodas, the opening holes are lit and reflected on the water and set an interesting scene with the reflections of the moon in the lake. Looking south from "Woxin Xiangyin Pavilion", one can see that the three pagodas are placed symmetrically at the central axis of the isle, making the isle and the pagodas an integrated whole.

杭州西湖小瀛洲、三潭印月鸟瞰图
Bird's Eye View of Xiao Yingzhou, Santan Yinyue, Hangzhou

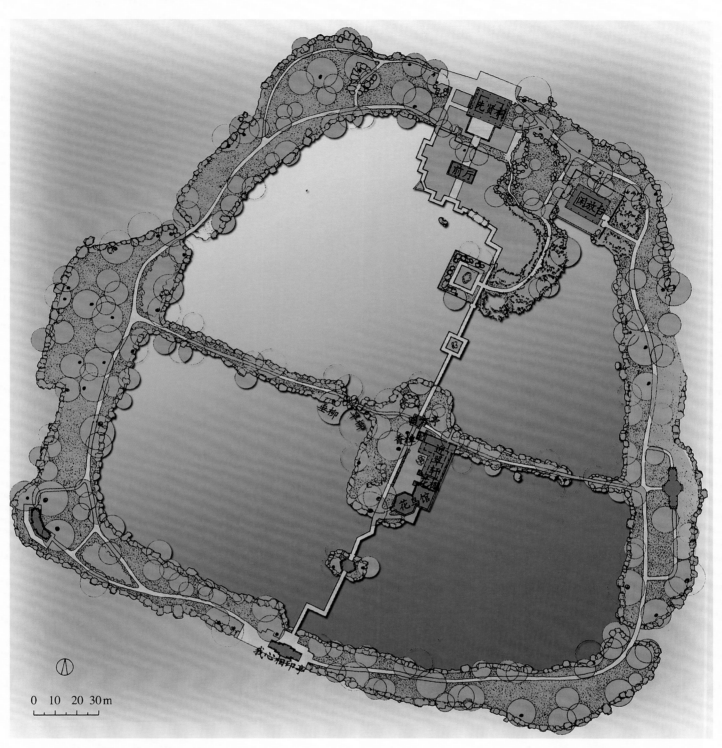

杭州西湖小瀛洲、三潭印月平面图
Plan of Xiao Yingzhou, Santan Yinyue, Hangzhou

扬州二十四桥景区 Twenty Four Bridge Scenic Zone of Yangzhou

瘦西湖长河如带，为接驾乾隆（弘历）皇帝南巡，两岸皆建名园，形成长达十余里的古典园林群，不论规模之大，艺术之高均首屈一指。随着时代变迁，瘦西湖已失去昔日光彩，但就其现状，亦堪称国内现存古典园林群之最。

"之"字形的瘦西湖，景点的精华均集中在河道转折处，尚存最好的为小金山一带。小金山高踞湖中，东边与南边由四桥烟雨、徐园（原名桃花岛）环抱，别具风韵。小金山之西由钓鱼亭、五亭桥、白塔环绕开阔水面，其高低错落的优美画面，已成为扬州的象征，给观光者留下极深刻的印象。五亭桥以西与大明寺南水面交错处，就是二十四桥景区（原为农田）。此景区于1989年完成复建，并获江苏省优秀设计二等奖。其规划设计旨意综述如下：

1. 尊重历史

瘦西湖建设应遵照国家规定，恢复清代最盛期的精粹，所以首先需尊重历史，规划设计之初首先是按历史原貌进行复原工作。该景区是由"春台祝寿"、"玲珑花界"、"望春楼"三部分组成。据《扬州书舫录》记载并故宫博物院所藏乾隆南巡盛典图示，春台祝寿位于湖西岸，由熙春台为主景，集楼、阁、亭、廊、层台于一体，与五亭桥遥遥相对。由于系朝廷所封奉宸卿汪廷璋所建，颇具皇家园林格局：规模恢宏，气魄雄伟。在其湖对岸为白塔晴云景区尾声："小李将军书本"与"望春楼"一组。湖南岸是以栽植芍药为主，以存"扬州芍药天下冠"之誉，园名玲珑花界。依据古园，参照文字作补充，绘制成复原图。

2. 面对现实

鉴于历史原因，以上三景点各自为政，自成体系，共同空间为水体。根据旅游观光者的需要，三景点除水面贯通外，需要有桥梁连系，故玲珑花界与熙春台之间布置为九曲桥。鉴于熙春台与望春楼之间湖面宽达40米，一桥相连必比例失调，故中间设计为圆拱桥，即新"二十四桥"，西侧连以山涧（以通画舫），东侧以临水

Shouxi Lake runs long and in order to prepare for Qianlong (Hongli) emperor's southern tour royal visit, gardens along both banks were built, forming a long chain of garden zone running for more than ten Lis. Not only the extent was immense, the level of artistry was also top notch. With the changing times, Shouxi Lake gradually lost its prime time glory; however, judging from its current condition, it still represents the best in China.

The lake is zigzag shaped and the cream of the scenes are all concentrated around the place where the river turns. The best of what still exists is around Xiaojinshan area. Xiaojinshan perches high in the lake and its east and south is surrounded by Siqian Yanyu and Xu Yuan (originally Taohua Wu), which is full of charm. West of Xiaojinshan is Diaoyu Tai (Note: Fishing Terrace), Five Pavilion Bridge and the white pagoda that surround broad water area. The silhouetted picturesque scenes become symbolic of the Yangzhou city, leaving lasting impression upon visitors. At the intersection of west of Five Pavilion Bridge and south water area of Daming Temple, is the Twenty Four Bridge Scenic Zone (originally farm land). It was rebuilt in 1989 and has won the second place in Jiansu Province's Outstanding Design. Here are its planning and design guidelines.

1. Respect History

The reconstruction of Shouxi Lake should abide by the national regulations to restore the peak time appearance of the Qing dynasty; therefore, we must first respect the history. At the beginning of the planning and design, our work was to restore what they used to be. The scenic zone consists of three parts: "Chuntai Zhushou", "Linglong Huajie", and "Wangchun Lou". According to "Yangzhou Huafang Lu" and the illustrations about the southern tour of Qianlong preserved in the Forbidden City, Chuntai Zhushou is located at the west shore of the lake. Xichun Tai was the main scene and it integrated Lou, Ge, Ting, Lang and Cengtai into one and faced the Five Pavilion Bridge from afar. Since it was built by Wang Tingzhang, a royal gardener, it has an air of royal gardens: it is grand and majestic. Opposite at the other side of the lake shore, is a conclusion of the Baita Qingyun Scenic Zone: "General Xiao Li Huaben" and "Wangchun Lou". South of the lake is mainly planted with peonies and there is a saying that goes "Yangzhou peony tops the world". The garden name is Linglong Jie. Based on the old gardens with the aid of literature, a restoration rendering was made.

2. Face the Reality

Because of historical reasons, the above three scenic spots are self-contained but shared the water body as their common space. Dictated by the needs of tourism, the three scene spots needs to be connected by bridges over water; therefore, between Linglong Huajie and Xichun Tai a nine-turn bridge was built. Since the distance between Xichun Tai and Wangchun Lou is as long as 40 meters, if a bridge connects the two, the scale would not be appropriate. So an arched bridge was built, and that is "Twenty Four Bridges". It connects to the west's mountain gullies (linking to the boat house) and connects to the east on a low and flat bridge. This basically resolved the mutual connection issue.

The three scenic spots in the scenic zone are all residence gardens, which can also be referred to as private villa. Because of this, the garden owners' everyday needs must have been met. According to historical records, the auxiliary building volume was large, so

平桥相贯。基本解决相互联系的交通。

该景区组成的三个景点均为宅园（也可以称为私人别墅）。故需满足园主人诸方起居需要，按记载附属用房面积很大，故按造园艺术的要求给予删减。

3. 基调统一

瘦西湖园林群为接驾之用，故园主均标新立异，以求御览为宠。标新立异产生多样化的园林，然而亦存在基调不一之弊。突出的是熙春台，记载为："飞甍反宇，五色填漆，上覆五色琉璃瓦"，"一片金碧"，"如榄仑山五色云气变成五色流水，令人目迷神恍。"如按此记载进行设计，必于整个瘦西湖的基调不统一，故仅屋面用绿琉璃瓦，其体量尺度介于皇家于私家之间，以求得整体的统一。为突出奉宸苑卿建造的接驾楼阁，故柱窗用紫红色油漆，露台为大块麻石铺装，固以较大尺度白岩石栏板，以增加色彩的鲜艳与对比度，其他景点仍用灰瓦赭色油漆。望春楼用灰绿色矮石栏，铺地踩灰砖拼砖砌，砖缝植草，降低色彩的对比度，以烘托熙春台，产生官式与民式的对比。设计加强翘角之翼然效果，也是为加强熙春台浑厚端庄的气氛；

now we cut it according to the needs of recreational gardens.

3. Consistent Style

Since the garden cluster at Shouxi Lake is for royal reception purpose, all gardens strived for uniqueness in order to get the attention of the emperor. This effort might be good to build diversity but the downside is that it lacks consistent style. A typical example is Xichun Tai, whose annals recorded as using "flying and inverted roof with five colors lacquer filled, on top of which there are five-colored glazed tiles", "magnificent golden color", and "as if Denglun Mountain's five-colored fog turning into five-colored flowing water, making one dizzy". If we followed this historical record for the re-design, the general style of Shouxi Lake would have been inconsistent. Therefore, we used the chromium green glazed tile on the roof and its size is between the royal palace and private residence. This is for achieving general consistency purpose. In order to give prominence to the royal reception building built by the Royal Gardener, the columns between windows were painted purple red, and the terrace was built using sizable granite pavers, surrounded by larger scale white alum stone made railings. This is to enhance the color and brightness contrasts. Other scenic installations still used gray tiles and red brown colored lacquer. Wangchun Lou was built using grayish green short stone railings, gray tiles pavers, and grass was planted between brick seams. This is to lower color contrast so that Xichun Tai could stand out more, making comparison between the governmental vs. folk styles. Its eave corners stick out higher and this was also intended to contrast Xichun Tai's deep and dignified feeling. A big tree by the lake shore was preserved and it hides the Wangchun Lou complex. This is also to ensure the imposing manner of Wangchun Lou so that horse

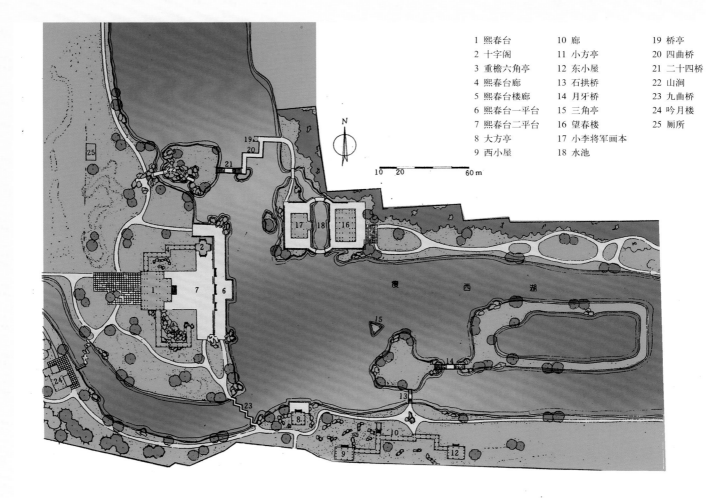

1 熙春台	10 廊	19 桥亭
2 十字阁	11 小方亭	20 四曲桥
3 重檐六角亭	12 东小屋	21 二十四桥
4 熙春台廊	13 石拱桥	22 山涧
5 熙春台楼廊	14 月牙桥	23 九曲桥
6 熙春台一平台	15 三角亭	24 吟月楼
7 熙春台二平台	16 望春楼	25 厕所
8 大方亭	17 小李将军画本	
9 西小屋	18 水池	

扬州二十四桥平面图　Plan of Twenty-Four-Bridge Scenic Zone, Yangzhou

保留湖边大树，隐蔽望春楼一组，以保证熙春台"横可跃马，纵可方轨"的气势。

4. 借古开今

杜牧的二十四桥诗已成为讴歌扬州的千古绝唱，加之《扬州画舫录》记载，二十四桥即是熙春台后的吴家转桥，扬州人妇孺皆知，国际上知名度亦甚高，加之此景区在乾隆水上游览线中起着承前启后的作用，故以二十四桥命名景区，同时这也是扬州人的美好愿望。

既以二十四桥景区命名，景区就要充分体现杜牧诗中的意境。从造园角度来看，要有具象的形象思维，其做法如下：

（1）二十四桥

采用半圆拱桥，长24米，宽2.4米，踏步24阶24根汉白玉栏柱，栏板上雕刻云月图。桥为东西向，登桥北眺蜀冈，远峰呈现"青山隐隐水迢迢"的画境；南望即由熙春台、望春楼、玲珑花界三景点环绕的大面水域，似可领略"清风明月本无价，近水远山皆有情"的诗情与画意。

（2）听箫亭

四方形听箫亭设在与新"二十四桥"相连的平水曲桥北隅，增加桥的分量与气氛。

（3）山涧、天桥

在熙春台与新二十四桥之间，在山涧上架天桥（名"落帆栈道"），以增加高低错落和色彩变化，使之层次丰富。

（4）熙春台

熙春台的平台与建筑是以古园为蓝本的，为使有梦幻境界，在楼南增计"云梯"（室外假山楼梯）直连二楼；自古誉石为云根，故登假山寓"腾云驾雾"之意。熙春台内装修装饰，底层迎面巨幅磨漆壁画"玉女吟月图"，集扬州漆器工艺于一体，用浪漫主义处理手法，再现杜牧诗中的画意。画前陈设月牙形乐台，上置两面琴几一张，上面为演奏古筝、古琴，下面供置香炉，拟产生在烟雾缭绕的朦胧中，玉女吹弹，筝柔琴脆，加之民间丝绸扎染藻井，云纹漏窗以及伴奏的凤凰夹鼓、夹锣器乐，怎能不使人飘飘然有欲仙之概，顿生吊古之情。二楼迎面屏立铜编钟，北墙竹简壁饰，大竹简上融书法、篆刻于一气，均为现代文人墨客吟诵扬州诗句。竹编天棚，悬吊三百余只无影灯笼，色彩各异，如天空星辰熠熠生辉。陈设的琴几、鼓凳为黑

racing and two carriages can be run simultaneously.

4. Seeking Inspirations from the Past

The twenty four bridge poem by Du Mu has been a singing praise for Yangzhou. With additional reference of "Yangzhou Huafang Lu", we know that the twenty four bridge is the Wu family's brick bridge behind Xichun Tai. It is well known by all in Yangzhou, as well as internationally. In addition, since this scenic zone served as a link between past and future as a water route for Qianlong's royal visit, using the twenty four bridge's name as that for the scenic area is also what the local people's fond wish.

If we decided to use the twenty four bridge scenic area for scene names, they should fully reflect what is expected in Du Mu's poems. From garden design's perspective, it should have some figurative imaginations. Here are the implementation details:

(1) Twenty Four Bridges

It is a semi-circular arch bridge with a length of 24 meters and width 2.4 meters. There are 24 stairs and 24 railing columns. The panels between railings have moon and cloud drawings. The bridge is east-west oriented. Looking to the north, one sees Shu Gang, and the distant peaks shows the artistic scene as "looming verdant mountains and distant waters". Looking to the south, one sees Xichun Tai, Wangchun Lou, and Linglong Huajie three scene spots surrounded by large water area, and could savor the poetic and artistic state reflected in the poem "The clear wind and bright moon are priceless; nearby water and faraway mountain have feelings and emotions."

(2) Tingxiao Ting

The square-shaped Tingxiao Ting (Note: Flute Appreciation Pavilion) is located at the north side of the low and flat zigzag bridge that connects to the new "Twenty Four Bridge". This adds to its weight and enhances its ambience.

(3) Gullies and Flyovers

Between Chunxi Tai and the new Twenty Four Bridge, a flyover bridge was built across the mountain gully (named as "Luofan Zhandao"), to enrich the changes in elevation and color, adding scene depths.

(4) Xichun Tai

The platform and architecture of Xichun Tai were built based on blueprints in the old times. To help create its fantasy state, a "cloud stairway" (outdoor rockery stair) was added to the south of the building leading directly to the second floor. Since the ancient times, rocks have been praised as the roots of clouds; therefore, mount a rockery is like riding on the clouds. As for the internal decoration, on its base floor, facing the entrance there is a large grinding lacquer mural "Yunǔ Yinyue Tu". It is a collective craft art of lacquer work that re-creates the idyllic scenes in Du Mu's poem using romantic touches. In front of the painting a crescent shaped band stand is set, on which there is a two-storied instrument table. The upper level is for playing ancient Zheng and instrument; the lower part is for placing an incense burner. The purpose is to create a hazy cloudy ambient atmosphere where a lady plays beautiful music with an instrument under decorative silk caisson ceiling by decorative lattice windows. With Phoenix folder drums and folder gongs playing at the side, how can you not be carried away with this sense of nostalgia? On the second floor facing the entrance a set of bronze bells – a concussion instrument – was erected and on the north wall, there is a bamboo decoration, on which Chinese calligraphy and seal cutting were integrated into one. The inscriptions are all the historical poems in praise of

色漆器银朱描线，各种色彩彩均统一于紫红色的地毯，古色古香。台北十字阁装修，以青龙、白虎、朱雀、玄武四方之神灯和仿古铜镜为特色。以上装修以烘托杜牧诗中的音乐境界为主题，亦宣扬扬州自唐以来的古代文化艺术。

(5) 小李将军画本、望春楼

该两座建筑除按古园恢复其外貌，在设计上避免重复扬州已有的楼房。重檐形式的小李将军画本除增加东面小卷与望春楼呼应外，重点在产生唐代画家李昭道（又名小李将军）册页的效果，故西边设扇面景窗，收入熙春台景观，东边六角景窗、南北边矩形景窗，有"画幅窗"（又名"无心画"）的效果，证其联句"万井楼台疑绣画"。五开间望春楼的布局，不同凡响，下层南北两间分别为水院、山庭，将室外自然环境引入室内（即现代的室内外空间渗透手法）野趣横生。登楼西眺，柱楣中收入玲珑花界、熙春台。构成的画面俗称"夹景"，东望五亭桥，白塔尽收眼底。楼上皆为活动门窗，无论是炎夏纳凉，还是中秋赏月，卸去门窗则形成露台，方便娱乐活动。

(6) 水上游程

二十四桥景区是乾隆水上游览线的精华部分，故需注意水上观景效果。水上游览若从五亭桥岛南西行，至玲珑花界折而北，过明月桥即进入该景区，迎面为"浅黄轻绿映楼台"，折向西，毗邻吹箫亭而过。前面则"碧瓦朱甍照城郭"；再北行，落帆栈道、二十四桥、曲桥之字形屏列，听箫亭点缀其间，过二十四桥圆拱，"北榭远峰闲目望"；向北纵览则"青山隐隐水迢迢"的画面飘然而至，"宽窄收放、高低错落"的变化尽在其中，每个转折对景都似一幅山水条画屏，若联系起来则组成一幅洋洋大观的山水长卷了。

(5) General Xiaoli's Huaben, Wangchun Lou

The two buildings were not only restorations based on the ancient drawings, on its design, an intentional effort was made to avoid duplications of existing buildings in Yangzhou. For the double-eaved General Xiaoli's Huaben, besides the fact that it added an east Xiaojuan to echo Wangchun Lou to bring about a scroll page effect characterized by Tang dynasty's painter Li Zhaodao's (also known as General Xiaoli) work, it also added at the west side fan-shaped scene windows, taking in scenes from Xichun Tai, and at the east hexagonal scene windows as well as rectangular scene windows in the south and north sides. This has a "scroll window" (also known as "hollow painting") effect, which confirms with the poem "thousands of buildings and scenes thought to be embroidery paintings". The layout of the five-unit Wangchun Lou is an extraordinary design. The lower floor's south and north units are water court and mountain court respectively, introducing the outdoor natural landscape indoors (same as modern architecture's interchanging of external and internal spaces) and thus full of natural interests. Getting onto the second floor and looking west, a picture containing Linglong Huajie and Xichun Tai is framed between the columns. The picture formed is commonly known as "sandwiched scene". Looking east toward the Five Pavilion Bridge, the white pagoda is within sight. Upstairs there are detachable doors and windows. When there is a need for shade in hot summer or moon appreciation at the Mid-Autumn Festival, it is convenient for leisure activities to have the doors and windows removed.

(6) Scenic Route on Water

The Twenty-Four Bridge scenic zone is the core part of Qianlong's water route and special waterscapes need to be maintained. For a water tour, if one departs from the Five Pavilion Bridge to southwest, turning north at Linglong Jie, passing Bright Moon Bridge to enter the zone, he sees "light yellowish green water reflecting the buildings". Turning west, one passes Chuixiao Ting. Going forward, one sees "town houses with blue tiles and red roof". Further north, Luofan Zhandao, Twenty-Four Bridge, and the winding bridge are placed in a zigzag way, punctuated by Tingxiao Ting. Passing the Twenty-Four Bridge's arch, one sees "the building in the north against distance peaks". Looking to the north it is "looming green mountains over distant water". Changes in "narrowness, openness, and elevations" are blended together, making every turn blessed with a view like natural landscape paintings. Linking them together a long scroll of natural landscape paintings is unfolding consecutively.

扬州二十四桥景区鸟瞰图　Bird's Eye View of Twenty-Four-Bridge Scenic Zone, Yangzhou

园冶
图释

Yuan Ye Illustrated
——Classical Chinese Garden Explained

二 立基

2 Base

凡园圃立基，定厅堂为主，先乎取景，妙在朝南。倘有乔木数林，仅就中庭一二。筑垣须广，空地多存，任意为持，听从排布，择成馆舍，余构亭台，格式随宜，栽培得致。选向非拘宅相，安门须合厅方。开土堆山，沿池驳岸。曲曲一湾柳月，濯魄清波；遥遥十里荷风，递香幽室。编篱种菊，因之陶令当年。锄岭栽梅，可并庾公故迹。寻幽移竹，对景莳花。桃李不言，似通津信。池塘倒影，拟入鲛宫。一派涵秋，重阴结夏。疏水若为无尽，断处通桥；开林须酌有因，按时架屋。房廊蜒蜿，楼阁崔巍，动"江流天地外"之情，合"山色有无中"之句。适兴平芜眺远，壮观乔岳瞻遥；高阜可培，低方宜挖。

（一）厅堂基

厅堂立基，古以五间三间为率。须量地广窄，四间亦可，四间半亦可，再不能展舒，三间半亦可。深奥曲折，通前达后，全在斯半间中，生出幻境也。凡立园林，必当如式。

（二）楼阁基

楼阁之基，依次序定在厅堂之后，何不立半山半水之间，有二层三层之说，下望上是楼，山半拟为平屋，更上一层，可穷千里目也。

（三）门楼基

园林屋宇，虽无方向，惟门楼基，要依厅堂方向，合宜则立。

（四）书房基

书房之基，立于园林者，无拘内外，择偏僻处，随便通园，令游人莫知有此。内构斋、馆、房、室，借外景，自然幽雅，深得山林之趣。如另筑，先相基形：方、圆、长、扁、广、阔、曲、狭，势如前厅堂基余半间中，自然深奥。或楼或屋，或廊或榭，按基形式，临机应变而立。

（五）亭榭基

花间隐榭，水际安亭，斯园林而得致者。惟榭只隐花间？亭胡拘水际！通泉竹里，按景山颠；或翠筠茂密之阿，苍松蟠郁之麓；或借濠濮之上，入想观鱼；倘支沧浪之中，非歌濯足。亭安有式，基立无凭。

（六）廊房基

廊基未立，地局先留，或余屋之前后，渐通林许。蹑山腰，落水面，任高低曲折，自然断续蜿蜒，园林中不可少斯一断境界。

All garden bases are determined by buildings after confirmed with the location of scene objects. One must first find a view, better oriented towards the south. If there are already a few trees, one or two need to be preserved in the central court. Walls are built to divide spaces and scenic zone extent should be as broad as possible. More vacant slots should be kept to maintain the original design concept of the garden and also to help reasonably arrange scenic layout. Appropriate locations should be chosen for buildings, and pavilions and platforms should be scattered around. The architectural style should be compatible with the general landscape design layout. All plants should be grown to display rich and varied interests. The orientation of buildings should not be affected by Fengshui factors and (buildings near the main entrance of the garden) should line up. Mounds should be piled up using the dirt dug out from the pond and rocks should be installed along the pond banks. Bright moon and willow branches should be reflected in the winding pond water and the moonlight should be mingled with water ripples to create twinkling interests. Clear breeze blows through miles of lotus pond, bringing fragrant aroma into quiet residence homes. Chrysanthemum can be used to make garden fences, just like what Tao Yuanming did. Paths can be created and plum flowers can be grown, just like what General Geng did in the past. If one wants tranquility, a bamboo grove can be grown; if one wants scenic spots, flowers can be planted. Winding paths under quiet peach and plum trees can lead people to cross; clear pond water with the reflection of buildings can trick people into thinking whether they can enter them. Water scenes may carry a sense of autumn and heavy shady areas can help alleviate heat in the hot summer. If you want to have a sense of infinity, you need to add a bridge at the end of a water body. To get seasons and special interests, it should be under consideration when to grow garden plants, which appropriate garden buildings should match. Buildings and corridors need to be tortuous, extended, built up higher to add a feeling of "rivers flowing beyond" and blended in a scene of "looming landscape". To solicit a magnificent mind, one can look far towards the high mountain peaks. One may add more dirt to make a high place much higher and may dig deeper at the place where it is already low to make it even deeper.

1) Base of Halls

To build a base of a hall, in ancient times, it was typically three or five units as standard. However, to build a garden hall, one must consider the actual site's size. A broader site can allow four units, or four and a half. If it is narrow and hard to expand, three and a half is also fine. The depths and complexities of a garden and its connections throughout all depend on this half unit, and all the spatial variations and illusions are from here. All garden designers must master this principle.

2) Base of Multi-Story Buildings

According to the traditional spatial sequence, the base for multi-stories must be behind a hall. Why not set it at a mountain slope or by the water? There is a kind of building that can be either one story or two stories. Looking up from down under it, there is a two-story building; however if you look from the hillside, it is a single story building. When you step in from there, you are already at the second story, and now you could enjoy "a grander view by ascending to another story".

3) Base of an Entrance Gate

Garden buildings have no fixed orientation rules, except for those at the garden entrance. Its orientation must be consistent with the main hall, and this is in accordance with the general garden layout.

4) Base of Studies

The base location of a study in a garden, whether it is inside or outside a scenic zone, should be a little remote, yet well connected to other parts of the garden to make visitors unsure about the exact location of the study. A study can be built in the form of a Zhai, Guan, Fang, Shi etc. Leveraging outside scenes, it should be natural and elegant and full of natural interests. If a study is built outside a garden, a survey of the site is necessary. Whether it is a square or round shape, an elongated or broad shape, a tortuous or narrow shape, its form must be based on the "leftover half unit" in the previously described "Base of Halls". Only through this manner can a hidden, quiet, private and natural space be formed. A study can be built in the form of a multi-storied

（七）假山基

假山之基，约大半在水中立起。先量顶之高大，才定基之浅深。掇石须知占天，围土必然占地，最忌居中，更宜散漫。

building or a regular one, or made into a Fanglang or Tingxie, which all depends on the shape of the site and varies according to the context.

5) Base of Pavilion and Xie

Generally, a Xie is privately built among flowers and a pavilion is built by the water. This is a common practice to generate interests. However, every rule has exceptions and a Xie does not have to be built among flowers and a pavilion does not have to be by the water. For example, a pavilion can be built by a stream flowing under bamboo grove or on a scenic mountain peak. It can also be built on a hill covered with thick vegetation or a foothill with verdant pines. A pavilion can also be built on a bridge for people to enjoy the fish by the rails. If a pavilion is built inside a pond, then it can be used to purify one's mind. The structure of a pavilion has a fixed format but there is no strict rule to build its base.

6) Base of Corridor Building

Before the base of a corridor is set, one must reserve sufficient pathways. One may plan for it in front of or behind a building and it could gradually lead to a forest. Climb up the mountain or step down to water, and follow the terrain just like the movement of a snake. This is an indispensable experience with garden design.

7) Base of a Rockery

Most of the bases of a rockery are built from the bottom of a water body. One must first measure the peak's size to be able to determine the base's depth. To pile up rocks one must know how to take advantage of space. When adding dirt, one must realize that land is occupied. The location of a rockery should avoid the center spot and should spread apart and around.

早在 20 世纪 70 年代，笔者被学科泰斗汪菊渊老先生选为《中国古代园林史》编委，重点编撰江南古典园林部分。当时汪老强调力求寻找到计成造园的实物，经多年史料稽考和实地考察，只找到计成在扬州建造的"影园"遗址。并为此依据众多史料进行了影园复原研究，发表"计成与影园兴造"论文于中国建筑工业出版社《建筑师》第 23 期（1982 年）。2000 年江苏省建委偕同扬州市园林局申报并获批准《恢复性重建影园》科研项目计划，并成立以园林大师朱有玠先生为组长，孟兆祯院士、潘谷西教授为副组长的国家级专家组。在研究、制定复建影园的大纲同时，决定由笔者执笔绘制规划设计方案，历时一年。2000 年 12 月由江苏省建委主持，扬州市政府及相关的园林局、规划局、文管会、文物考古所以及国家级专家共计三十余人论证笔者设计方案。（附：《影园》总体规划方案专家论证会评审意见）

根据评审意见，笔者继续完成总体规划设计，包括图解、文字、模型三部分。并将成果发表论文《借长洲镂奇园——影

As early as 1970s, the author was selected by the profession's authority Mr. Wang Juyuan as an editorial board member for "Chinese Ancient Garden History", focusing on classical Chinese gardens south of the Yangtze River. At that time, Mr. Wang stressed the need to look for real relics of Ji Cheng's gardens. After years of research and field study, only "Ying Yuan", a garden built by Ji Cheng in Yangzhou, was found. Based on this finding, many historical researches on Ying Yuan's restoration were performed. A research paper, "Ji Cheng and the Construction of Ying Yuan", was published in "The Architect" Volume 23 (1982) by Chinese Construction Industry Press. In 2000, the Construction Commission of Jiangsu Province and Bureau of Gardens and Parks of Yangzhou submitted a research project "Restorative Construction of Ying Yuan" and received approval. A garden design masters' team was formed and this national level group was led by Master Zhu Youjie, and Academician Meng Zhaozhen and Professor Pan Guxi served as deputy directors. At the same time when the research and establishment of reconstruction outline were being worked on, it was decided to task the author to make the planning and design drawings, and that took a year. In December 2000, under the direction of the Construction Board of Jiangsu Province, a group of 30 people participated in the project's planning and design presentation and they are from Yangzhou's government and its associated Bureau of Gardens and Parks, Planning Administration, Cultural Relic Management Committee, Institute of Cultural Archeological Research and other relative national institutes. (Attached is "Ying Yuan" overall plan project after expert evaluation meeting.)

园复建设计识语》于《中国园林》2005年第九期。文章首先考证出影园是计成设计并主持施工的，另影园位置通过史料记载与文物部门考古勘查、试掘报告，证实笔者确认的位置是准确的。文中影园恢复性重建规划设计包括：总体规划原则、山景、水景、建筑、植物以及影园恢复性重建设计的启示。论文采用图文并茂的方式。

文章重点是准确复原，但笔者从复原的影园实体发现诸如：选址、借景、延山引水、地形、建筑等，均与计成《园冶》中的描述惟妙惟肖，由此可悟出：园冶书中的论述均是其实践与实例的总结。

本章节"立基"是谈园林的规划与设计，笔者觉得用复原后影园实例来比对，是读懂立基篇最佳方法，其一，影园是计成的作品；其二，图文并茂更易理解与读懂原文。

Per the evaluations, the author continued to finish the master planning and design, which include three parts: illustrations, texts, and architectural models. The result was published as a paper in "Build an Exquisite Garden on a Long Isle – Comments on Ying Yuan's Restoration" in "Chinese Landscape Architecture" magazine issue number 9 in 2005. The paper first indicated that Ying Yuan was designed by Ji Cheng and the construction was also directed by him. In addition, according to historical records, archeological survey and explorative excavation work by the department of cultural archeology, it is proved that the location confirmed by the author was accurate. In the paper, the restorative construction planning and design of Ying Yuan include principles of master planning, mountain view, water view, architecture, plant design and inspirations and revelations of the restorative project. The paper combines texts with graphics.

The focus of the paper is on how to precisely restore the original design. From the restored physical installations, the author found that the siting, scene leveraging, mountain extending and water channeling, terrain treatment and building construction and so on all corresponded extremely well to what Ji Cheng's "Yuan Ye" describes. From this we could realize and conclude that all the discourses in the book are the summaries of practices and real examples.

The Base section in this chapter is about the planning and design of a garden. The author argues that using the real restored Ying Yuan as an example is an effective way to understand the Base. It is because that 1) Ying Yuan is the work of Ji Cheng and 2) Ying Yuan's illustrations contain both texts and graphics and they help understand the original scripts.

2000年12月11日—13日在扬州市荷花池公园、扬州金陵西湖山庄召开了"影园"总体规划方案专家论证会。会议由江苏省建设厅俞惠珍处长主持，扬州市园林管理局具体组织。参加单位及人员有：朱泽民副市长、宋建国副秘书长、市建委、市园林局、市规划局、市文管会、唐城文物考古队等单位负责人、专家30余人，均对方案发表了意见。一致认为这是一个高水平的总体规划。现将国家园林界顶级专家评审结果如下：

《影园》总体规划方案专家论证会评审意见

评委认为：影园总体规划方案是依据考古发掘资料、文献材料，并融入造园哲师计成《园冶》专著的理法、董其昌山水画意、扬州园林风格。故依据翔实，定性、定位准确，思路清晰。化文字资料为具象的总平面设计，基本上将历史文献对影园的描写具体地表述出来。

这是一个高水平的总体规划方案。现提出进一步完善方案的意见如下：
一、编制影园外围环境的规划，并与城市规划协调。
二、综合专家意见，继续完善规划，并制作出模型，在模型上进行推敲。
三、总体规划应包括图解、文字、模型三部分。
四、总体规划的进一步完善，应提高到国家水平的学术研究的高度来完成。
五、建议科学研究部分也要出一个高水平的具体成果。

中国工程院院士	北京林业大学教授	
清华大学	教授	
中国风景园林学会	常务副理事长	
华南理工大学	教授	
东南大学	教授	

二〇〇〇年十二月十三日

借长洲镂奇园——影园复建设计

Building an Exquisite Garden on a Long Isle – Ying Yuan Restorative Design

"影园"与计成有什么关系呢？据《扬州揽胜录》记载："在西门外花山涧南，扫垢山麓，与瘦西湖通，湖中长屿上南段，为明末清初郑元勋先生影园故址。"可知影园是郑元勋的宅园，而郑元勋是计成的挚友，计成是否参与影园的建设，从郑元勋的墨迹中可以看出。郑元勋在为《园冶》一书的题词中写道："即予卜筑城南，芦汀柳岸之间，仅广十笏，经无否略为区画，别现灵幽"。又曰："予自负少解结构，质之无否，愧为拙鸠，宇内不少名流韵士，小筑卧游，何不问途无否"？在郑元勋的《影园自记》中又写到："又以吴友计无否善解人意，意之所向，指挥匠石百不失一，故无毁画之恨。"从郑元勋的说法，可以得到以下结论：

（1）影园的位置是在扬州城南湖中长屿上，也就是现荷花池以北、西门桥之南。东、西为内、外城河夹。

（2）影园不但是计成规划设计的，而且还是他亲自指挥施工的，园主人对其高超的造园艺术十分钦佩，评价很高。

（3）影园建造的时间，据《五架书屋文苑实录》载："……七年（即崇祯七年，1634年）得京口造园名家计无否赞助，经年园成。"从目前史籍稽考来看，可能是计成晚年最后建的一个园林，所以造园艺术更趋成熟。

2000年12月由江苏省建委主持，扬州市政府及相关的园林局、规划局、文管会、文物考古所以及国内知名专家召开"影园"总体规划方案专家论证会，并形成"专家论证会评审意见"：肯定了影园复原设计是准确的。

1 影园恢复性重建规划设计
1.1 总体设计原则

关于影园原有的概貌，茅元仪有评价："于尺幅之间，变化错综，出人意料，疑鬼疑神，如幻如蜃……"，是"内含深远意境，外具自然风韵、情景交融、绘声绘色的形象化的诗篇和立体画卷。"影园造园的目的就是于山影、水影、柳影之间

What does "Ying Yuan" have to do with Ji Cheng? According to "Records of Yangzhou Attractions", "Out of the West Gate and to the south of Flower Mountain gully at Saogou foothills where it is connected with Shouxi Lake, on the south part of Changyu in the lake, is located the ruins of Ying Yuan, a property of Mr. Zheng Yuanxun in the end of the Ming dynasty and early Qing dynasty." From this we know that Ying Yuan is Zheng Yuanxun's residence garden and Zheng Yuanxun is Ji Cheng's close friend. Whether Ji Cheng participated in Ying Yuan's construction we could tell from Zheng Yuanxun's calligraphic work. In the dedication that Zheng Yuanxun wrote for "Yuan Ye", "As for my small residence in the south town, it is between an isle with reeds and a bank with willow and the size is very small. With some design by Wupi (Note: Zi, or another name of Ji Cheng), it suddenly became exquisitely scenic." Also, "I thought I knew how a garden works; however, compared to Wupi, I'm almost like a clumsy bird unable to build a nest. There are a lot of people with good taste, but if one wants to build a recreational garden, how can he not consult Wupi?" Zheng Yuanxun wrote again in "Ying Yuan Notes", "My friend Wupi is very understanding and thoughtful. He directed many workers and craftsmen at garden construction sites and never lost once. He didn't leave regret in his design work." According to what Zheng Yuanxun said, we could come to this conclusions:

(1) The location of Ying Yuan is at Changyu isle in the lake south of Yangzhou city, about the same location north of the lotus pond and south of Ximen Bridge. It is sandwiched in the east and west by the inner and the outer city moats.

(2) Ying Yuan was not only designed by Ji Cheng but the construction work was also directed by him. The owner of the garden gave high compliments on his garden design art.

(3) The date of Ying Yuan's construction, according to "Wujia Shuwu Wenyuan Records", "…seventh year, i.e. Chongzhen 7th year (1643), we had sponsorship from the well-known garden designer Ji Wupi from Jingkou and it lasted for a year to be completed." From what we had found from our historical research, it may be the last garden design work that Ji Cheng had done in his late years; therefore, his garden design skills tended to be maturer.

In December, 2000, under the auspices of Jiangsu Province Construction Board, Yangzhou city government and associated Bureau of Gardens and Parks, Bureau of Planning, Cultural Relics Management Committee, Archeological Institute and some well-known national experts convened to discuss and evaluate Ying Yuan's general planning and formed "Expert Discussion Review Comments", which gave approval to the restoration project.

1. Ying Yuan Restorative Rebuilding Planning and Design
1.1 General Design Principle

Regarding Ying Yuan's original look, Mao Yuanyi commented, "Within a small extent of space, there are so many variations or changes. A lot of them are really unexpected and illusionary…" It "contains deep artistic concept internally and possesses natural charm externally; people's inner emotions intermingle with the outer scenes, creating vivid poetic visualizations and real-life landscape scrolls." The goal of Ying Yuan is to

"是足娱慰"，塑造山水环境的艺术空间，使园主人"得闲即诣，随与携游。"

从文献资料来看，影园面积不足10亩（0.67公顷，不包括菜园和花圃部分），在江南属中小型规模的宅园。其布局特点大致可归纳为以下几点：

（1）"虽由人作，宛自天开"

影园是以水为中心、山为衬托的大山水环抱中的园林境地。南借江南连绵的宁镇山脉，北眺"苍莽浑朴，欲露英雄本色"的蜀冈，东隔内城河和南门城墙，西面"柳外长河，河对岸，亦高柳，阎氏园、冯氏园、员氏园皆在目，园虽颓而茂，竹木若为吾有"，达到了极目所至，佳景"靡不物私其所有"。通过借景突破自身在空间上的局限，延伸视野的深度和广度，使人作的园林与天开的山水巧妙融合，天开的山水是骨架，而精神则在园子。正如计成说："借者：园虽别内外，得景则无拘远近，晴峦耸秀，绀宇凌空；极目所至，俗则屏之，嘉则收之，不分町疃，尽为烟景，斯所谓'巧而得体'者也。"好的布局来源于对园林整体空间的各种环境的丰富想象和高度概括，可谓"景以境生"。

（2）"多方景胜，咫尺山林"

被内、外城河环抱，影园位于湖中的岛上，形成了"湖中有岛"的格局，而岛中内湖上的玉勾草堂，又是"湖中有岛"的设置，这样从总体上就形成了"湖中有湖"、"岛中有岛"、步步深入、层层叠叠的空间布局，显得格外深邃、含蓄。

（3）"略成小筑，足征大观"

全园建筑量较少，为使建筑融入大自然之中，故采用散点式布置，建筑因景而生，体现了一种疏朗、质朴的自然情调。

（4）"涉门成趣，得景随形"

依自然之势，巧妙安排观景路线，并有节奏地串联起园中大小空间，于曲折变化之中求得空间上的深度、广度，极大增加了园子本身的层次，凡人之所处，目之所极，皆能感受到诗情画意。

1.2 山水关系

位于苏北平原扬州的影园，要塑造出山环水抱的景观确实不易。记载中的影园是一幅"列千寻之丛翠"、"浚一派之长源"的山水构图主题画，山与水创造了声色俱全的园林景观，令人玩味。笔者在研究有关影园史料的基础上，再现出影园山水、

"enjoy to the fullest" among images of mountains, waters, and trees. It is to create artistic space of natural landscape, enabling the garden owner to have a "leisurely tour".

According to historical records, the garden is less than 10 Mus (or 0.67 hm2, excluding vegetable and flower gardens), and is among mid or small sized residence gardens in Southern China. Its layout has the following characteristics:

(1) "Although man-made, it appears as natural"

Ying Yuan is centered around water and uses surrounding mountains as its general background setting. In the south, it leverages continuous Ningzhen mountain ranges. To the north, it looks in the distance at the "vast, crude, and genuinely simple" Shugang. To the east it is separated by the inner city moat and the South Gate city wall. To its west, "It's the long river beyond the bank of willows. At the opposite bank, there are more tall willows, and Yan's garden, Feng's garden, and Yuan's garden are all in sight. Although the gardens are derelict, the plants are luxuriant and they seem to be owned by me." Whatever is in sight, great scenes "all seem belong to me". By scene leveraging, the spatial limit has been broken through and the depth and breadth of the scene have been much broadened. This way, the man-made garden has been cleverly mixed with the natural landscape. The grand structure is the natural landscape while the spirit is within the garden. Just like what Ji Cheng said, "On leveraging, although there is a divide between the inside and outside of a garden, there is no difference in obtaining a scene from afar or close by. Green mountain or red architecture, whatever is visible, we adopt the best and screen out the worst. It doesn't matter whether it's a house or farmland, as long as it can form a scene. This is so-called 'clever and appropriate'". A great composition lies in rich imagination and a high level generalization of the garden space and various environments. This is what is referred to as "scene created out of imaginative situations".

(2) "Multi-Faceted Scenes, Microscopic Landscapes"

Surrounded by the inner and outer moats, Ying Yuan is situated on an islet in the lake, forming an "isle in lake" situation. And Yugou Caotang in the inner lake of the islet is also a "isle in lake" composition, which creates a general layout that has "lake in lake", "isle in isle" composition that is step after step and layer after layer spatial interest, making it reserved with depths.

(3) "Limited Construction Makes Unlimited Landscape"

The amount of buildings in the garden is very limited. In order to blend the architecture into nature, all buildings are scattered around. A building is erected there because of the scene, reflecting a sparse and simple natural taste.

(4) "Creating Scenes at Key Points on Context"

Based on natural terrain, a visitor's route was cleverly designed to link rhythmically to all garden spaces, big and small. The garden scene layer depths were greatly enhanced by seeking spatial depths and breadths through directional changes. Wherever one reaches or sees, poetic and artistic feelings can be felt.

1.2 The Relationship between Mountains and Waters

It is not easy to create a landscape scene that seems to be surrounded by mountains and waters at Yangzhou's Ying Yuan, whose location is on the Northern Jiangsu plain. According to historical records, Ying Yuan enjoyed a landscape of "green mountains after mountains" and "long flowing waters" with rich mountain and water scenes. Based on the author's research on its history, to restore its mountain and water scene, there are a few key points

其中妙处主要有如下几点：

1.2.1 山景

（1）借山　影园北面隔水是蜿蜒起伏的蜀冈。登读书楼则蜀冈、迷楼皆在项臂；南面是江南诸山，青山隐隐，若可攀跻，正可谓"晓起凭栏，六代青山都到眼；晚来把酒，二分明月正当头。"这也正是计成在《园冶·园说》中强调的"远峰偏宜借景，秀色堪餐"，"山楼凭远，纵目皆然"，充分体现出园林虽有内外之别，但得景则并无远近的限制。晴山耸立的江南秀色，古寺凌空的平山胜景，此时皆化为己有了。

（2）掇山　影园"入门山径数折，松杉密布，间以梅杏梨栗，山穷……"又在一字斋"隔垣见石壁二松，亭亭天半。"在媚幽阁"一面石壁，壁立作千仞势，顶植剔牙松二……""涧旁皆大石如斗，石隙俱五色梅。"以上数段记载是园内大的地形起伏，也就是人工的堆山，园子的东南部是土山，松杉成林，北部是石包土山，以陡壁处理。为什么要在东南边与北边堆山呢？从影园位置图来看，这两边远处都有山体，园内的掇山可以视为园外山体向园内的延伸，是主山的余脉在平地又起，中间以丛林相连，这样使园内的掇山与天然山势气韵相连，似与真山一脉相承，在这样大的自然空间内，园内的山让人感受到一种强烈的气势，这种自然主义与浪漫主义相结合的造景手法，可称得上是"形神兼备"。江南淡淡的云山，园内以平冈小阪、漠漠平林陪衬，显得远山更为平远，从而构成一幅山水的长卷；北面较近的蜀冈，园内以色泽苍古的千仞峭壁和虬曲古松相呼应，尽显山势宏大高远，又恰似一幅山水立轴。

影园内的掇山，即"以真山为依据。又融合在真山之中"，这也正是学习运用计成"以假为真"、"做假成真"手法的实例。

（3）叠石　园内掇山是改造地形的大动作，而园内采用野趣盎然的叠石，增加了内部小空间的起伏变化。计成在掇山篇中详细地总结出园内不同环境地点的掇山叠石手法，如"凡掇小山，或依嘉树卉木，聚散而理。或悬岩峻壁，各有别致，书房中最宜者。"又如"以予见：或有嘉树，稍点玲珑石块；不然，墙中嵌理壁崖，或顶植卉木垂萝，似有深境也"。仅从史料记载中就可以看到，影园在书房的位置"庭

here:

1.2.1 Mountain Scenes

(1) Mountain Leveraging

North of Ying Yuan across the river is the meandering and undulating Shugang. Once you get onto the study building, then Shugang and Milou both seem to be at your arms reach. To the south are several green mountains that appear to be so close to the mount. This is like "Rising in the morning, against the railings, all green mountains are in sight; Drinking at night, the bright moon is right shining above me." This is also what Ji Cheng stressed in "Yuan Ye – On Garden", "Distant peaks are good for scene leveraging, as they are really beautiful to enjoy." "Looking afar from my mountain building, all distant scenes are within my sight." This fully reflects the rule that although there is a divide between inside and outside of a garden, there is no difference in scene leveraging. Beautiful southern scenes with verdant mountains and broad and far landscape scenes with ancient temples become ours at this point.

(2) Mountain Building

Entering Ying Yuan, "there is a mountain path with several turns with thick pine and cedar trees, interrupted with some plum, apricot, pear and chestnut trees. When the mound ends…" At Yizi Zhai, one sees "two pine trees growing out of rack wall, with their canopies covering half of the sky". At Meiyou Ge, "with one side of stone wall like a cliff, there are two pine trees planted on top of it…", and "there are big foot-long rocks with five-colored plum growing out of their cracks." The above commented the garden's general terrain, which is the artificial mound. In the southeast of the garden is a dirt mound with pine and cedar groves. North of it is a dirt-inside-rock mound, treated like a steep cliff. Why is it necessary to create mounds in the southeast and the north? From Ying Yuan's location, there are mountains in both these directions. So creating mounds in these directions can be viewed as an extension inside the garden of the outside mountains, or the remaining mountain ranges rising again at flat land. There are tree groves in between them, making the artificial mounds inside the garden connect with the natural mountains in spirit, extending the same strain. In this grand natural space, the artificial mound inside the garden makes one feel a great momentum. This integration of naturalism with romanticism as a means of garden design can be complimented as "well-presented in both the form and the spirit". The light cloudy mountain in Southern China, relative flat hillock and foothill inside the garden, and coupled with uneventful groves, make the mountains seem farther away, forming a broad landscape painting scroll. The closer Shugang in the north, with the ancient looking rock cliff and tortuous and curling old pines inside the garden, make the mountain grand, tall and far, forming a vertical landscape painting scroll.

The artificial mound at Ying Yuan, is "based on real mountains and well-integrated into the real mountains". This is a real example of applying Ji Cheng's "using fake as real" and "making fake as real".

(3) Rockery Construction

It is a big move to make mound inside the garden; however, using rockeries of wild interests could add richness and changes of inner smaller spaces. Ji Cheng elaborated in his mound making chapter as they apply to different contexts. For example, "If you make a small mound, you may add pretty flowers and trees and scatter them around. Or you may make them like steep cliffs with unique design, which is particularly suitable for study buildings." Another example, "in my opinion, one can plant some pretty trees and place some exquisite rocks; otherwise, making cliff-like walls, or growing some plants on its top or

前选石之透漏秀者，高下散布，不落常格，而有画理。室隅作两岩，岩上多植桂，缘枝连卷，溪谷嵌岩，似小山招隐处。岩下牡丹……备四时之色。而以一大石作屏。石下古桧一，偃蹇盘壁"，如此详尽的描绘亦为复建设计提供了资料。计成还提倡池山，"池上理山，园中第一胜也。若大若小，更有妙境。就水点其步石"，"穿岩径水，峰峦缥缈，漏月招云；莫言世上无仙，斯住世之瀛壶也。"影园中也布置有池山，史料中记载有："影园池中多石磴"，"大者容十余人，小者四五人，人呼为小千人"。恢复性重建中影园的石壁是采用计成所总结的"理者相石皴皱纹，仿古人笔意。植黄山松柏、古梅……"的手法，而园中的围墙则是"甃以乱石，石取色，斑似虎皮者"，也是计成认为的"聚石叠围墙。居山可拟"。只要胸有丘壑并多方细致营造，就能"得自然之理，呈自然之趣"。

总之，影园掇山是着力于追求宛若天成的效果，即所谓"山林意味深求"，所以只有"有真为假，做假成真"，才能"深意画图、余情丘壑"。

1.2.2 水景

"山要环抱，水要萦回"，水体处理贵在萦回，只要水体萦回了，就会产生"脉流贯通，全园生动"的效果。影园是东、南、西三面环水长屿，环绕的水把园地包围起来，园界自然也就形成了。而内外城河北通瘦西湖并直达蜀冈，南流入古运河，这就形成了水在园外围第一层大的萦回。《影园自记》中有载："通古邗沟、隋堤、平山、迷楼、梅花岭、茱萸湾皆无阻……盖从此逮彼，连绵不绝也。"简练、苍劲、自然的笔触，描绘出的是"山高水长"的意境，视觉空间与联想境界都无限地扩展了。

影园内"四方池，池外堤，堤高柳"，借助堤上高大柳树分割，使岛内视线统一起来，并与外部广阔的湖面产生隔离，从而得到一片如镜的内湖风景。茅元仪《影园记》中载："水一方，四面池，池尽荷。远翠交目。近卉繁殖，似远而近，似乱而整……"，这就是"岛中有湖"；当通过小石桥看到中央是玉勾草堂小岛，则又成了"湖中有岛"。层层叠叠，空间也随之逐步展开。当进入"淡烟疏雨"，穿曲廊登上读书楼，周围景色尽收眼底，显得空

hanging vines, appear to be a scene with depths." We could see only from the historical record that, in the study area, "the selected rocks in the yard are hollow and good looking, scattered up and down with unconventional setting but abide by the artistic landscape painting rules. At the corner of the room there are two rocks, on top of which there are osmanthus trees whose branches mingle with rocks. Rocks and creeks, seemingly to have formed a place for secluded living. The peonies by the rocks…… have different colors in four seasons. A large rock, standing tall and erect, serves as a screen with a Chinese juniper tree down." Such detailed descriptions helped the restoration work. Ji Cheng also advocated rockery in a pond, "To build a rockery in a pond is most attractive in a garden. It can be a wonderful setting no matter how large it is. Stepping stones can be used by the water", and "passing through rocks over water among ethereal peaks with leaking moon light and lingering clouds, don't tell me that there are no immortals around, since I'm an earthly one living as immortals right here." There is also rockery in the pond at Ying Yuan, as goes a historical record, "There are many stepping stones in the pond of Ying Yuan", "with larger ones that can hold more than ten people and smaller ones can only hold four or five. People refer to them as Xiaoqian Ren." During the restorative reconstruction of Ying Yuan, the rock wall work adopted a method that Ji Cheng summarized, "making rocks have cracks and other textures to imitate ancient landscape painting. Pines and Chinese junipers from Mount Huang, old plum trees were planted…" The garden fence wall, "used assorted stones selecting their different colors to imitate tiger's fur", is compatible with his view that "by using stones to build fences, a mountain living life style can be mocked up". As long as there is a commanding idea in your mind and a meticulous management method, you can achieve the status of "getting the nature's rule and showing the nature's cool". In a word, the mound or rockery building of Ying Yuan is to strive for its natural appearance, or so-called "mountain living lifestyle pursuit". Therefore, only "using real as fake and making fake as real" can "distill nature's image in design work and evoke nature's feeling from them".

1.2.2 Water Scenes

"Mountains should give a sense of embracement and water encircling." The handling of water bodies lies in its encircling feeling. As long as it encircles, it would bring "liveliness for the whole garden because of great circulation". The east, south, and west three sides of the garden face water and the garden is surrounded almost all by water and the boundary is naturally defined. The inner and outer moats connect to the north to Shouxi Lake and reach directly to Shugang. To the south it flows into the ancient canal. It forms the first layer of the circling water outside the garden. According to "*Ying Yuan Records*", "it links to Hangou, Suidi, Pingshan, Milou, Meihua Ling, and Zhuyu Wan without obstruction…therefore, from here to there, they are all connected." Simple, powerful and natural strokes realized a "high mountain and long river" artistic conception and so the visual space and the imaginative world are both expanded indefinitely.

Inside Ying Yuan, there are "square pond, dykes at the perimeter of the pond, and tall willows on the dyke". Via the tall willows on the dyke, the visual extent on the isle is united and separated from the outside vast lake, and a calm inner lake scene is created. According to Mao Yuanyi's "*Ying Yuan Annals*", "There is a piece of water that is a four-sided pond. At the end of the pond there are lotus plants. Faraway green scenes and nearby luxuriant

间格外广阔，气势宏大。然步履园内，见珍奇花木、亭台楼阁、假山水池的曲折起伏变化无穷，与简单的外部构图产生了强烈的对比，更觉园中深邃含蓄的美。

以上是通过岛、堤、桥、岸的分割处理，从大处着眼，使水体产生趣味不同、余意不尽的感觉。园内长形水池的经营亦得章法。整个水面以聚为主，聚中有分。特别是在池面的中间部位，曲桥东部的石磴与一字斋西南的土山相应地向池中伸展，收缩成夹峡之势，使水面像个葫芦，在此之北，又以"荣窗"和亭桥"湄荣"再一次收缩水面，空间划分为"展—缩—展—缩—展"3个大层次，成了南北之间夹景式构图。"似隔非隔，水意连绵，反而显出水面的弥漫、深远"。茅元仪评价："水狭而若有万顷之势矣。"水面上散点的水景山石"石磴"，既丰富了水景，也增加天然山水朴质的野趣。

影园中的水体处理还注意了深邃幽奇、变化莫测的意境塑造。媚幽阁三面临水，一面石壁，"壁下石涧，涧引池水入，哇哇有声，涧旁皆大石，恕立如斗……至水面穷不穷也"。得"巨石峡细流潺缓如琴之韵，因有是名。其间奇石嶙峋，古木蓊蔚，有天然林泉之致。"入影园大门后，开门见山，山径数折。"右小涧。隔涧疏竹百十竿。"往前走是"窄径隔垣，梅枝横出，不知何处，水来柳近，疑若已穷，而小径忽横，苔华上下者，其折入草堂之路也。有水一方，四面荷……"真可谓"山重水复疑无路，柳暗花明又一村"了，这均是计成所说的："门引春流到泽"，"看竹溪湾，观鱼濠上"，水贵在源泉，故应"疏源之去由，察水之来历。"由于溪口、河湾、石岩之巧于安排，以假乱真，层层相属，使园内水体又组成若干小的萦回，体现出"来去无踪，弥漫无尽"的妙境。

影园十分注意随园内地形起伏塑造幽朴的涧溪，在"媚幽阁"石壁下的涧流，利用水面高差，形成淙淙有声的流水，使水有蔓延流动的神态。园东南部分溪流的处理，是自然界山水的缩影。"溪水因山成曲折，山蹊随地作低平"，这样水就仿佛是劈山凿岩造成的湍流、曲涧等多种水景，寓意"水源"，使全园山水景观符合天然山水的模式。

影园水景设计与山紧密结合，"水随

flowers meet the eye. They look far but also appear close, unordered but actually orderly…" This is "lake inside an isle". After passing the small stone bridge one sees in the middle Yugou Caotan islet, it becomes again "an islet inside a lake". Layer after layer, the spaces are gradually and sequentially exposed. When entering "Danyan Shuyu" (Note: light fog and scattered drizzle) and passing the corridor and getting up onto the study building, you see all surrounding scenes at a glance with wide open space with a grandeur feeling. However, when you walk into the garden and see the rare flowers and plants, architectures, ponds and rockeries with all the variations and unlimited changes, they seem to form a sharp contrast against the external simplicity. This further enhances the garden's beauty in its depths and reservation.

The above is about how to use islet, dyke, bridge, and bank to create interests with infinite feeling at a higher level. The inside elongated pond also has its own design methodologies. The whole water body mainly uses concentration approach but with concentration, there is also dispersion. Especially at the center location of the pond, the stepping stone east of the zigzag bridge and the dirt mound southwest of Yizi Zhai correspondingly extended into the pond, pressed from both sides, making the water body a gourd shape. North of here, "Rong Chuang" and the pavilion bridge "Mei Rong" shrink the pond again and so the space sequence has three layers showing "expansion – contraction – expansion – contraction – expansion", forming from north to south a flanking composition. "It seems a separation but actually not. The continuous on and off water scene adds a widespread and distant feeling." Here Mao Yuanyi commented, "although narrow, the water here gives an impression of vastness." The scattered scene making stepping stones on water add the richness to the waterscape while at the same time, it brings forth wild interests of natural landscape.

The water scene handling in Ying Yuan also includes ever-changing tranquil wonder scene creation. Meiyou Ge faces water at three sides and a stone wall at one side. "At the gully under the stone wall, gurgling water is introduced from the pond. Large rocks are by the gully standing at the water … showing no sign of their depths." "The water passing the rocks creates a melody similar to that from a musical instrument and hence the name. Here rare rocks accompany old trees, which creates some quiet natural wilderness scenes." You see the mountain right after entering Ying Yuan's gate. After walking a few sections of mountain paths, you see "a small gully on your left and across the gully, there are about a hundred bamboo trees." If you keep walking forward, you see "narrow pathways separated by walls. A plum branch sticks out from nowhere and you don't know where you are. Coming near to water and willow trees, you might think it's the end; however, you bump into a mossy path up and down leading into a thatched house, and see another water body with all the lotus plants…" This is indeed a "silver lining or pleasant surprise in a seemingly dead end". This is all what Ji Cheng referred to as "channeling spring stream to lake", and "enjoying bamboos by the brook and watching fish frolicking from above the gully". The most precious thing about water is its source; therefore we need to "make the water source clear" to arrange stream head, river bend, and rockery to have a tight integration of the fake with the real, making the garden water body system form several sub circling systems, to reach a wonderful state where the water "comes and goes without a trace to unlimited extent".

The designer paid much attention to creating natural looking and quiet

山转、山因水活"。园中湖面收柳影、山影、山林之影倒入池中，愈增深邃、含蓄之感，令人心怡。

1.3 建筑

影园内建筑以散点式布置为主，数量不多，给人以朴实无华，疏朗淡泊之感。园内建筑多"因景而生"，如取李太白"浩然媚幽独"诗句的媚幽阁等。茅元仪说："水狭而若有万顷之势矣，媚幽所以自托也。"媚幽阁是借景而成的，使人们见景生情，意味无穷。

园林建筑毕竟不同于一般性的建筑物，除了满足居住、休息或娱乐等实际需求外，往往是园景的构图中心，即"按时景为精"。影园内"当正向阳之屋"的"玉勾草堂"，是全园的活动中心，与媚幽阁"彼此相望。可呼与语，第不知径从何达？"在近水远山之中，远翠交目，近水繁花，与周围的景色合成一体。园内东北面的一字斋，隔曲桥与媚幽阁为一组读书楼，其间以花木、蹊径和小桥串联，又形成多种景观。

计成在《园冶·立基》中说："房廊蜒蜿、楼阁崔巍。动'江流天地外'之情，合'山色有无中'之句。适与平芜眺望，壮观乔岳瞻遥；高阜可培，低方宜挖"。影园中的"淡烟疏雨"正是如此由廊、室、楼构成的一组院落，藏书室在下，而读书楼峙于上，"能远望江南诸峰，可近收林树翠色"，真可谓是"清风明月本无价，近水远山皆有情"。"花间隐榭，水际安亭，斯园林而得致者"。影园内设有淳翠亭，园主人曾说："盛暑卧亭内，凉风四至，月出柳梢，如濯凉壶中。薄暮望岗上落照，红沉沉入录。"可知此亭安排得妙趣横生。在桥亭的东面临水还置有一小阁，名曰"半浮"，好似漂在水面上，旁有供水上游玩的小舟"泳庵"，此处"专以候鹕"，使人能即兴吟诵古诗"两个黄鹂鸣翠柳"。

据史料记载，将园内诸个建筑进行复原设计，并安排在特定的位置上，影园内的建筑是有主次之分的，主景是玉勾草堂，其他建筑尽管散漫，但总给人有秩序、有条理的感受。同时，在秩序之中也不乏活泼、多变的布置，林林总总皆是"藉景而成"。

1.4 植物

植物是构成园林的重要因素，也是组成园景的重要素材。园林中的树木花草不

gullies on the undulating terrains in the garden. At the stream under the stone wall by the "Meiyou Ge", taking advantage of different water levels, a gurgling water sound is formed, posing a flowing air of the water. The handling of water in southeast of the garden can be viewed as a miniature of the natural landscape. "A stream water path becomes tortuous and winding because of mountains and a mountain stream exhibts low and flat depending on the terrain." In this way, the water seems to be turbulent currents or tortuous streams formed by the chiseled rocks, alluding "water source", making the garden landscape in compliance with the forms of the natural landscape.

The water scene at Ying Yuan is tightly integrated with the mountain scene. "Water follows mountains and mountains become alive because of water." The reflections on the lake water include that of the willow trees, the mountains, and the mountain forests, and these make it rich in its depths and reservation. What a wonderful scene!

1.3 Architecture

The buildings inside the garden are scattered around and the number of buildings is limited. This gives people a plain and simple feeling. The buildings in the garden mostly were named and built "because of a scene", e.g. the building Meiyou Ge's name is derived from Li Taibai's poem "Hao Ran Mei You Du". Mao Yuanyi said, "Although the water body is narrow, it appears to be a vast one; therefore we got to create a closed, quiet and self-dependent micro-environment." Meiyou Ge was created by the scene and its purpose is to evoke people's emotions by seeing the scene and bringing about infinite imaginations.

Buildings in a garden are different from general purpose buildings after all. Besides its goal to meet living, resting and recreational and other practical needs, they often serve the purpose of being a composition center of a garden, and "take what's in vogue as the essence". Ying Yuan's "Yugou Caotang", "a building that directly faces the sun", is the whole garden's activity center, and it corresponds to Meiyou Ge. "They are so close and face each other, but there is no apparent path to connect them." With nearby water and faraway mountains, what one can see is distant greens and near flowers, and they mingle well with the surrounding scenes. In the northeast part of the garden, Yizi Zhai, Gequ Bridge and Meiyou Ge form a set of study buildings, and between them there are flower trees, paths and small bridges for connection, forming multi-variant scenes.

In "Yuan Ye – Base", Ji Cheng said, "Buildings and corridors need to be tortuous, extended and built higher to add a feeling of "rivers flowing beyond" and be blended in a scene of "looming landscape". To solicit a magnificent mind one can look far towards the high mountain peaks. One may add more dirt to make a high place much higher and may dig deeper at the places where it is already low to make it even deeper." Ying Yuan's "Danyan Shuyu" (Note: Thin Fog Scattered Rain) is exactly like this. It is a courtyard made up of corridors, rooms, and buildings, and the library is at the lower level and the study is at above. From there one can "look at mountains in the distance and enjoy beautiful trees nearby." This is indeed like what the poem goes, "The clear wind and bright moon have no price tags; both nearby water and distant mountain have emotions." "Generally a Xie is built privately among flowers and a pavilion by the water. This is a common practice to generate interests." There is a Chuncui Pavilion at Ying Yuan and the owner once said, "Resting in the pavilion in the hot summer, when the cool breeze comes from all directions and the moon climbs up on the tree canopy, it feels like having a cool shower. At dusk watching

仅是为了使山水"得草木而华"，或是为陪衬园林建筑而点缀其间，其本身也常组成群体，成为园林中的景。影园进门后，山径几曲，松杉密布高下，垂荫间以梅、杏、梨、栗，隔涧溪土丘陵上丛竹一片，观赏植物使园内东南面土丘上形成了古木交柯、雄健挺拔的气势，浓荫如盖，更增添了山林的浑厚苍劲和园内深邃幽奇的情趣。园内西部堤畔水际遍植垂柳，间种桃李，柳枝柔丝，洒落有致，有"轻盈袅袅占年华，舞榭妆楼处处遮"的韵味，一派垂柳依依、柔丛千缕的含蓄媚态。这些植丛除了自身的景观外，还与园外景色联系起来，"造成园外有园，山外有山，树外有树的自然气氛"。

汪菊渊教授早有论著："各种植物均有一种性格或个性，也就是所谓'自然的人格化'，然后借着这种艺术的认识，以植物为题材，创作艺术的形象来表现所要求的主题。这是我国园林艺术上处理植物题材的优秀传统，是客观通过主观的作用"。影园建造的时代和园主人封建士大夫的社会地位决定了他们"幽雅、冷洁、宁静"的基调。今影园得以复建，基本遵照史料，如树种选择就少不了清标韵高的梅花、节格刚直的竹子、幽谷品逸的兰花、操介清逸的菊花。蜡梅之标清、杏花之繁灼、桃花之夭冶、木樨之香胜、紫荆荣而入、芙蓉丽而升，以及梨之韵、李之洁等。此外种植这类植物，也有因物取祥的原因，如紫薇象征高官；萱草忘忧；紫荆和睦；石榴多子；玉兰、海棠、牡丹、桂花齐栽，象征玉堂富贵等。通过植物配景，更进一步了解了计成的种植艺术。影园主人曾对计成的花木点缀赏识至极，说"一花、一竹、一石，皆适其宜，审度再三，不宜，虽美必弃。"从记载来看，的确很有章法，"俨然画幅也"。具体做法是沿堤垂柳，夹岸桃李；千百本芙蓉坐趾水际，池内尽荷；高梧流露，荫以梨栗；历年久苔之华。梅枝横出，壁立千仞，苍松挺立；窗外置石，芭蕉荫翳，修篁弄影；岩上植桂，岩下奇卉，阶下石榴，台上牡丹。可谓高低参差，天然成趣。

此外，设计中还注意到植物群体布置不但要疏密有致，偃仰适宜，更重要的是"顾盼生情"。《影园记》中载："……夹其傍，如金谷合乐；玉兰、海棠、绯白

sunset, a weighty sun sinks behind the mountain." From this we could know that the design of this pavilion did take into account some fun factors. East of the bridge pavilion by the water there is a small cabinet named "Ban Fu" (Note: semi-floating), which seems to float on water. By its side, there is a small boat "Yuan An" for recreational use on water. This is a "dedicated place for waiting for orioles", which prompts people recite the old poem "Two orioles chirp among willow trees."

Based on historical records, all garden buildings were re-designed and placed on appropriate locations. The buildings in the garden have a hierarchy and the main building is Yugou Caotang. The rest of the buildings, although seemingly unorganized, do contain an order and are disciplined. At the same time, with this order they are also lively and not rigid; they were "all built with certain scenery taken into account".

1.4 Plants

Plant is an indispensable part of a garden and is also an important scene making component. The plants in a garden not only serve the purpose of making landscape "more flowery because of plants", or as a background or decorations of garden buildings, they in their own right become a scene in a garden. After entering Ying Yuan, passing through a few sections of mountain paths, we see thick and tall pine and cedar trees, whose shadows covers plum, apricot, pear, chestnut trees. Across the gully on the hill there are bamboo groves. Ornamental plants make the hill in southeast part of the garden into a luxuriant vegetation area. Powerful and towering posture and large shadow area added the natural forest's vigorous air and an appeal of depths and tranquility. In the west part of the garden by the bank, willow trees were planted everywhere with occasional peach and plum trees. The billowing willow in the breeze has a lingering charm of "a light and wavy young lady who is a bit bashful" and appears coquettish. These plant clusters not only create scenes with themselves but also form a large system by connecting to the outside scene, contributing "an organic atmosphere that makes people feel there are gardens outside the garden, mountains outside the mountain, and trees outside of the trees".

Professor Wang Juyuan wrote this long time ago, "Every plant has a character or personality, and this is so-called 'nature's personification'. Based on this artistic recognition, we could use plant materials to show artistic image that represents a main theme. This way of handling plant materials is an excellent tradition in our garden art; it is an objective effect influenced by subjectivity." When Ying Yuan was built, the social status of the feudal literati garden owner determined their fundamental style of "quiet and elegant, cool and clean, serene and tranquil". Ying Yuan's restoration was basically guided by historical records. For instance, plant selection cannot live without elegant and tasteful plum, upright and outspoken bamboo, serene and tranquil orchid, and light and nonchalant chrysanthemum. The clarity of wintersweet, the luminous apricot flower, the coquettish peach flower, and the fragrance of melilot are all indispensable. The glory of bauhinia makes it a great candidate; the beauty of hibiscus elevates its status. The lingering charm of pears and the purity and cleanness of plums make them great garden plants. Another reason for using these plants is for propitiousness. For instance, crape myrtle represents senior officials and day lily is for forgetting sorrow. Bauhinia is for harmony, and pomegranate is for prolificacy. Planting magnolia, crabapple, tree peony, and osmanthus represents wealth and so on. Through these, Ji Cheng's plant design art can be understood further.

桃护于石，如美人居闲房；良姜、洛阳、虞美人、曰兰、曰蕙，俱如媵婢盘旋，呼应恐不及"。生动活泼的艺术形象能引起人们的情感和想象，一字斋北的墙面开圆形洞窗，框入桂花的景色，构思不凡。"……留一小窦，窦中见丹桂，如在月轮中"，激起观赏者对嫦娥奔月这一民间传说的美好遐想，景意无限开阔。园内的种植，还考虑安排禽鸟的栖息场所，园西南面均为柳，"鹂性近柳，柳多而鹂喜，歌声不绝，故听鹂者，往焉。"并且筑半浮阁，"传以候鹂"听其鸣唱。为增加园内自然之趣，在菰芦中水际，佳苇生之，"芦花白如雪，雁鹭家焉，书去夜来，伴予读。"这又形成无数的画幅，咏不尽的诗篇。

影园绚丽多姿的花木使硬直静止的山石、屋宇变得生动活泼，树木花草的容貌、色彩、芳香又对园内季相变化起到重要的标志作用，显示出春、夏、秋、冬的基调。大量的观赏植物在烘托、渲染、陪衬影园景观的气氛上起到了决定性作用。

2 影园恢复性重建设计的启示

影园的恢复性重建是令人振奋的事。对于这样一个计成《园冶》理论的实践作品，同时又是集大成的江南山水园，笔者在恢复性重建设计时得到了很多启示，特别有以下3点值得我们借鉴。

2.1 巧于因借

"因"是处理地形条件的技法原则。影园地处苏北平原，又是三面环水，所以要在山水景上下工夫。从大的方面看是园子南部大的湖面和北面带状萦回的水体产生强烈对比，各尽其趣；而南面大的湖面又通过山丘和亭桥的两次收缩，产生了"放－收－放－收－放"的空间变化。所谓"一勺则江湖万里"。

"借"是处理环境条件的空间透视构图原则。影园成功地借北面的蜀冈和南面的青山为自己环境中的构景，这是空间的构图，也是游人视野的境域，借山影、水影、柳影成园，从而突破了园林本身的范围和限制，达到"虽由人作，宛自天开"的艺术效果。

2.2 以简寓繁，以少总多

造园前辈朱有玠先生曾比喻说："江南山水园很像绘画中的白描，只有抛开彩色，皴擦和渲染，才能达到'墨寓五色、

The owner of Ying Yuan showed extreme appreciation of his masterful use of plants and commented, "A flower, a bamboo, a rock seem all good; however, after repeated assessment, he still gave up their uses, although they're all beautiful." According to the written records, the art of composition is of high standard, "as if they're artistic paintings." The specific implementations are, planting willows along a dyke and growing peaches and plums on both banks. Hundreds of hibiscus were planted by the water and there are many lotuses planted in the pond. Tall Chinese parasol trees cast over pear and walnut trees with years of shade covering. Plum tree branches grow out sideways against high vertical stone walls, sided by old pines. Rocks stand outside by the windows shaded by banana trees and bamboo groves. Osmanthus trees were planted on top of rocks and colorful flowers cluster its base. Pomegranate trees were planted below terraces and above it, there are tree peony flowers. High and low with variant heights, all plants appear to be natural.

Besides paying attention to the density and openness variations of plant design, another important factor is "working in concert". According to "Ying Yuan Annals", "…by its side, just as a symphony. Magnolia, Crapapple, red and white peach by a rock, are just like beautiful ladies living in a leisure house. Lesser galangal, Luoyang peony, corn poppy, Yuelan, and Yuehui, are all like maids working around the master." Lively artistic image can evoke people's emotions and imaginations. For example, on the wall north of Yizi Zhai, there is a round-shaped window used as a picture frame framing osmanthus flowers. What an excellent design! "…leaving a small opening, through which orange osmanthus can be seen, just as if it's displayed against a full moon." This alludes to the legend of the Goddess Chang E's flying to the moon.

What a beautiful scene with unlimited imaginations! The planting design in the garden also took birds' perching sites into consideration. The southwest part of the garden is filled with willows, since "orioles love willows and when there are a lot of willows, orioles are happy and enjoy chirping. As such, people who like hearing orioles chirping would go over there." And Banfu Ge was built there, "which was said to have been built for people who're waiting there" to hear orioles' chirping. To add more natural interests, reeds were planted in water. Great reeds grew there, and "the reed flowers are white as snow and they became the home of water fowls. The sound from the reeds rubbing each other served as a soothing companion for my reading at night." There are countless images and poetic verses forming the scenes.

Ying Yuan's colorful garden plants contrast with static rocks and buildings and help making them livelier. The forms, colors, and fragrances of the flowers play an important role in marking seasonal changes, showing basic tones of spring, summer, autumn and winter. Large amount of ornamental plants act decisively as foils to render and contrast Ying Yuan's landscape ambience.

2. Revelations on Ying Yuan's Restorative Reconstruction

Restorative reconstruction of Ying Yuan is a very exciting event. Based on the theories from "Yuan Ye", this real work by Ji Cheng is an epitome of all southern Chinese gardens, and the author received much revelation in the process of the restorative reconstruction. There are especially three aspects that we could learn from.

2.1 Artful Leveraging

"Context" is the rule for terrain handling. Ying Yuan is located on the northern Jiangsu plain; therefore, more work needs to be done at the handling of terrains, or mountains

笔几刚柔'，朴素蕴藉，雅淡耐看的效果。"也就是绘画中常说的"惜墨如金"。影园正是以简练的布局，疏朗的建筑点缀，古树、土丘、水面、堤、岛的自然过渡，在山水境地里"略成小筑"而"足征大观"。借用清·沈复的话来看影园，即："大中见小，小中见大；虚中有实，实中有虚；或藏或露，或浅或深"。这也就是美学上所说的"绚烂之极归于平淡"。

2.3 情景相融，意趣耐寻

影园设计中的意境安排是出类拔萃的，给人一种意远情深，品味不尽的艺术享受。朱有玠先生对于意境的创作有过总结，他说："江南山水园林在设计过程中，形象思维的核心是意境。"又说："意境可以说是通过艺术形象而深化了的思想感情，同时意境又是运用形象以直抒情意的表现方式。"影园的恢复性重建设计是"意在笔先"，以诗情画意写影园，志在画中游。意境确定后以泉石为皴擦，以花草为点染，随着游程的空间流动和季相、晨昏、气候的变化，呈现出静态和动态的综合效果，完全是一幅鲜活的立体画卷。而此时人们的"胸中之竹，已不是眼中之竹"，胸中之景已不是眼中之景，山水的艺术形象变成了深化的思想感情，正像王国维在《人间词话》中说的："词家多以景寓情……一切景语多情语。"影园的恢复性重建设计"首先是创造自然美和生活美的'生境'，然后进一步上升到艺术美的'画境'，进而升华到美的'意境'，最终达到三者互相渗透、情景交融的高潮。"

3 结语

《园冶》是计成毕生造园理论和实践的总结，由于《园冶》的刊行在影园筑成之后，故影园乃计成《园冶》的实景蓝本。在影园的恢复性重建设计工作中，《园冶》一书的理论从渺茫到逐渐清晰再到最终从图纸上再现，笔者领悟到：影园的恢复性重建也就是计成《园冶》的实景再现。

影园恢复性重建是园林界的大事，也是众望所归之举。

鞠躬致谢：

在影园复建规划设计的过程中，吾师孟兆祯院士不辞辛劳关注着复建设计的学术水准，亲自画图致信于笔者，对地形、

and waters in Chinese garden design jargon. A general examination of the existing situation shows a large lake area in the south of the garden which contrasts sharply with the linear and meandering water bodies in the north, each with their own unique interest. Moreover, the large water body area in the south has been contracted twice by a hill and a pavilion bridge, forming an "open-close-open-close-open" spatial sequence. This is so-called "a spoonful of water represents thousands of miles of rivers and lakes".

"Leverage" is a spatial perspective composition principle for handling each individual situation. Ying Yuan successfully leveraged Shugang in the north and the green mountains in the south, embracing them in its own scenery framework. This is a spatial composition. It is within people's vision extent and it leverages mountain, water and willow scene to form a garden. It breaks the bound and limit of the garden proper, and achieves artistic state that "although man-made, it appears to be natural".

2.2 Simplicity over Complexity - Less is More

Senior garden master Mr. Zhu Youjie once used an analogy, "Southern Chinese gardens are like line drawing in traditional ink and brush style. Only if we avoid color, texture and rendering can we achieve what's stated in 'Ink only can represent five colors and brush stroke alone can express rigidity and flexibility'. This is the effect that is simple, suggestive, elegant and engaging." This is also what artists often refer to as "using one's ink as if it were as precious as". It is here at Ying Yuan that simple composition was applied and sparse buildings were built. Ancient trees, dirt mounds, water bodies, dykes and islets transition naturally. In such a landscape, "a little construction" can "achieve a great deal". To borrow the Qing dynasty Chen Fu's words to describe Ying Yuan, "Seeing small among big, big among small. There is void in solid and solid in void. It may be hidden or it may be exposed; it may be shallow or it may be deep." This is what the saying "extreme gorgeousness resides in plain and ordinary" goes in aesthetics.

2.3 Fusion of Garden Scenes and Human Emotions - Room for Rumination

The artistic conception management at Ying Yuan is marvelous, as it gives you an artistic enjoyment of deep meaning and emotion and lingering taste and engagement. Mr. Zhu Youjie once summarized about the creation of an artistic conception and he said, "In the design process of Chinese Southern landscape gardens, the core of image thinking is artistic conception." He then went, "An artistic conception is thoughts and feelings deepened through artistic images. At the same time, it is also a direct expression of emotions by using images." The restorative reconstruction of Ying Yuan is "plot before drawing", creating a garden using poetic emotions and picturesque thoughts, intending for a tour in pictures. After an artistic concept is set, then springs and rocks can be used and served as textures created by brush strokes, and flowers can be placed here and there as decorations. The changing parts are spatial perceptions along a path, varying seasons in a year, different hours in a day, and the ever-changing weathers of a place. All these combined make complex effects with both static and dynamic scenes, creating a lively three dimensional picture scroll. At this stage, "the bamboo in people's mind is no longer the one in their eyes", meaning what people perceive is no longer what they see. The artistic image of landscape is elevated into people's thoughts and feelings, just as what Wang Guowei said in "Renjian Cihua", "Men of letters tend to express their emotions through physical scenes, … and there are more language of

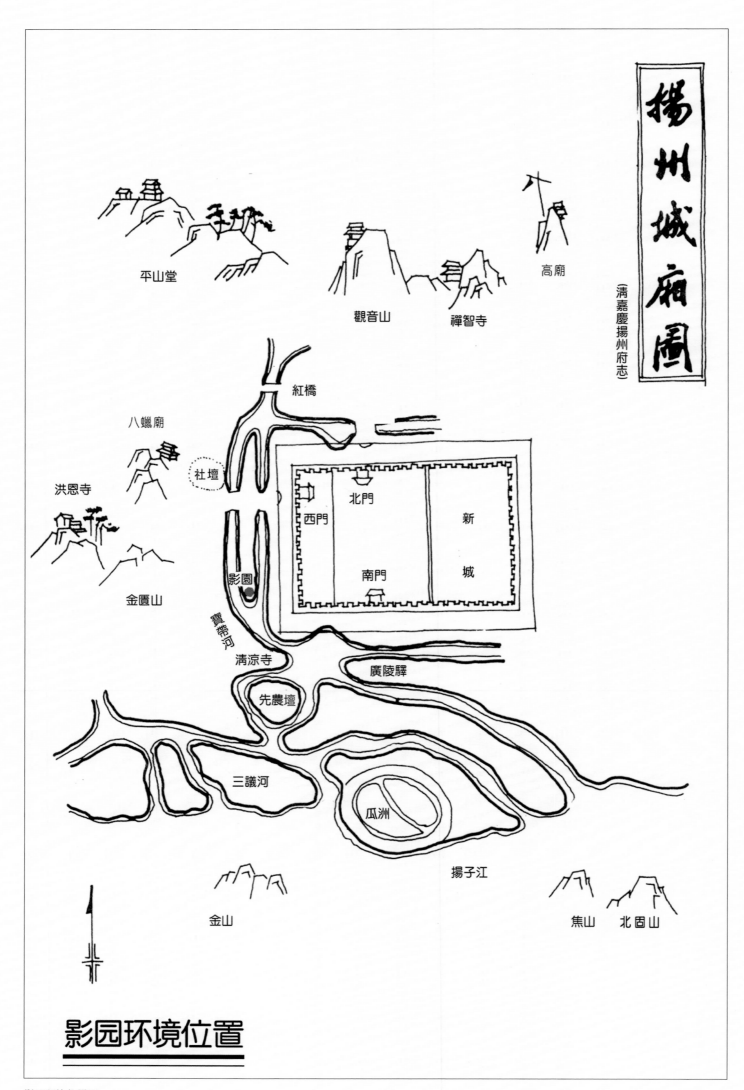

影园环境位置图
Locational Map of Ying Yuan

景点等提出指导。并亲自操刀督导影园实景模型制作，学生感激不尽，在此只有鞠躬，以表谢意（附上孟院士的亲笔书信和图纸，以飨读者）。

scenes than that of emotions." The restorative reconstruction of Ying Yuan is "first to create a 'state of nature' that has beauties of life and nature, then elevate to a 'state of art' that has artistic beauty, and finally sublime to a 'state of mind' that has a sublimed beauty. The ultimate climax is when the three states intermingle and bind the subjective emotion with the objective scene."

3. Conclusion

"Yuan Ye" is the summary of Ji Cheng's garden design theory and practice over his whole life time. Since the publication of "Yuan Ye" happened after the construction of Ying Yuan, the garden became Ji Cheng's materialized blueprint of "Yuan Ye". Through the restorative reconstruction of Ying Yuan, the theory in "Yuan Ye" went from uncertainty to clarity and finally restored on paper. The author realized that the restorative reconstruction of Ying Yuan is actually the materialization of Ji Cheng's "Yuan Ye".

The restorative reconstruction of Ying Yuan is a great event in China's garden and landscape architecture's community, and it was also much expected and widely welcomed.

Acknowledgement and Appreciation:

During the restorative reconstruction of Ying Yuan, my mentor Meng Zhaozhen paid tireless attention to the academic standards of the restorative design, and personally wrote letters and drew sketches for the author and offered guidance to terrain and scenic attractions' handlings. He was also personally involved in supervising the garden's physical model's construction, for which I'm greatly indebted. I really appreciate the help and advice. (Attached please find Academician Meng's personal letter and drawing.)

影园遗址－现状地形图
Ruin Site of Ying Yuan—Current Topographic Map

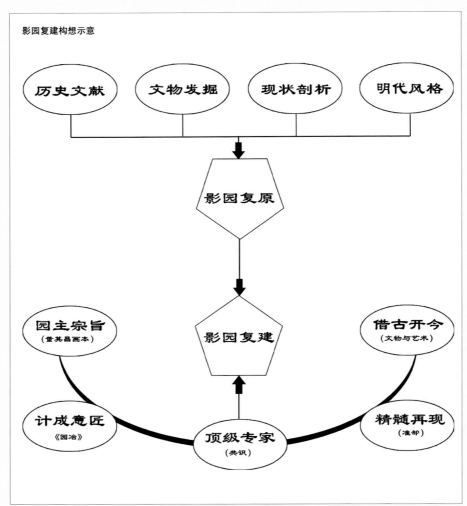

影园复建构思　Restorative Reconstruction Concept of Ying Yuan

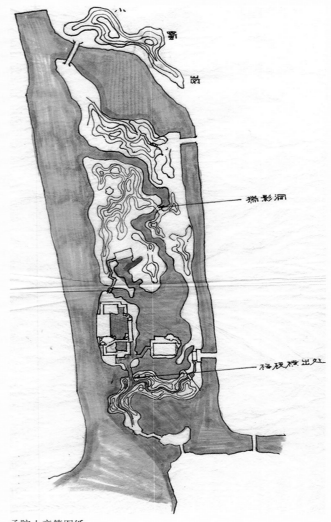

孟院士亲笔书信
Academician Mr. Meng's Handwritten Correspondence

孟院士亲笔图纸
Academician Mr. Meng's Handwritten Map

影园——竖向规划图
Vertical Design Map of Ying Yuan

影园地形处理——延山引水 "列千寻之丛翠" "浚一派之长源"
Handling of Terrains of Ying Yuan—Mountain Extension and Water Channeling.

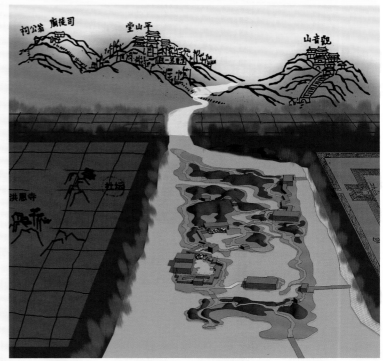

影园建筑布局——因景而生，融入自然
Overall Architectural Layout of Ying Yuan – Derived from the Scene and Blend Well into the Natural Surrounding.

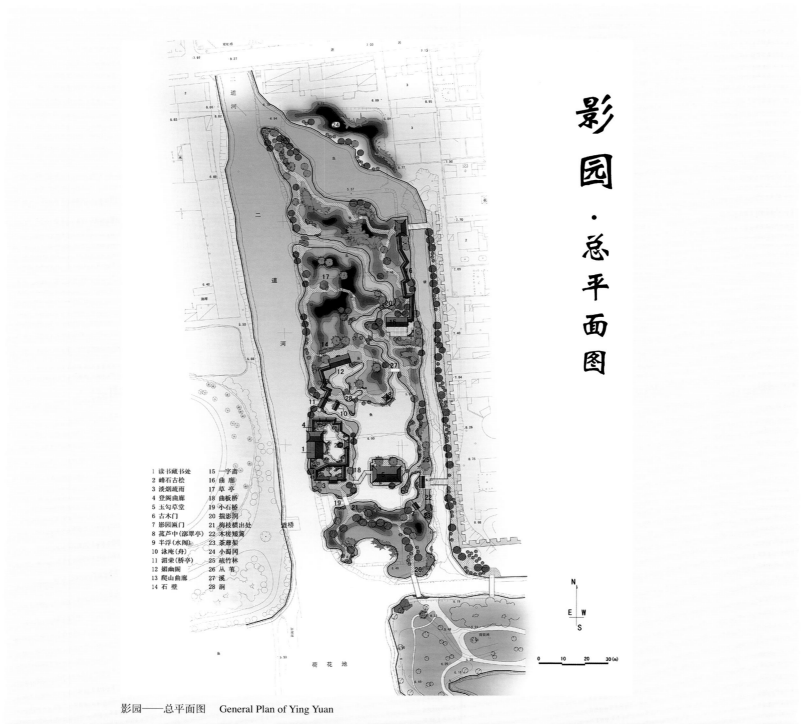

影园——总平面图　General Plan of Ying Yuan

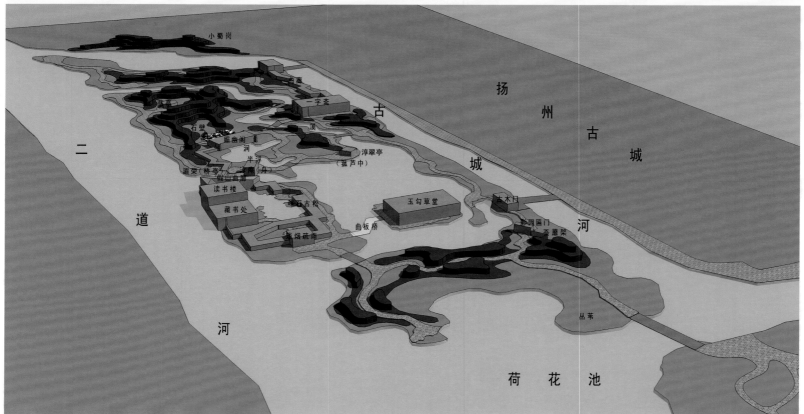

影园主要建筑位置示意图　Locational Map of the Main Buildings in Ying Yuan

影园——鸟瞰图　Bird's Eye View of Ying Yuan

影园——模型（局部）　Portions of Model of Ying Yuan

影园——模型
Model of Ying Yuan

影园——玉勾草堂效果图　　Rendering of Yugou Caotang

影园——读书藏书处、泳庵（舟）、湄荣（亭桥）效果图
Rendering of Study/Library, Yongan (Boat) and Meirong (Pavilion Bridge)

影园——淡烟疏雨效果图
Rendering of Danyan Shuyu

影园——菰芦中、漷翠亭效果图
Rendering of Gulu Zhong and Huocui Pavilion

园冶

Yuan Ye Illustrated
— Classical Chinese Gardens Explained
II

吴肇钊　陈艳　吴迪　著
By Wu Zhaozhao, Chen Yan and Wu Di
马劲武　译（英文）
English Translation by Jinwu Ma

图释

【中卷】

中国建筑工业出版社

目 录

序一——孟兆祯............ 6
序二——陈晓丽............ 9
序三——王绍增............ 11
序四——刘秀晨............ 13
序五——甘伟林　王泽民..... 15
序六——王金满............ 18
自序一................... 20
自序二................... 22
《园冶图释》缘起.......... 24
《园冶》冶叙——阮大铖..... 26
《园冶》题词——郑元勋..... 26
《园冶》自序——计成....... 28
兴造论................... 001
园说..................... 013

一　相地................. 025
（一）山林地............. 045
（二）城市地............. 055
（三）村庄地............. 081
（四）郊野地............. 084
（五）傍宅地............. 095
（六）江湖地............. 109

二　立基................. 121

三　屋宇................. 141
（一）门楼............... 152
（二）堂................. 153
（三）斋................. 154
（四）室................. 157
（五）房................. 158
（六）馆................. 159
（七）楼................. 161
（八）台................. 169
（九）阁................. 174

（十）亭................. 176
（十一）榭............... 190
（十二）轩............... 191
（十三）卷............... 194
（十四）广............... 195
（十五）廊............... 199
（十六）五架梁........... 201
（十七）七架梁........... 202
（十八）九架梁........... 203
（十九）草架............. 204
（二十）重椽............. 204
（二十一）磨角........... 205
（二十二）地图........... 207

四　装折................. 209
（一）屏门............... 210
（二）仰尘............... 211
（三）户槅............... 212
（四）风窗............... 222

五　栏杆................. 225
栏杆图式................. 226

六　门窗................. 245
门窗图式................. 246

七　墙垣................. 253
（一）白粉墙............. 254
（二）磨砖墙............. 254
（三）漏砖墙............. 254
（四）乱石墙............. 258

八　铺地................. 259
（一）乱石路............. 260
（二）鹅子地............. 260

（三）冰裂地............. 260
（四）诸砖地............. 261
砖铺地图式............... 261

九　掇山................. 265
（一）园山............... 272
（二）厅山............... 276
（三）楼山............... 279
（四）阁山............... 281
（五）书房山............. 282
（六）池山............... 283
（七）内室山............. 284
（八）峭壁山............. 285
（九）山石池............. 286
（十）金鱼缸............. 286
（十一）峰............... 289
（十二）峦............... 292
（十三）岩............... 293
（十四）洞............... 294
（十五）涧............... 295
（十六）曲水............. 296
（十七）瀑布............. 297

十　选石................. 301

十一　借景............... 311
《园冶》自识............. 329
海外交流................. 331
主要参考文献............. 365
跋一..................... 366
跋二..................... 369
译后记................... 370
鸣谢..................... 374

CONTENTS

Preface 1, by Meng Zhaozhen 7
Preface 2, by Chen Xiaoli 10
Preface 3, by Wang Shaozeng 12
Preface 4, by Liu Xiuchen 14
Preface 5, by Gan Weilin,
Wang Zemin 16
Preface 6, by Wang Jinman 19
Preface by Author 1, by Wu Zhaozhao 21
Preface by Author 2, by Chen Yan,
Wu Di 23
The Origin of *Yuan Ye Illustrated* 25
Preface of *Yuan Ye*, by Ruan Dacheng ... 26
Inscription of *Yuan Ye*,
by Zheng Yuanxun 27
Preface of *Yuan Ye*, by Ji Cheng 28
On Design and Construction 001
On Garden Design 013

1. Siting 025
(1) Hilly Forest Land 045
(2) Urban Land 055
(3) Village Land 081
(4) Suburban Land 084
(5) Residence Side Land 095
(6) River and Lake Land 109

2. Base 121

3. Buildings 141
(1) Men Lou 152
(2) Tang 153
(3) Zhai 154
(4) Shi 157
(5) Fang 158
(6) Guan 159

(7) Lou 161
(8) Tai 169
(9) Ge 174
(10) Ting 176
(11) Xie 190
(12) Xuan 191
(13) Juan 194
(14) Guang 195
(15) Lang 199
(16) Five-Frame Girder 201
(17) Seven-Frame Girder 202
(18) Nine-Frame Girder 203
(19) Cao Jia 204
(20) Chong Chuan (Double Rafter) 204
(21) Mo Jiao 205
(22) Floor Plan 207

4. Decorations 209
(1) Screen Doors 210
(2) Yang Chen 211
(3) Hu Ge 212
(4) Feng Chuang 222

5. Railings 225
Railing Patterns 226

6. Doors & Windows 245
Door & Window Illustrations 246

7. Walls 253
(1) White Plaster Wall 254
(2) Ground Brick Wall 254
(3) Lattice Brick Wall 254
(4) Random Masonry Wall 258

8. Pavings 259

(1) Random Masonry Path 260
(2) Cobblestone Path 260
(3) Ice Crack Path 260
(4) Brick Path 261
Brick Paving Patterns 261

9. Rockery 265
(1) Yuan Rockery 272
(2) Ting Rockery 276
(3) Lou Rockery 279
(4) Ge Rockery 281
(5) Study Rockery 282
(6) Chi Rockery 283
(7) Indoor Rockery 284
(8) Cliff Rockery 285
(9) Rockery Pond 286
(10) Gold Fish Pond 286
(11) Peaks 289
(12) Luan 292
(13) Yan 293
(14) Caves 294
(15) Jian 295
(16) Meandering Streams 296
(17) Waterfalls 297

10. Rock Selection 301

11. Scene Leveraging 311
Postscript of *Yuan Ye* 329
Overseas Exchange Activities 331
Major Reference 365
Postscript 1 368
Postscript 2 369
Postscript by
the English Translator 372
Acknowledgement 374

目　次（日本語）

序一　孟兆禎
序二　陳暁麗
序三　王紹増
序四　劉秀晨
序五　甘偉林　王沢民
序六　王金満
自序一
自序二
「園冶図釈」縁起
冶叙——阮大鋮
題詞——鄭元勲
自序——計成
興造論
園説

一　相地
(一)山林地
(二)城市地
(三)村荘地
(四)郊野地
(五)傍宅地
(六)江湖地

二　立基

三　屋宇
(一)門楼
(二)堂
(三)斎
(四)室
(五)房
(六)館
(七)楼
(八)台
(九)閣

(十)亭
(十一)榭
(十二)軒
(十三)巻
(十四)広
(十五)廊
(十六)五架梁
(十七)七架梁
(十八)九架梁
(十九)草架
(二十)重椽
(二十一)磨角
(二十二)地図

四　装折
(一)屏門
(二)仰塵
(三)戸槅
(四)風窓

五　欄杆
欄杆図式

六　門窓
門窓図式

七　墻垣
(一)白粉墻
(二)磨磚墻
(三)漏磚墻
(四)乱石墻

八　舗地
(一)乱石路
(二)鵝子路
(三)氷裂路
(四)諸磚地

磚舗地図式

九　掇山
(一)園山
(二)廳山
(三)楼山
(四)閣山
(五)書房山
(六)池山
(七)内室山
(八)峭壁山
(九)山石池
(十)金魚缸
(十一)峰
(十二)巒
(十三)岩
(十四)洞
(十五)澗
(十六)曲水
(十七)瀑布

十　選石

十一　借景

自識
海外交流
参考文献
跋一
跋二
後書き
謝辞

园冶图释

Yuan Ye Illustrated
—— Classical Chinese Garden Explained

三 屋宇

3 Buildings

凡家宅住房，五间三间，循次第而造；惟园林书屋，一室半室，按时景为精。方向随宜，鸠工合见。家居必论，野筑惟因。虽厅堂俱一般，近台榭有别致：前添敞卷，后进余轩；必用重椽，须支草架；高低依制，左右分为。当檐最碍两厢，庭除恐窄；落步但加重庑，阶砌犹深。升栱不让雕鸾，门枕胡为镂鼓。时遵雅朴，古摘端方。画彩虽佳，木色加之青绿；雕镂易俗，花空嵌以仙禽。长廊一带回旋，在竖柱之初，妙于变幻；小屋数椽委曲，究安门之当，理及精微。奇亭巧榭，构分红紫之丛；层阁重楼，迥出云霄之上。隐现无穷之态，招摇不尽之春。槛外行云，镜中流水，洗山色之不去，送鹤声之自来。境仿瀛壶，天然图画，意尽林泉之癖，乐余园圃之间。一鉴能为，千秋不朽。堂占太史，亭问草玄，非及云艺之台楼，且操般门之斤斧。探奇合志，常套俱裁。

All residence buildings, no matter if it is three units or five, must be built in certain order. However, a garden house such as a study, no matter whether it is one room or a half, must be built according to seasonal sceneries. The orientation of a building can be flexible but a consensus between the owner and the garden designer must be reached. Attention must be paid toward the orientation of a residence house, but for other garden buildings, contextual adaptation is recommended. Although all hall oriented buildings are pretty much the same, a platform of pavilion type of architecture should be unique. At its front there should add open Juan and at its back there should be extra Xuan. At its top there must be double rafters and the structure must be supported by Cao Jia. The part manufacturing must be done from higher to lower and the construction must be done one side before the other. In front of a main hall there should not build two compartments because this would make the court appear small. If more rafters were added over steps, there should be an extension of the steps. There is no need to carve ornamentations on a Dougong (Note: a Chinese style building bracket) and neither is there a need to make a door pillow drum shaped. The style should be ancient and elegant and if we model after an old style, it should be proper and naturally poised. Although being colorful is nice, it would be more elegant if we paint the white wood with viridity. Hollow carving may turn philistine, such as when an immortal fowl is carved, which is inappropriate. If one would like to build a zigzag and meandering corridor, a prior careful design is a must, and some changes and variations must be pre-conceived. If one wants to build a small building with only a few units, he must think about how to arrange its door, and use some ingenuity for this small extent design. Fantastical and quaint pavilion and open hall buildings should be built separately among garden flowers and plants. Multi-storied buildings should be built at different elevations but at much higher places. Unlimited scenes can be created inconspicuously and indefinite spring scenes can be presented at full length if we apply this garden design rule. The scenes outside the railings are like moving cloud and the natural scenes in the pond are like flowing water. The verdant mountains cannot be washed off by rain water and the cranes' cry can be heard through the wind. It is like a wonderland and a natural landscape painting. I can enjoy natural landscape at any moment and entertain myself in the garden at my leisure time. If one can learn from all these and build a garden in this way, then it must be a masterpiece. A hall like garden building should show characters of integrity and nobility while pavilions and terraces should appear nonchalant about fame and wealth. Although my garden design and construction skills cannot match those of master Lu Yun, I just regard myself as a student of the great master Lu Ban. To study and savor the wonders of garden design art, like-minded people should get together to exchange ideas. Don't bother with conventional, uncreative and outmoded ideas.

计成强调园林中书屋"按时景为精",扬州个园中书屋面对"冬景",围墙北面堆叠宣石,寓意积雪未化;墙面四排圆洞,由于墙外窄巷高墙的负压作用,且洞小空气流动速度急增,风入洞口呼呼作响,四排洞口如口琴音孔式排列,音响各异,极富感染力,故书屋取名"透风漏月",可谓形神兼备。苏州鹤园书屋前为洞壑山池,山林之好,随意可得。

Ji Cheng stressed that garden studies should "best use seasonal scenes", and one example is the "winter scene" in Geyuan in Yangzhou. Quartzite rockery was constructed north of the wall, implying accumulated snow not yet melted. There are four rows of round holes on the wall. Due to the narrow alley tall wall negative pressure effect outside the wall, and fast air flow through small holes, wind blows through the holes making wuthering sound. The four rows of holes are like the pattern on a harmonica, and they each make different sounds with extreme verves; therefore the study was named as "Loufeng Touyue" (Note: Seeping Wind and Spilling Moon), a name carrying both form and spirit. Another example is the Donghe Shanchi in front of the study in Heyuan, Suzhou. There your love for natural landscape can be easily achieved.

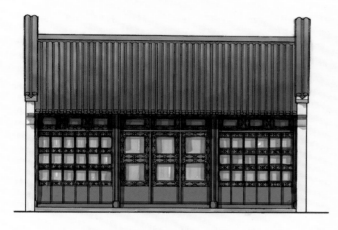

扬州个园透风漏月屋
Toufeng Louyue in Geyuan, Yangzhou

苏州鹤园屋前洞壑山池
Donghe Shanchi in Front of a Study in Heyuan, Suzhou

扬州个园冬景
Winter Scene in Geyuan, Yangzhou

家居建筑讲究朝向，而造园建筑应当因地制宜。拙政园绣绮亭建在庭园中，故强调花木环绕，开漏窗框景。天平山坡麓建亭则需顺应自然，力求"天坠地出"的艺术效果，四仙亭六个翘角飘然欲飞，似有仙人境界的畅想。

Much attention needs to be paid towards the orientation for residential buildings, while other garden buildings stress more on the context and adaptation. Xiuyi Pavilion of Zhuozheng Yuan was built inside a courtyard, emphasizing the fact that it's surrounded by garden plants, and it has decorative lattice windows for framing sceneries. A pavilion on the hillside of Mount Tianping was built with the idea of being in harmony with nature, striving for an artistic effect of "naturally made", with its six up-tilt building corners posing as flying, giving one an impression of an object in a fairy world.

拙政园绣绮亭
Xiuyi Pavilion of Zhuozheng Yuan

天平山四仙亭
Sixian Pavilion of Mount Tianping

园林中的厅堂高大开敞，最好朝南，堂檐前勿建两厢。诸多园林中厅堂造型大都相似。图中苏州拙政园远香堂与沁园一朝风月堂的柱网开间均一样，仅仅是唐代与清代的建筑风格的差异。

厅堂与台榭连接时，可以前添敞卷，后进余轩，顶上用重椽并立支草架。计成主张建筑装折应"时遵雅朴"，不要彩绘，雕镂也易俗气。

A hall building in a garden is tall and open, better facing south and should avoid adding two compartments. A lot of open halls like buildings in gardens have similar forms. The building units of Yuanxiang Tang (Note: Distant Fragrance Hall) in Zhuozheng Yuan in Suzhou and Yizhao Fengyue Tang of Qin Garden are the same, with the only difference being the architectural style of the Tang vs. the Qing dynasty.

When such a hall-like building connects with Taixie, a platform-like building, extensions such as Changjuan in front, or Yuxuan in back, could be added. Double rafters should be used on top and a Caojia should be erected. Ji Cheng argues that architectural elements such as doors and windows should be "simple and elegant" and they should avoid color painting and philistine carving.

拙政园远香堂
Yuanxiang Tang of Zhuozheng Yuan

沁园一朝风月堂
Yizhao Fengyue Tang of Qin Garden

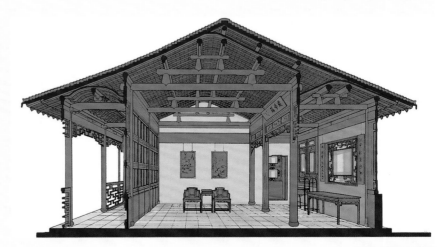

苏州网师园殿春簃
Dianchun Yi of Wangshi Yuan, Suzhou

一带迴旋长廊，当开始立柱之时，要妙于曲折变幻。苏州网师园廊分别与亭、轩组合，错落有致。下面三幅图分别展示：曲廊与廊桥的组合；曲廊与水榭的组合；折廊与亭台的结合；均为园林中的佳作。

The design of a long and zigzagging corridor should be well-conceived in its changes at turns. The garden corridor in Suzhou's Wangshi Yuan connects with pavilions and larger open buildings and is handled well with all the combinations. The following three figures show the combinations of corridor with a bridge, corridor with an open building, and corridor with an elevated pavilion respectively. They are all excellent works in garden design.

苏州网师园
Wangshi Yuan, Suzhou

园林中数间小屋的组合要委婉曲折，须研究安门的当否，并体察手法的精微。图中虎丘悟石轩、拥翠山庄月驾轩是利用山地巧于安置，达到事半功倍的效果；木渎羡园则是人工掇山叠石塑造起伏地形，使建筑产生起伏错落的韵律。

Combinations of small buildings in a garden should be tactful and should avoid directness. Careful study should be made to decide whether to have a door and where to place it, paying attention to implementation details. In the following figures, Wushi Xuan in Hu Qiu and Yuejia Xuan in Yongcui Shanzhuang show examples of skillful handlings of terrain to have achieved prominent effects. Muduxian Yuan shows an example of artificially creating a mound and rockery to have an undulating terrain and adding a rhythmic wavy movement to the garden buildings.

苏州虎丘悟石轩
Wushi Xuan in Hu Qiu, Suzhou

木渎羡园爬山廊
Climbing Corridor in Muduxian Yuan

拥翠山庄月驾轩
Yuejia Xuan in Yongcui Shanzhuang

奇亭巧榭，要分建于花木之中。拙政园雪香云蔚亭建在池山丛林中，体现"蝉噪林愈静，鸟鸣山更幽"的山林景色；荷风四面亭则在水边垂柳畔，池中荷花飘香，体现出园主人"出淤泥而不染"的造园意境。网师园濯缨水阁则是取"沧浪之水清兮，可以濯吾缨；沧浪之水浊兮，可以濯吾足"的诗意，在一定程度上表明园主人出世的情怀。

Exotic pavilions and artful open buildings should be built among trees and flowers. Xuexiang Yunwei Pavilion of Zhuozheng Yuan was built in a small forest, reflecting a natural forest scene described in the poem, "The forest feels even quieter with cicadas' noisy sound and the mountain seems even more secluded with birds' chirping." Hefeng Simian Ting (Note: Lotus Wind Pavilion) is by a bank with willows. With all the lotus flower fragrance in the air, it reflects the garden owner's aspiration of "emerging unstained from the filth". The Zhuoying Shuige of Wangshi Yuan alludes to the poem "Clear water can be used to wash my tassels and muddy water for my feet." This shows to a certain extent the garden owner's supra-mundane mindset.

苏州拙政园雪香云蔚亭
Xuexiang Yunwei Pavilion of Zhuozheng Yuan, Suzhou

苏州拙政园荷风四面亭
Hefeng Simian Ting of Zhuozheng Yuan, Suzhou

网师园濯缨水阁
Zhuoying Shuige of Wangshi Yuan

层阁重楼，像驾乎云霄之上，这样的造园章法，就借来无穷的景物，隐现不定、不尽的春光，招摇而来，从下面实例诸图就得知"探奇合志，常套俱裁"。

Multi-storied buildings may seem as high as reaching the cloud. This way of handling could draw unlimited sceneries, either apparent or not. The following are examples of "exploring wonders of a garden with like-minded people and avoiding conventions and formalities".

沧浪亭看山楼建于山顶印心石屋之上，前轩后楼，造型优美，昔日登轩可望苏州西南诸山，可谓"洗山色之不去，送鹤声之自来"

Kanshan Lou (Note: Mountain View Building) of Canglang Ting was built on top of Yinxin Shiwu, with Xuan front and Lou back beautiful format. In it, one could view several mountains in the southwest of Suzhou. "The beautiful mountain scene is still there after being washed; the cranes' singing can still be heard after they're let go".

御风楼是昔日杭州城南最高建筑，由于高居假山之顶，登楼看西湖众山环抱，湖光塔影历历在目，"隐现无穷之态，招摇不尽之春"

Yufeng Lou used to be the tallest building in south Hangzhou. Since it's high on top of a rockery, one can enjoy all the mountain, lake and pagoda scenes from there. "With its various forms visible or not, it embraces unlimited beautiful spring scenes".

豫园燕爽阁屹立假山之上，确有"栏外行云，镜中流水"的感受

Yuyuan's Yanshuang Ge is on top of a rockery, showing a feeling of "fleeting cloud beyond the railings and flowing water in the mirror".

扬州个园拂云亭在秋山上，是昔日城东北部最高点，其东北方向系京杭大运河，俯瞰可"隐现无穷之态"

The Fuyun Ting (Note: Cloud Brushing Pavilion) of Geyuan in Yangzhou was built on top of Qiu Shan (Note: Autumn Mountain) and it used to be the highest point in the northeast part of the city. The be Beijing-Hang zhou grand canal runs in its northeast side. An overlook at the direction can expose numerous various scenes.

瘦西湖熙春台运用"叠石停云"的景境，使层阁重楼有腾云驾雾的艺术魅力，且北面借景蜀岗大明寺、栖灵塔，晨钟晚鼓，真有"洗山色之不去，送鹤声之自来。境仿瀛壶，天然图画"的境界。
Xichun Tai of Shouxi Lake utilizes "Dieshi Tingyun" (Note: Rock Piling Cloud Stalling) scene, causing a giddy feeling. To its north, it leverages Daming Temple of Shugang, Qiling Pagoda, and morning drum beats and evening bell rings, immersing itself in a scene described in the poem, "The beautiful mountain scene is still there after being washed; the cranes' singing can still be heard after they're let go. Almost like Yinghu, it is a natural landscape painting."

瘦西湖"卷石洞天"峦崖洞壑与阁楼、庭院结为一体，相得益彰，"意尽林泉之癖，乐余园圃之间"
"Juanshi Dongtian" of Shouxi Lake integrates cliffs, caves with buildings and courtyard, with each part promoting each other. "With a persistent love in forest and springs, and unending enjoyment in gardens"

狮子林屹立在池畔假山上的修竹阁，集湖光山色于一体
In Shizi Lin, Xiuzhu Ge (Note: Slender Bamboo Building) was erected by pond side rockeries, collecting the natural beauty of lakes and mountains.

上海秋霞圃霁霞阁建在湖石堆叠的仙人洞上，可谓"意尽林泉之癖"
Jixia Ge of Qiuxia Pu in Shanghai was built on top of rockery made "fairy cave", showing intense interest in seeking natural enjoyment in forest and springs.

留园冠云楼南，以冠云峰为主题，可谓"栏外行云，镜中流水"
Taking Guanyun Feng (Note: Cloud Topping Peak) as the central theme at the south of Guanyun Lou in Liuyuan, it embodies the feeling of "fleeting cloud beyond the railings and flowing water in the mirror".

（一）门楼

门上起楼，像城堞有楼以壮观也。无楼亦呼之。

(1) Men Lou

A building added on top of a garden gate is like a battlement and that makes the gate spectacular. On the other hand, if no such building is added on top, it is still referred to as a Men Lou (Note: Gate Building).

无锡寄畅园磨砖雕花门楼，此乃江南古典园林最典型的做法
A ground brick and flower carving Men Lou in Jichang Yuan, Wuxi, is a typical type of architecture in southern Chinese gardens.

瘦西湖徐园门景为圆形，墙后为门房建筑，入园荷池映莲，石桥卧波
The gate scene at Xuyuan in Shouxi Lake is round shaped. Behind it is a gate building. Lotus pond greets people at the entrance, with stone bridges over water.

苏州耦园门楼，是采用粉墙与亭廊相结合的形式
Ouyuan's Men Lou, Suzhou, combines painted wall with pavilion corridor.

（二）堂

古者之堂，自半已前，虚之为堂。堂者，当也。谓当正向阳之屋，以取堂堂高显之义。

(2) Tang

Tang in ancient China is the front half of a building, usually without any obstruction or separations. The meaning of Tang is Dang, which means "facing front". It is on the central axis facing south towards the sun, suggesting it is dignified, impressive and with the central location.

正立面图　Front Elevation

横剖面图　Cross Section View

拙政园远香堂实测图
Plan – Actual Survey Drawing of Yuanxiang Tang in Zhuozheng Yuan

（三）斋

斋较堂，惟气藏而致敛，有使人肃然齐敬之义。盖藏修密处之地，故式不宜敞显。

(3) Zhai

Zhai is similar to Tang, but the difference is that it is a place for meditating and collecting Qi, as in Taichi. It is supposed to be filled with deep esteem. Since it should be in a secluded area for such a purpose, it should be a bit hidden from public access.

怡园画舫斋横剖面图
Cross Section of Huafang Zhai of Yiyuan

怡园画舫斋正立面图
Front Elevation of Huafang Zhai of Yiyuan

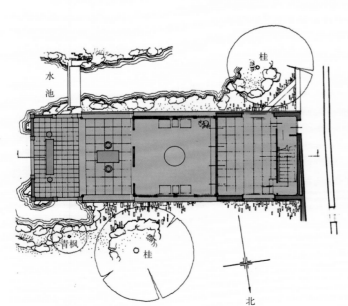

怡园画舫斋平面图
Plan of Huafang Zhai of Yiyuan

怡园画舫斋侧立面图
Side Elevation of Huafang Zhai of Yiyuan

拙政园香洲侧立面图
Side Elevation of Xiangzhou of Zhuozheng Yuan

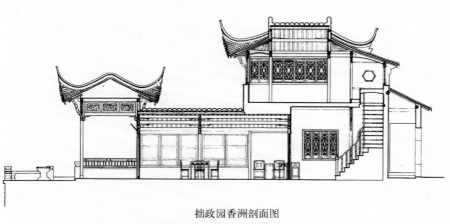

拙政园香洲剖面图
Cross Section of Xiangzhou of Zhuozheng Yuan

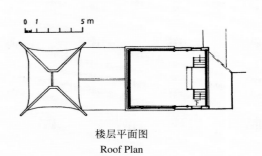

楼层平面图
Roof Plan

拙政园香洲正立面图
Front Elevation of Xiangzhou of Zhuozheng Yuan

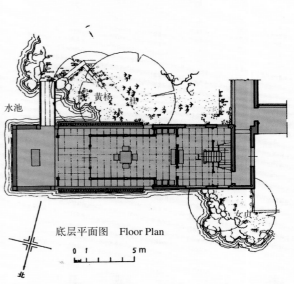

拙政园香洲平面图
Plan of Xiangzhou of Zhuozheng Yuan

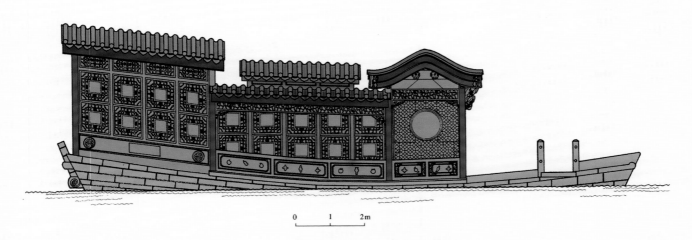

南京煦园不系舟立面图
Elevation of Buji Zhou in Xuyuan, Nanjing

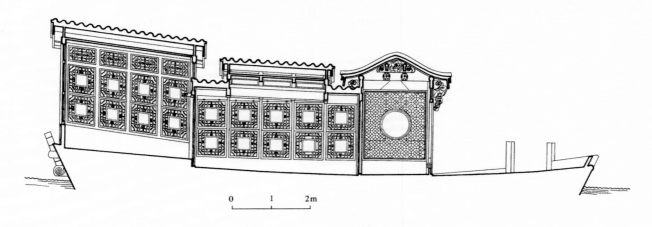

南京煦园不系舟剖面图
Cross Section of Buji Zhou in Xuyuan, Nanjing

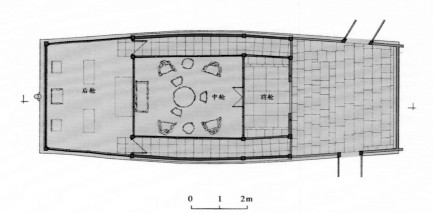

南京煦园不系舟平面图
Plan of Buji Zhou in Xuyuan, Nanjing

南京煦园不系舟南立面图
South Elevation of Buji Zhou in Xuyuan, Nanjing

南京煦园不系舟北立面图
North Elevation of Buji Zhou in Xuyuan, Nanjing

（四）室

古云，自半已后，实为室。《尚书》有"壤室"，《左传》有"窟室"，《文选》载："旋室媛娟以窈窕"指"曲室"也。

(4) Shi

According to what people said in ancient China, a Shi is the back half of a building where people live and it is a space enclosed with four walls. In "*Shang Shu*", there is a record for "Rang Shi", there is "Ku Shi" in "*Zuo Zhuan*", and in "*Wen Xuan*", a deep and quiet "Qu Shi" was recorded and it is supposed to be indirectly connected and secluded.

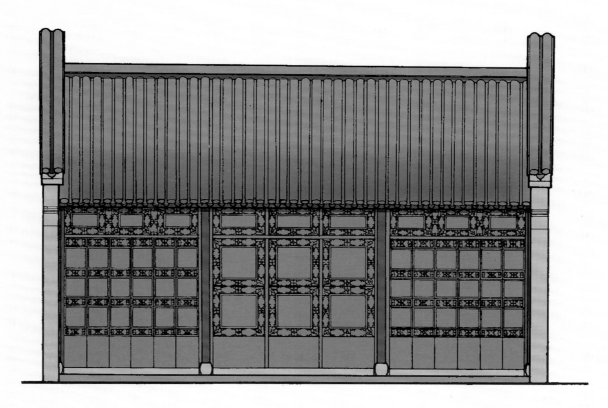

扬州个园透风漏月室南立面图
South Elevation of Toufeng Louyue Shi, Yangzhou

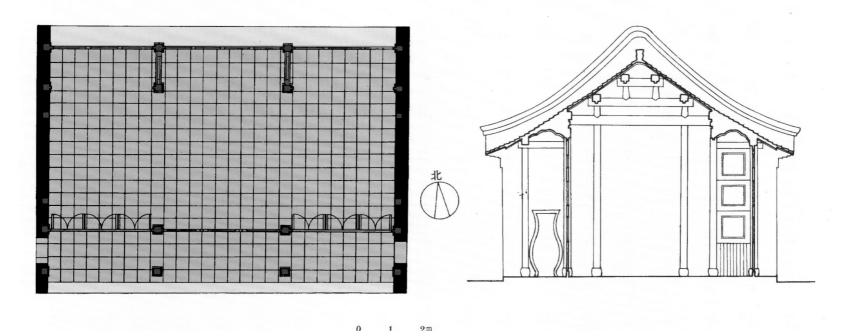

扬州个园透风漏月室平面图
Plan of Toufeng Louyue Shi, Yangzhou

扬州个园透风漏月室剖面图
Crosss Section of Toufeng Louyue Shi, Yangzhou

（五）房

《释名》云：房者，防也。防密内外以为寝闼也。

(5) Fang

According to "*Shi Ming*", a Fang means defense. This space should be concealed and divided into the inner and outer, and is supposed to be a bedroom for sleeping.

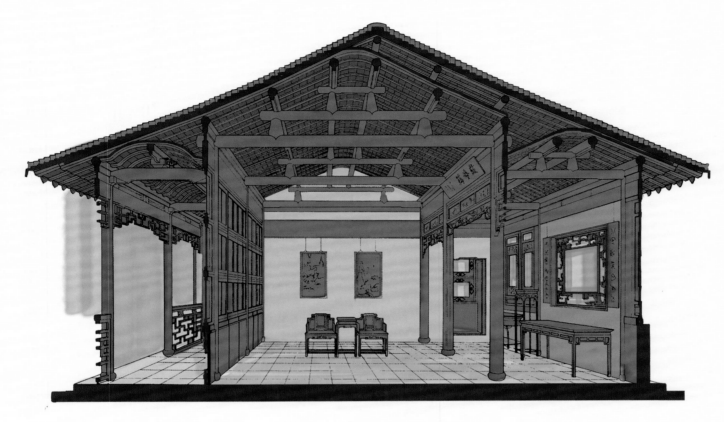

苏州网师园殿春簃剖视图
Cross Section of Dianchun Yi in Wangshi Yuan, Suzhou

苏州网师园殿春簃平面图
Plan of Dianchun Yi in Wangshi Yuan, Suzhou

苏州网师园殿春簃南立面图
South Elevation of Dianchun Yi in Wangshi Yuan, Suzhou

（六）馆

散寄之居，曰"馆"，可以通别居者。今书房亦称"馆"，客舍为"假馆"。

(6) Guan

A temporary dwelling is called Guan. It can also connect to another dwelling. Now a study can also be referred to as a Guan, and a guest house is called "Jia Guan" (Note: Holiday House).

拙政园三十六鸳鸯馆正立面图
Elevation of Thirty Six Yuanyang Guan in Zhuozheng Yuan

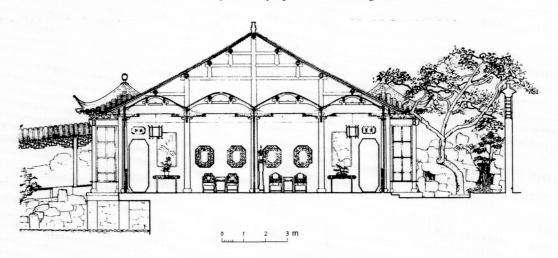

拙政园三十六鸳鸯馆横剖面图
Cross Section of Thirty Six Yuanyang Guan in Zhuozheng Yuan

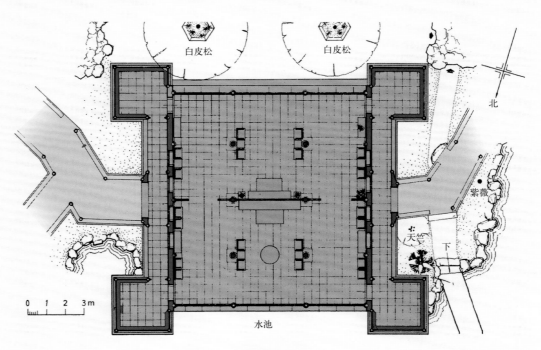

拙政园三十六鸳鸯馆平面图
Plan of Thirty Six Yuanyang Guan in Zhuozheng Yuan

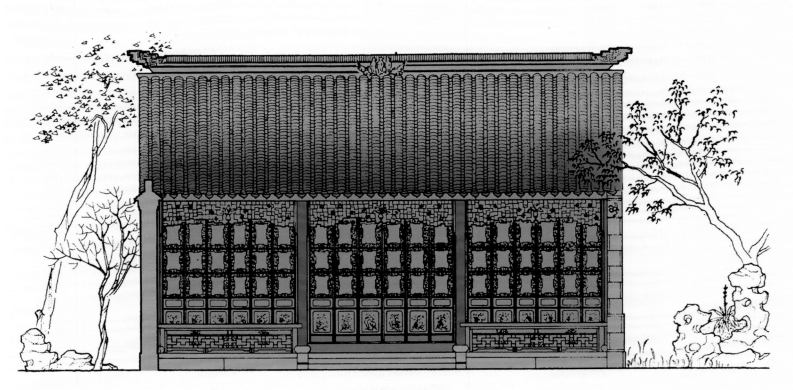

王洗马巷某宅花厅立面图
Elevation of a Flower Hall in Wangxima Lane

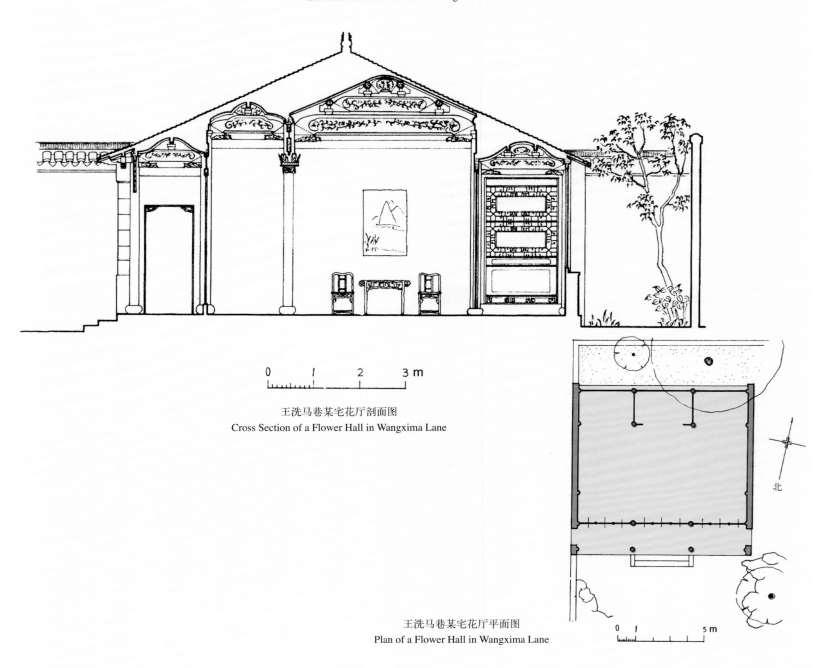

王洗马巷某宅花厅剖面图
Cross Section of a Flower Hall in Wangxima Lane

王洗马巷某宅花厅平面图
Plan of a Flower Hall in Wangxima Lane

（七）楼

《说文》云：重屋曰"楼"。《尔雅》云：陕而修曲为"楼"。言窗牖虚开，诸孔悽悽然也，造式，如堂高一层者是也。

(7) Lou

According to "Shuo Wen", a multi-storied building is a Lou. According to "Er Ya", a narrow and long building is called Lou, that is, there are many opened doors and windows, and the light through those openings makes the room appear open and lit. Its architectural structure is as adding a story on top of a Tang.

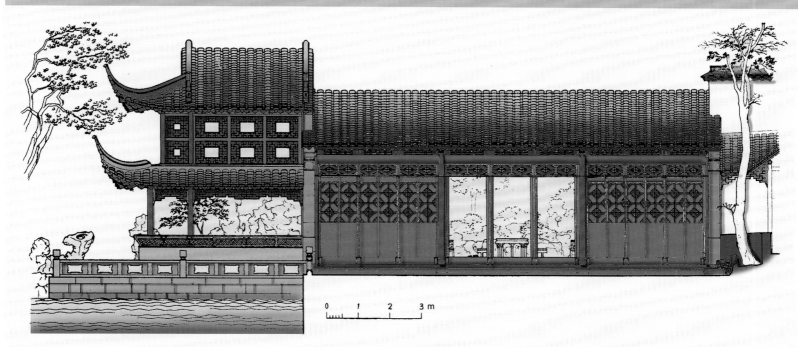

留园明瑟楼正立面图
Elevation of Mingse Lou in Liu Yuan

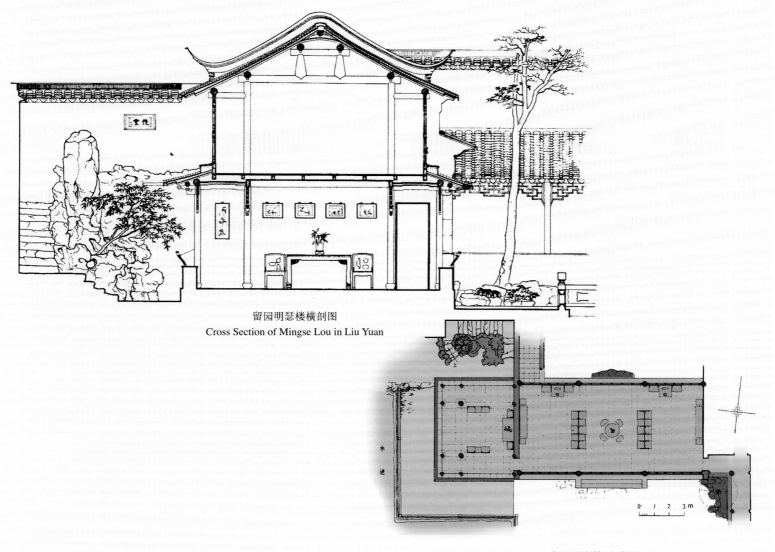

留园明瑟楼横剖图
Cross Section of Mingse Lou in Liu Yuan

留园明瑟楼平面图
Plan of Mingse Lou in Liu Yuan

拙政园倒影楼正立面图
Front Elevation of Daoying Lou in Zhuozheng Yuan

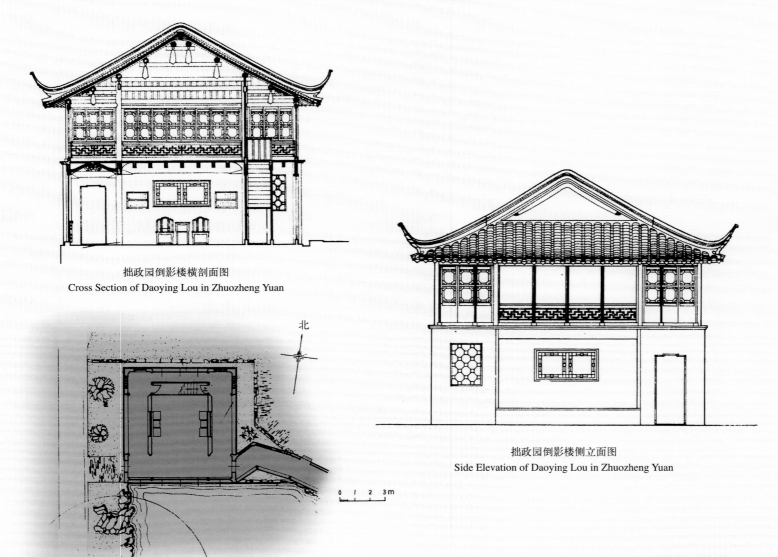

拙政园倒影楼横剖面图
Cross Section of Daoying Lou in Zhuozheng Yuan

拙政园倒影楼侧立面图
Side Elevation of Daoying Lou in Zhuozheng Yuan

拙政园倒影楼平面图
Plan of Daoying Lou in Zhuozheng Yuan

BUILDINGS

沧浪亭看山楼侧立面图
Side Elevation of Kanshan Lou in Canglang Ting

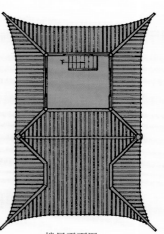

楼层平面图
Stair Plan of Kanshan Lou in Canglang Ting

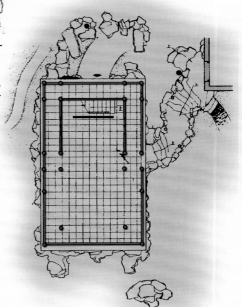

底层平面图
Ground Floor Plan of Kanshan Lou in Canglang Ting

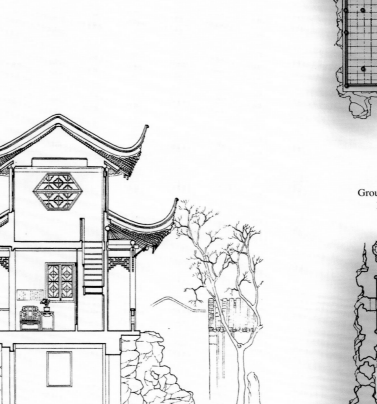

沧浪亭看山楼横剖面图
Cross Section of Kanshan Lou in Canglang Ting

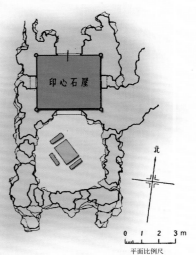

底层下石洞平面图
沧浪亭看山楼平面图
Below Floor Rock Cave Plan of Kanshan Lou in Canglang Ting

个园抱山楼壶天自春南立面图
South Elevation of Hutian Zichun of Baoshan Lou in Geyuan

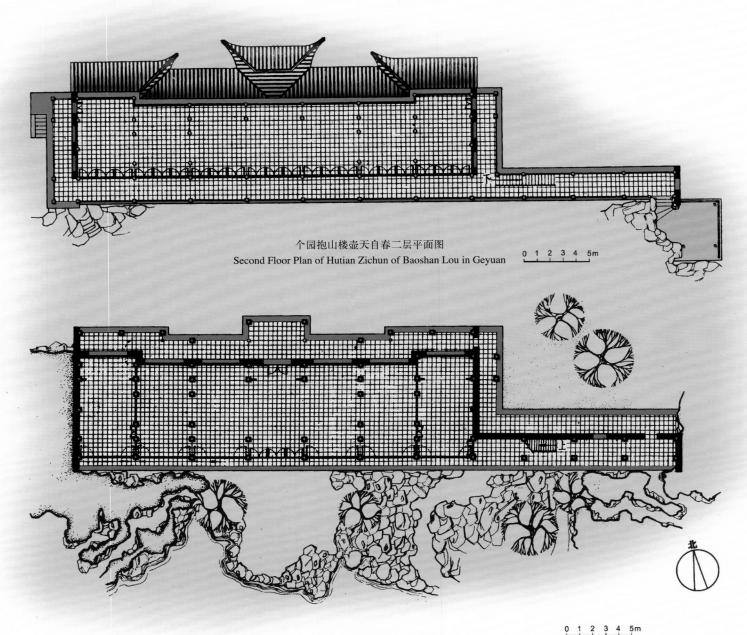

个园抱山楼壶天自春二层平面图
Second Floor Plan of Hutian Zichun of Baoshan Lou in Geyuan

个园抱山楼壶天自春底层平面图
Ground Floor Plan of Hutian Zichun of Baoshan Lou in Geyuan

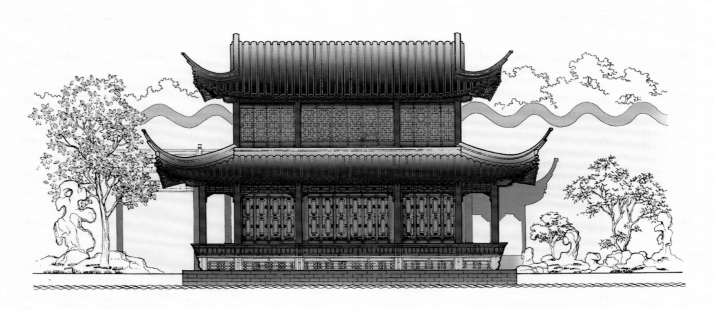

南京煦园夕佳楼东立面图
East Elevation of Xijia Lou in Xuyuan, Nanjing

南京煦园夕佳楼北立面图
North Elevation of Xijia Lou in Xuyuan, Nanjing

南京煦园夕佳楼二层平面图
Second Floor Plan of Xijia Lou in Xuyuan, Nanjing

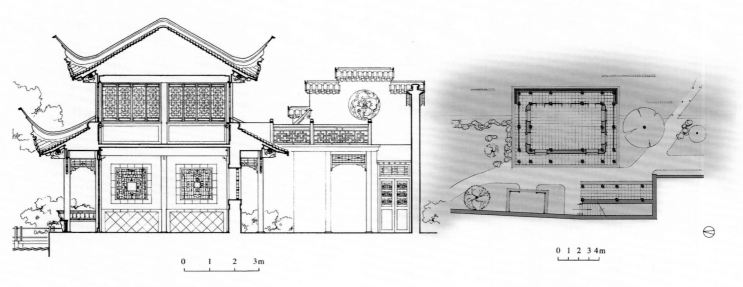

南京煦园夕佳楼剖面图
Cross Section of Xijia Lou in Xuyuan, Nanjing

南京煦园夕佳楼一层平面图
Ground Floor Plan of Xijia Lou in Xuyuan, Nanjing

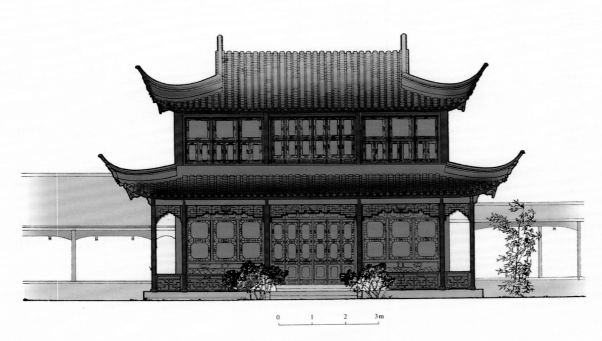

苏州寒山寺枫江第一楼西立面图
West Elevation of Fengjiang 1st Building in Hanshan Temple, Suzhou

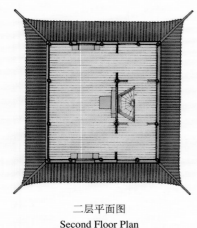

二层平面图
Second Floor Plan

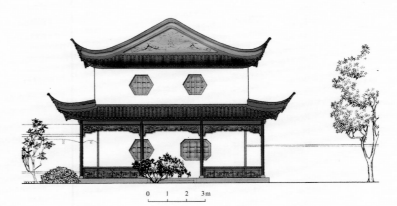

苏州寒山寺枫江第一楼南立面图
South Elevation of Fengjiang 1st Building in Hanshan Temple, Suzhou

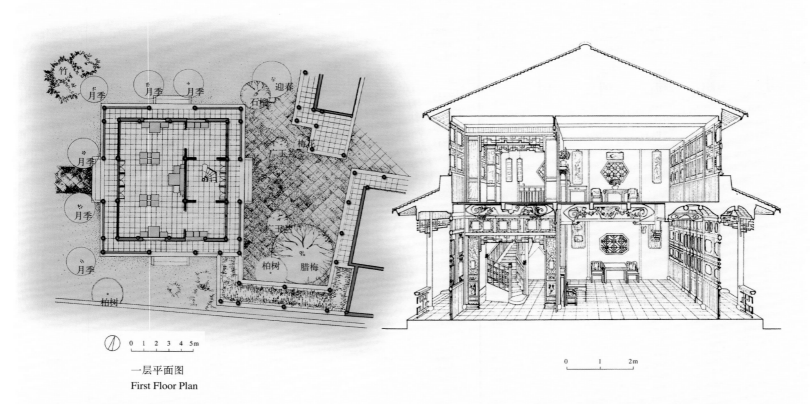

一层平面图
First Floor Plan

苏州寒山寺枫江第一楼一、二层平面图
Ground and First Floor Plan of Fengjiang 1st Building in Hanshan Temple, Suzhou

苏州寒山寺枫江第一楼剖面透视图
Cross Section Perspective View of Fengjiang 1st Building in Hanshan Temple, Suzhou

苏州拙政园见山楼东立面图
East Elevation of Jianshan Lou in Zhuozheng Yuan, Suzhou

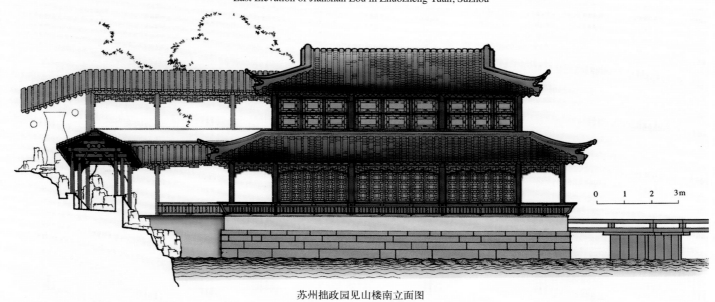

苏州拙政园见山楼南立面图
South Elevation of Jianshan Lou in Zhuozheng Yuan, Suzhou

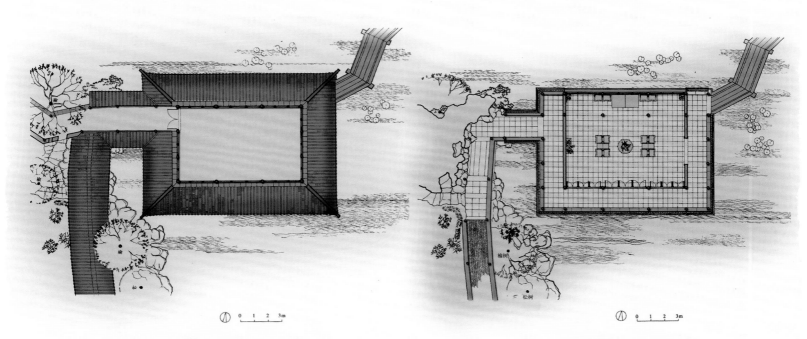

苏州拙政园见山楼上层平面图
Upper Floor Plan of Jianshan Lou in Zhuozheng Yuan, Suzhou

苏州拙政园见山楼底层平面图
Ground Floor Plan of Jianshan Lou in Zhuozheng Yuan, Suzhou

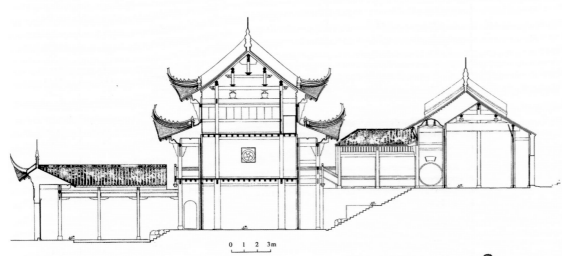

马鞍山采石矶太白楼剖面图
Cross Section View of Taibai Lou in Caishi Ji, Ma'anshan

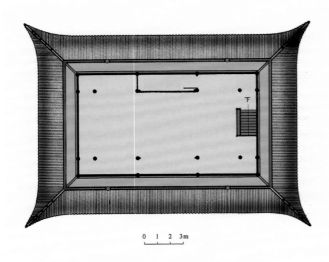

马鞍山采石矶太白楼三层平面图
Third Floor Plan of Taibai Lou in Caishi Ji, Ma'anshan

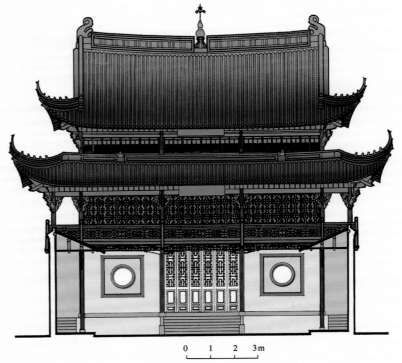

马鞍山采石矶太白楼正立面图
Front Elevation of Taibai Lou in Caishi Ji, Ma'anshan

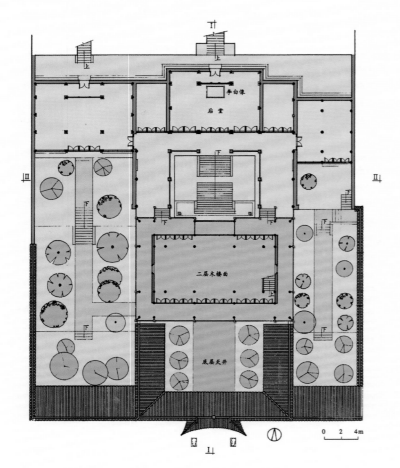

马鞍山采石矶太白楼二层平面图
Second Floor Plan of Taibai Lou in Caishi Ji, Ma'anshan

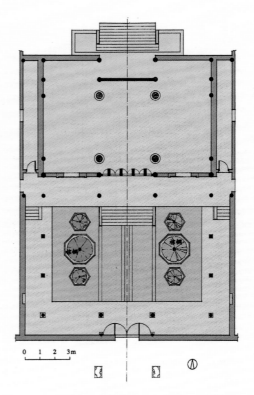

马鞍山采石矶太白楼一层平面图
First Floor Plan of Taibai Lou in Caishi Ji, Ma'anshan

（八）台

《释名》云："台者，持也。言筑土坚高，能自胜持也。"园林之台，或掇石而高上平者；或木架高而版平无屋者；或楼阁前出一步而敞者，俱为台。

(8) Tai

According to *Shi Ming*, Tai means support. It means piling up dirt to be high and sturdy and to be able to support itself. Generally a Tai in a garden is either constructed by stones piling up high with a flat top, or by wood structures with planks on top but with no further construction, or an extension, further away from a building, which is open and broad.

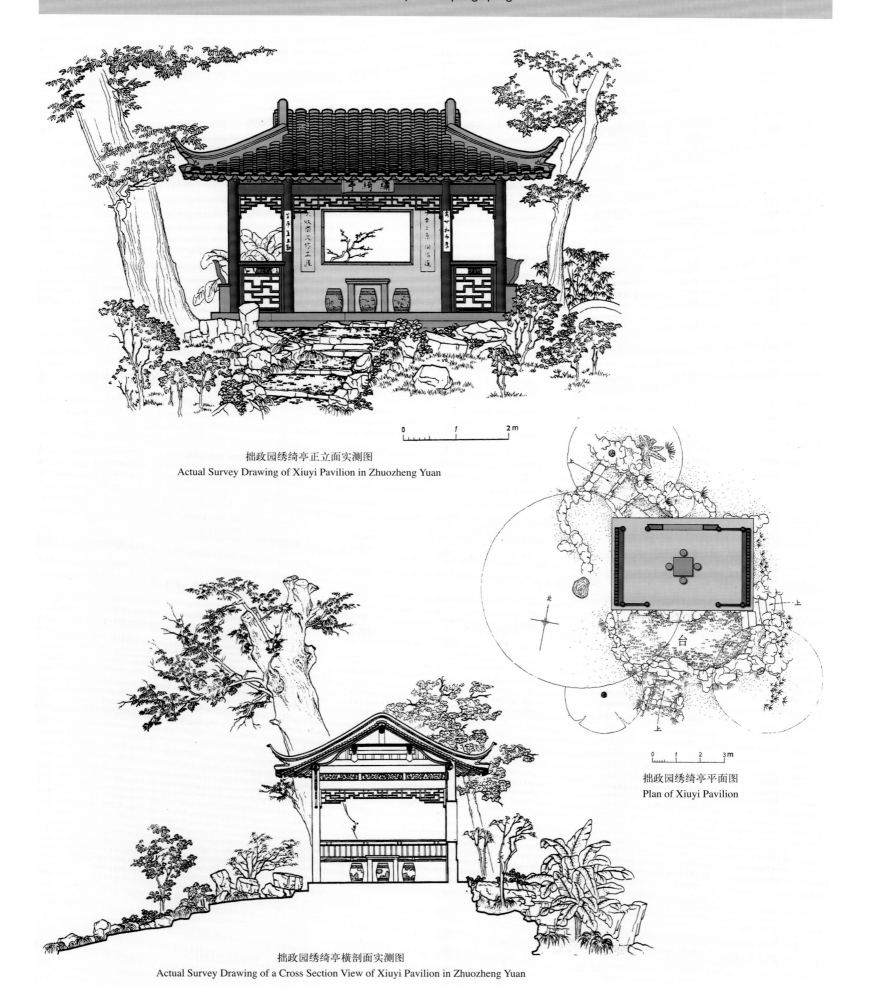

拙政园绣绮亭正立面实测图
Actual Survey Drawing of Xiuyi Pavilion in Zhuozheng Yuan

拙政园绣绮亭平面图
Plan of Xiuyi Pavilion

拙政园绣绮亭横剖面实测图
Actual Survey Drawing of a Cross Section View of Xiuyi Pavilion in Zhuozheng Yuan

苏州拙政园雪香云蔚亭南立面图
South Elevation of Xuexiang Yunwei Pavilion in Zhuozheng Yuan, Suzhou

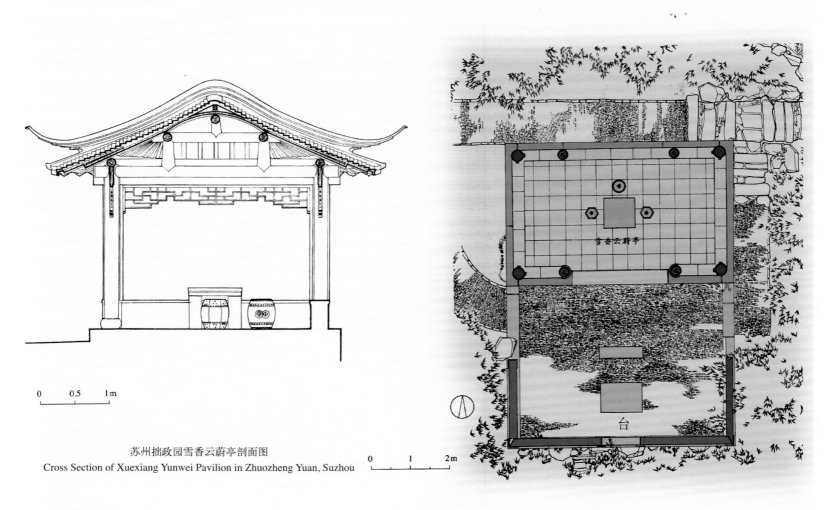

苏州拙政园雪香云蔚亭剖面图
Cross Section of Xuexiang Yunwei Pavilion in Zhuozheng Yuan, Suzhou

苏州拙政园雪香云蔚亭平面图
Plan of Xuexiang Yunwei Pavilion in Zhuozheng Yuan, Suzhou

屋宇
BUILDINGS

扬州何园月亭立面图
Elevation of Moon Pavilion in Heyuan, Yangzhou

扬州何园月亭剖面图
Section View of Moon Pavilion in Heyuan, Yangzhou

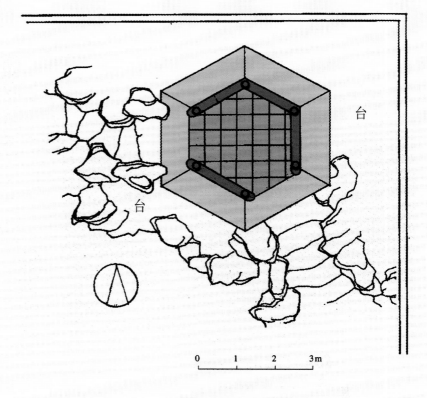

扬州何园月亭平面图
Plan of Moon Pavilion in Heyuan, Yangzhou

天平山四仙亭正立面图
Elevation of Sixian Pavilion Survey Drawing, Mount Tianping

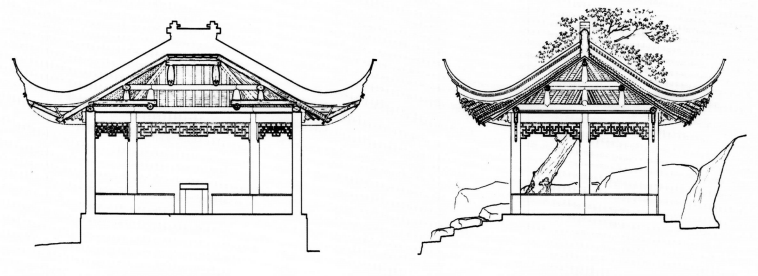

天平山四仙亭纵剖面图
Vertical Section of Sixian Pavilion Survey Drawing, Mount Tianping

天平山四仙亭横剖面图
Horizontal Section of Sixian Pavilion Survey Drawing, Mount Tianping

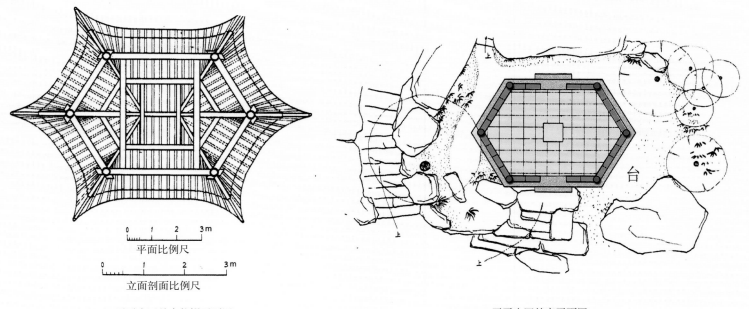

天平山四仙亭仰视平面图
Up-Looking View of Sixian Pavilion Survey Drawing, Mount Tianping

天平山四仙亭平面图
Plan of Sixian Pavilion Survey Drawing, Mount Tianping

屋宇
BUILDINGS 173

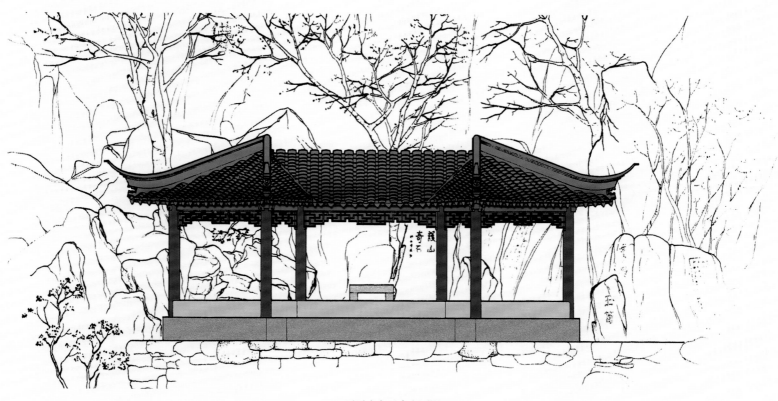

天平山白云亭立面图
Elevation of Baiyun Ting (Note: White Cloud Pavilion), Mount Tianping, from Actual Survey Drawing

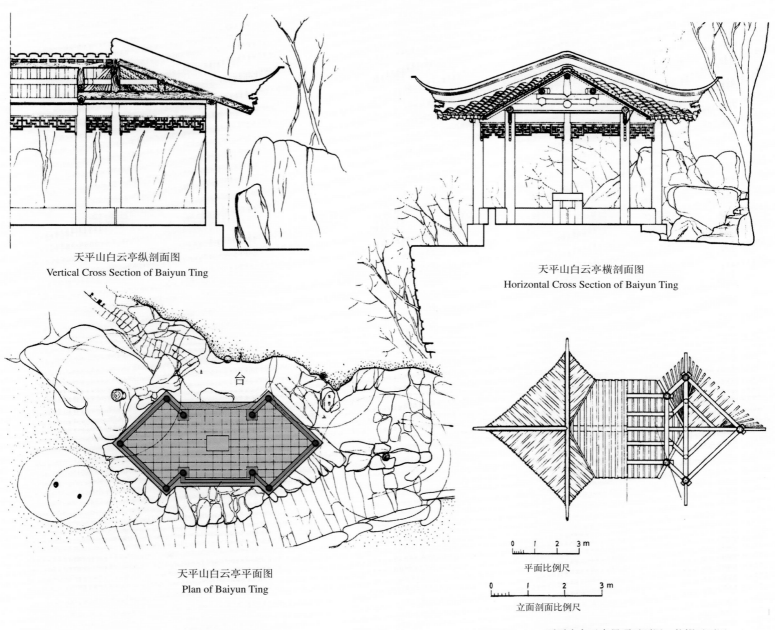

天平山白云亭纵剖面图
Vertical Cross Section of Baiyun Ting

天平山白云亭横剖面图
Horizontal Cross Section of Baiyun Ting

天平山白云亭平面图
Plan of Baiyun Ting

平面比例尺

立面剖面比例尺

天平山白云亭屋顶平面图、仰视平面图
Roof and Ceiling Views of Baiyun Ting

(九) 阁

阁者，四阿开四牖。汉有麒麟阁，唐有凌烟阁等，皆是式。

(9) Ge

A Ge is a building with four-sided sloped roof and windows on all its four sides. The Qilin (Note: An unreal animal imagined in Chinese mythology) Ge in the Han dynasty and the Lingyan Ge in the Tang dynasty are all such examples.

留园远翠阁正立面图
Elevation of Yuancui Ge in Liuyuan

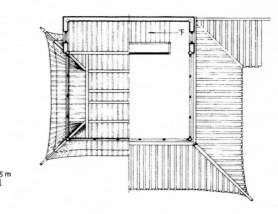

楼层仰视平面图、楼层平面图
Up-Looking View and Stair Floor Plan

留园远翠阁横剖面图
Cross Section View of Yuancui Ge in Liuyuan

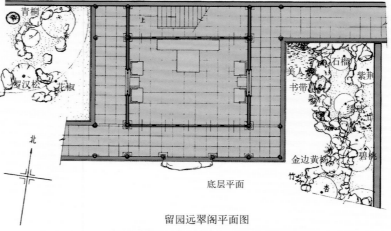

留园远翠阁平面图
Ground Floor Plan of Yuancui Ge in Liuyuan

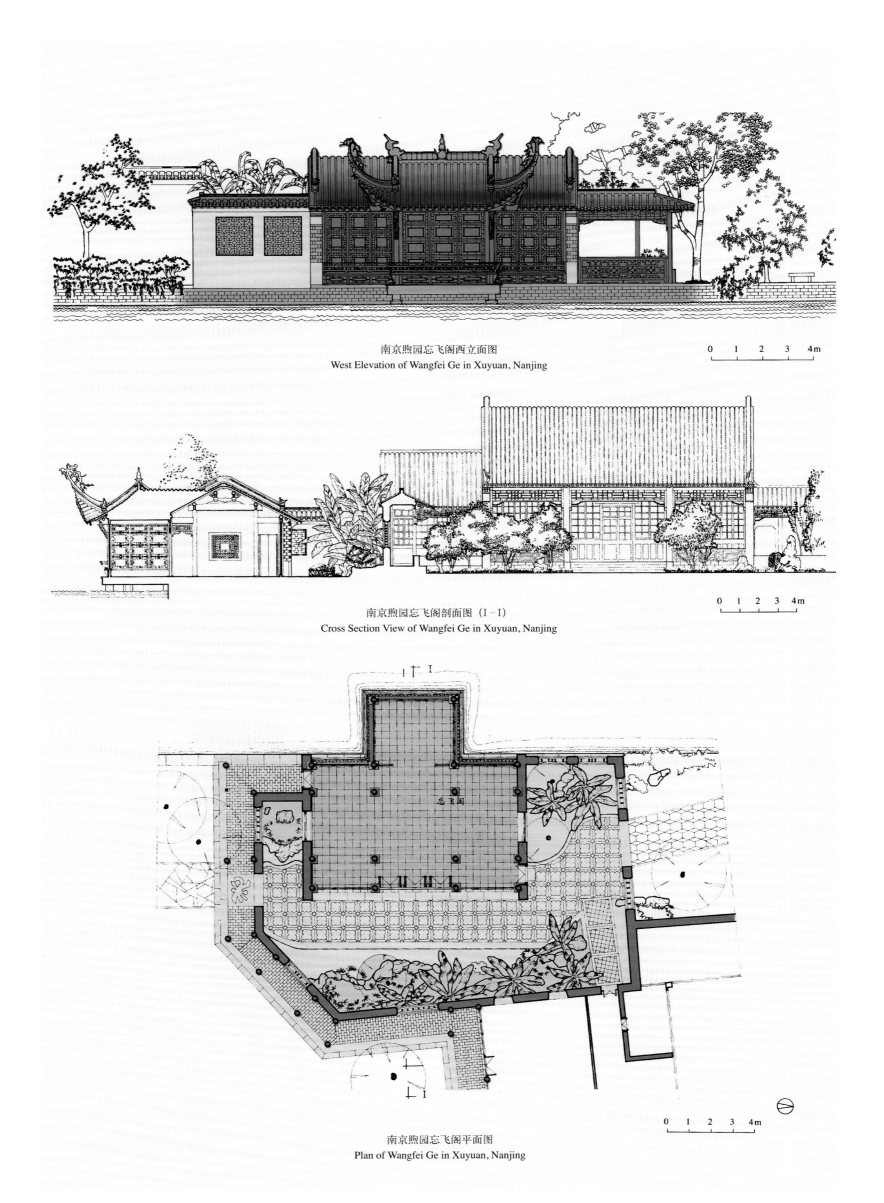

南京煦园忘飞阁西立面图
West Elevation of Wangfei Ge in Xuyuan, Nanjing

南京煦园忘飞阁剖面图（I-I）
Cross Section View of Wangfei Ge in Xuyuan, Nanjing

南京煦园忘飞阁平面图
Plan of Wangfei Ge in Xuyuan, Nanjing

（十）亭

《释名》云："亭者，停也。人所停集也。"司空图有休休亭，本此义。造式无定，自三角、四角、五角、梅花、六角、横圭、八角至十字，随意合宜则制，惟地图可略式也。

(10) Ting

According to *Shi Ming*, a Ting means a stop, that is, a building that is used for visitors to have a temporary rest. Sikong Tu of the Tang dynasty, based on this meaning, built a "Xiuxiu" Ting in his estate. There is no fixed format or architectural style to follow. A triangular, four cornered, pentagon shaped, plum flower shaped, horizontal shaped, octagonal shaped, or cross shaped one, as long as it is appropriate to the context, can be built. If we have a plan map, we could easily illustrate its architectural form.

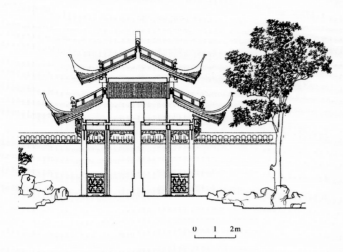

杭州文澜阁碑亭剖面图
Section View of Wenlan Ge Stele Pavilion, Hangzhou

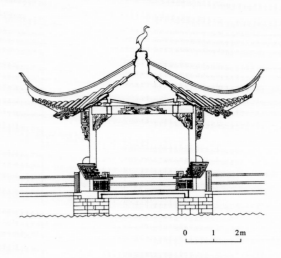

杭州小瀛洲三角亭剖面图
Section View of Xiao Yingzhou Triangle Pavilion, Hangzhou

杭州文澜阁碑亭立面图
Elevation of Wenlan Ge Stele Pavilion, Hangzhou

杭州小瀛洲三角亭立面图
Elevation of Xiao Yingzhou Triangle Pavilion, Hangzhou

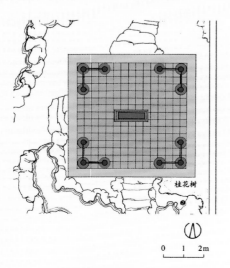

杭州文澜阁碑亭平面图
Plan of Wenlan Ge Stele Pavilion, Hangzhou

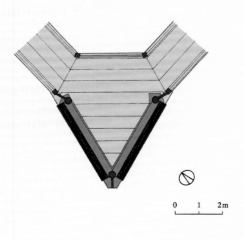

杭州小瀛洲三角亭平面图
Plan of Xiao Yingzhou Triangle Pavilion, Hangzhou

BUILDINGS 177

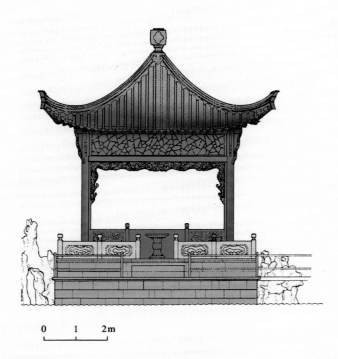

扬州何园水心亭立面图
Elevation of Shuixin (Note: Water Center) Pavilion in Heyuan, Yangzhou

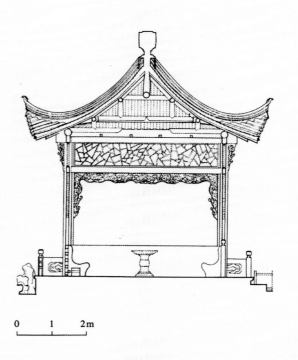

扬州何园水心亭剖面图
Section View of Shuixin Pavilion in Heyuan, Yangzhou

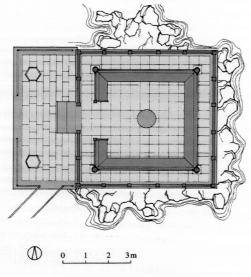

扬州何园水心亭平面图
Plan of Shuixin Pavilion in Heyuan, Yangzhou

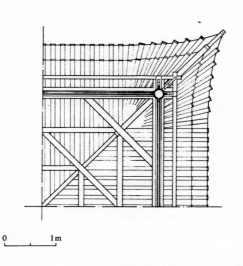

扬州何园水心亭仰视图
Ceiling View of Shuixin Pavilion in Heyuan, Yangzhou

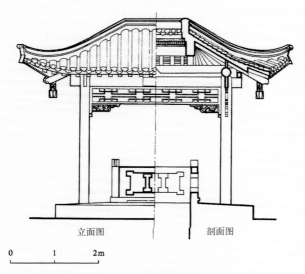

扬州平山堂西园池中第五泉井亭立、剖面图
Elevation/Section Views of 5th Spring Well Pavilion in West Pond in Pingshan Tang, Yangzhou

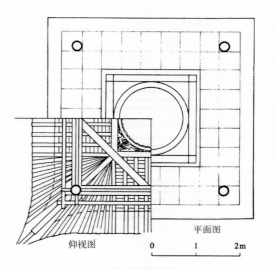

扬州平山堂西园池中第五泉井亭平面图
Ceiling/Plan View of 5th Spring Well Pavilion in West Pond in Pingshan Tang, Yangzhou

178 园冶图释
YUAN YE ILLUSTRATED—CLASSICAL CHINESE GARDENS EXPLAINED

真趣亭正立面图
Front Elevation of Zhenqu (Note: True Interest) Pavilion

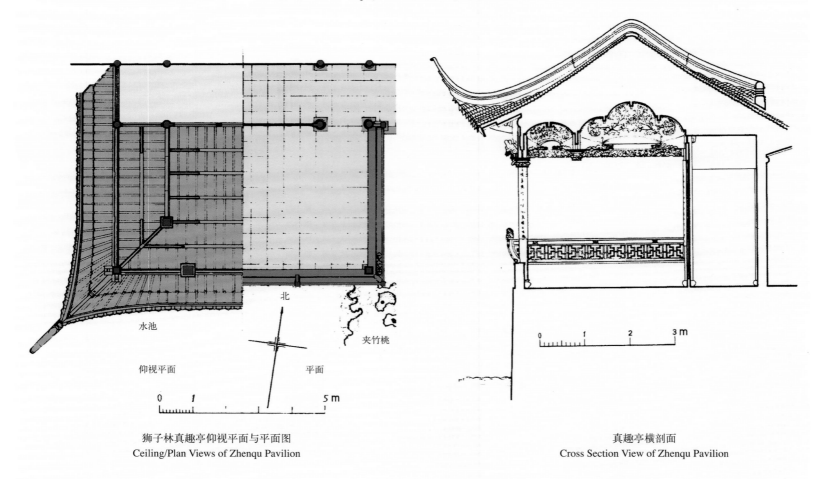

狮子林真趣亭仰视平面与平面图
Ceiling/Plan Views of Zhenqu Pavilion

真趣亭横剖面
Cross Section View of Zhenqu Pavilion

苏州拙政园梧竹幽居亭西立面图
West Elevation of Wuzhu Youju Pavilion of Zhuozheng Yuan, Suzhou

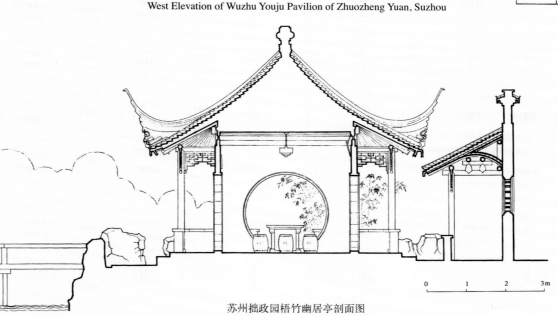

苏州拙政园梧竹幽居亭剖面图
Section View of Wuzhu Youju Pavilion of Zhuozheng Yuan, Suzhou

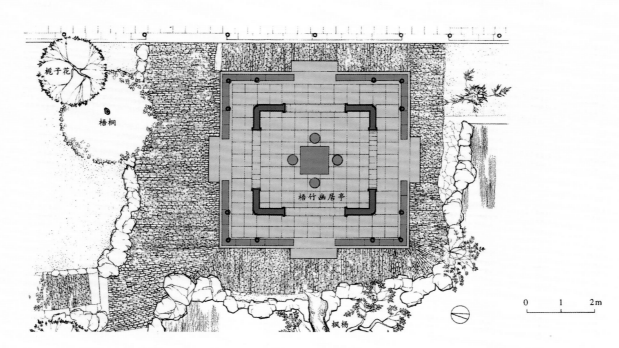

苏州拙政园梧竹幽居亭平面图
Plan of Wuzhu Youju Pavilion of Zhuozheng Yuan, Suzhou

沧浪亭正立面图
Front Elevation of Canglang Pavilion

怡园小沧浪亭正立面图
Front Elevation of Xiao Canglang Pavilion of Yiyuan

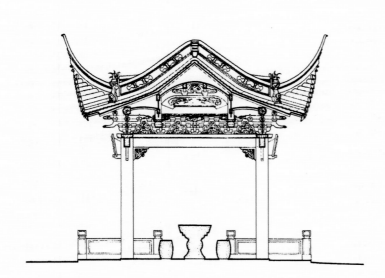

沧浪亭横剖面图
Section View of Canglang Pavilion

怡园小沧浪亭横剖面图
Section View of Xiao Canglang Pavilion of Yiyuan

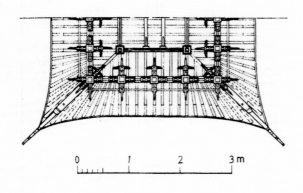

沧浪亭仰视图
Ceiling View of Canglang Pavilion

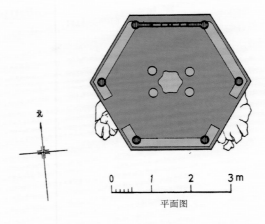

怡园小沧浪亭实测图
Plan of Xiao Canglang Pavilion of Yiyuan Actual Survey Drawing

绍兴兰亭小兰亭立面图
Elevation of Xiaolanting of Lanting, Shaoxing

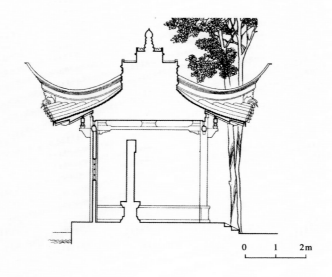

绍兴兰亭小兰亭剖面图
Section View of Xiaolanting of Lanting, Shaoxing

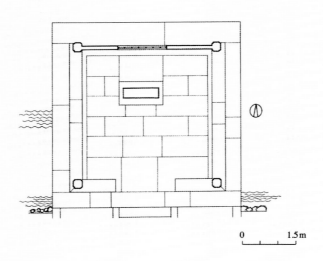

绍兴兰亭小兰亭平面图
Plan of Xiaolanting of Lanting, Shaoxing

苏州拙政园荷风四面亭立面图
Elevation of Hefeng Simian Pavilion in Zhuozheng Yuan, Suzhou

苏州拙政园荷风四面亭平面图
Plan of Hefeng Simian Pavilion in Zhuozheng Yuan, Suzhou

拙政园塔影亭正立面图
Elevation of Taying (Note: Pagoda Reflection) Pavilion

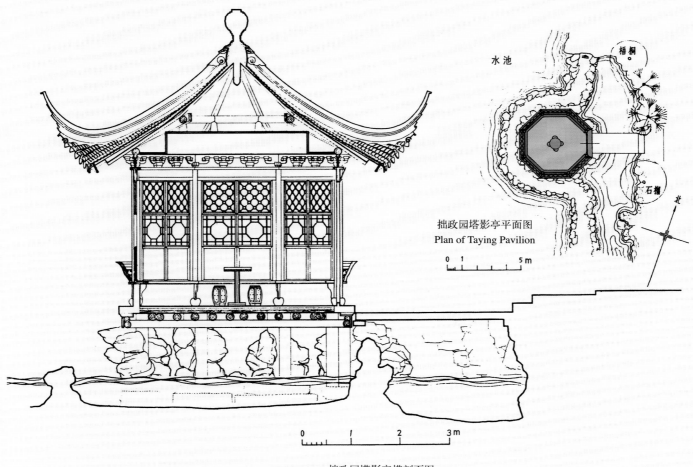

拙政园塔影亭横剖面图
Cross Section View of Taying Pavilion

拙政园塔影亭平面图
Plan of Taying Pavilion

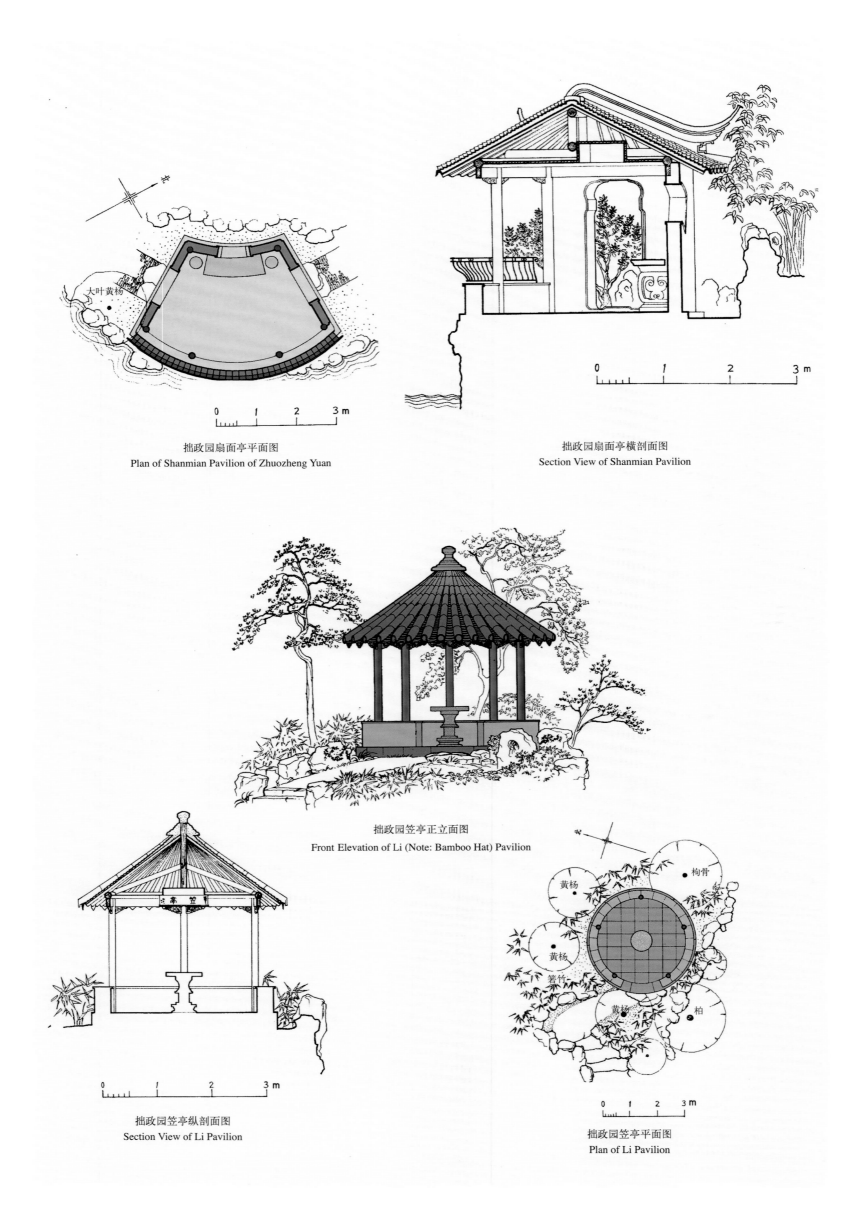

拙政园扇面亭平面图
Plan of Shanmian Pavilion of Zhuozheng Yuan

拙政园扇面亭横剖面图
Section View of Shanmian Pavilion

拙政园笠亭正立面图
Front Elevation of Li (Note: Bamboo Hat) Pavilion

拙政园笠亭纵剖面图
Section View of Li Pavilion

拙政园笠亭平面图
Plan of Li Pavilion

屋宇 BUILDINGS 185

苏州天平山御碑亭立面图
Elevation of Royal Stele Pavilion in Mount Tianping, Suzhou

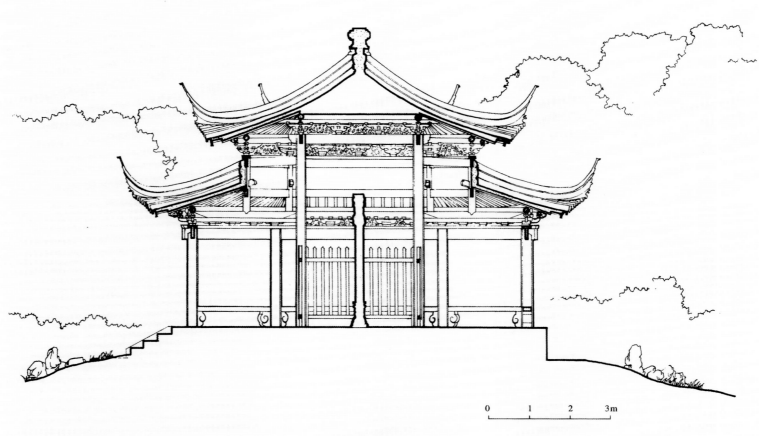

苏州天平山御碑亭剖面图
Section View of Royal Stele Pavilion in Mount Tianping, Suzhou

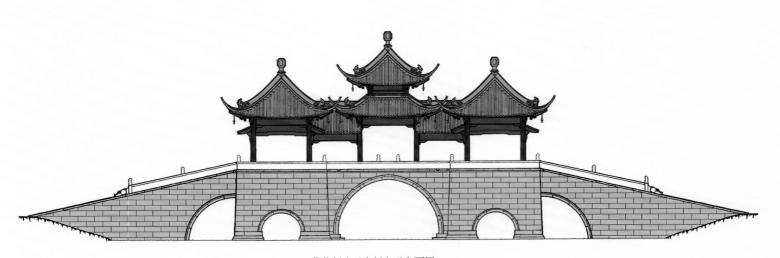

莲花桥（五亭桥）正立面图
Front Elevation of Wuting Bridge (Five Pavilions Bridge) in Yangzhou

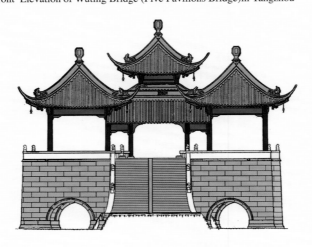

莲花桥（五亭桥）侧立面图
Side Elevation

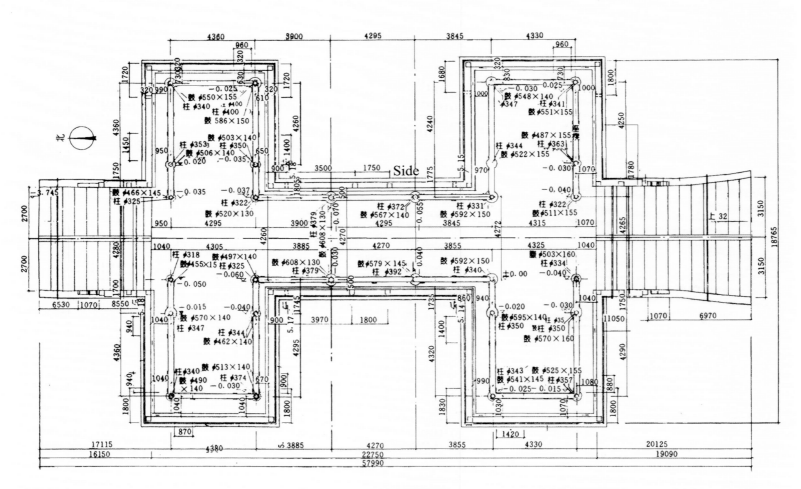

实测平面图
Actual Survey Plan

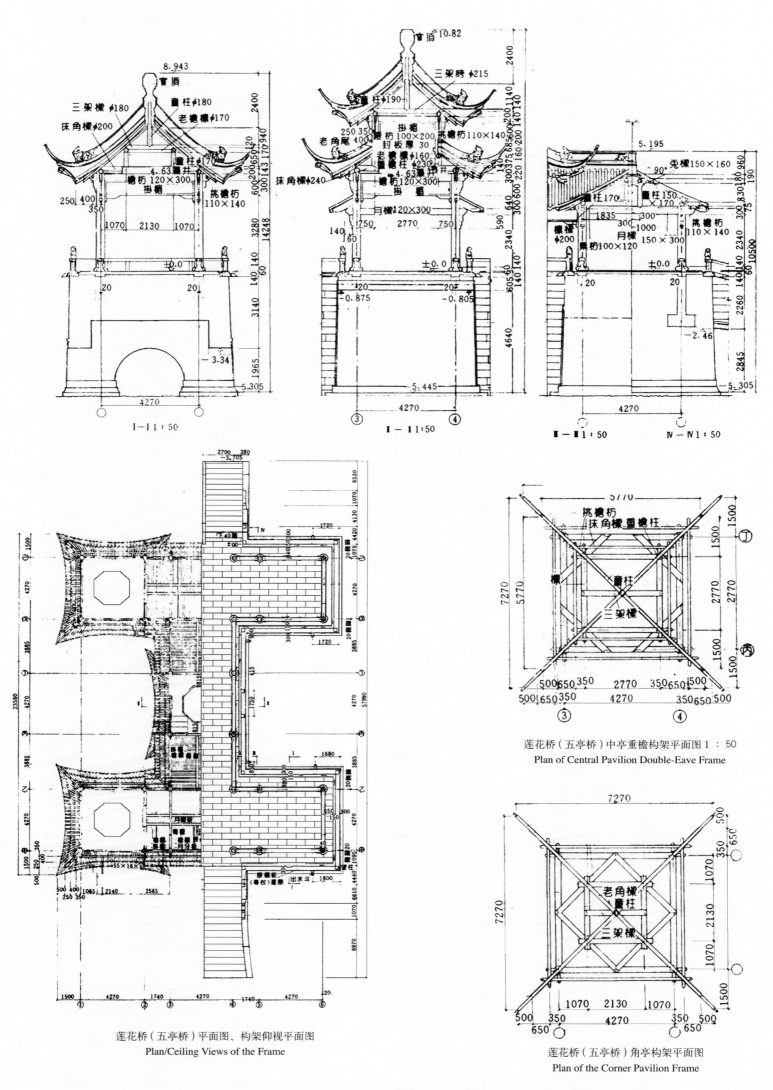

莲花桥（五亭桥）平面图、构架仰视平面图
Plan/Ceiling Views of the Frame

莲花桥（五亭桥）中亭重檐构架平面图 1：50
Plan of Central Pavilion Double-Eave Frame

莲花桥（五亭桥）角亭构架平面图
Plan of the Corner Pavilion Frame

Ⅰ-Ⅰ、Ⅱ-Ⅱ、Ⅲ-Ⅲ、Ⅳ-Ⅳ剖面，角亭梁架平面图，中亭梁架平面图

绿漪亭侧立面实测图
Actual Survey Drawing of Side Elevation of Lüyi Pavilion

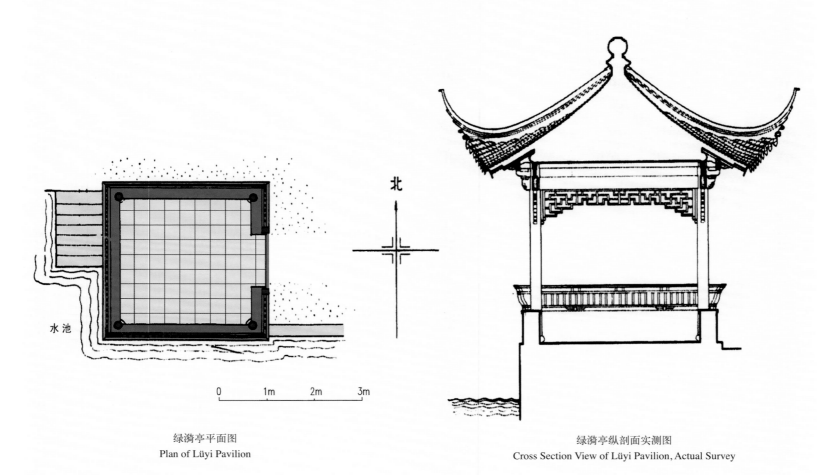

绿漪亭平面图
Plan of Lüyi Pavilion

绿漪亭纵剖面实测图
Cross Section View of Lüyi Pavilion, Actual Survey

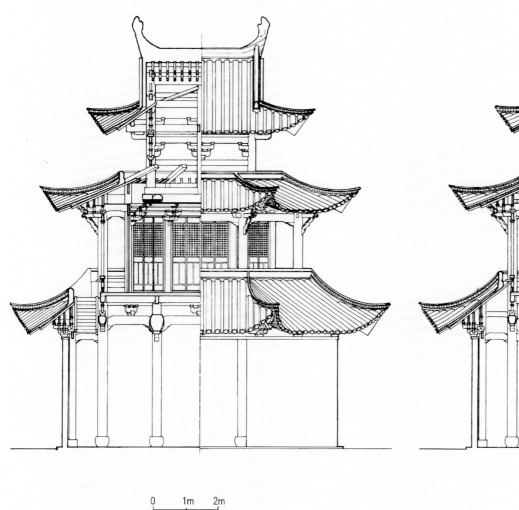

歙县许村大观亭侧立面及横剖面图
Side Elevation and Section View of Daguan Pavilion of Xucun, She County

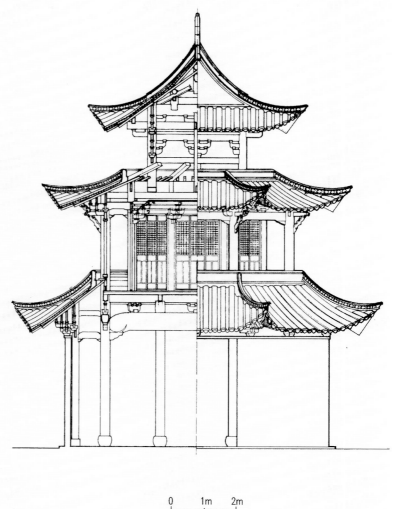

歙县许村大观亭立面及纵剖面图
Front Elevation and Section View of Daguan Pavilion of Xucun, She County

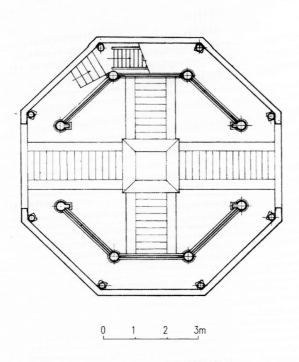

歙县许村大观亭一层平面图
1st Floor Plan of Daguan Pavilion of Xucun, She County

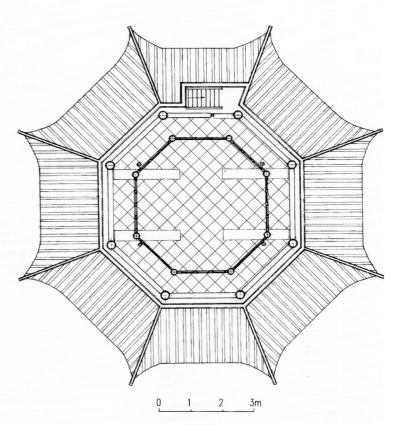

歙县许村大观亭二层平面图
2nd Floor Plan of Daguan Pavilion of Xucun, She County

（十一）榭

《释名》云：榭者，藉也。藉景而成者也。或水边，或花畔，制亦随态。

(11) Xie

According to *Shi Ming*, Xie means leveraging, meaning it is built by leveraging a landscape artistic concept. It can be built by water, or hidden among flowers, whose form and style are all dependent upon context and artistic conception.

网师园濯缨水阁正立面图
Front Elevation of Zhuoying Shuige in Wangshi Yuan

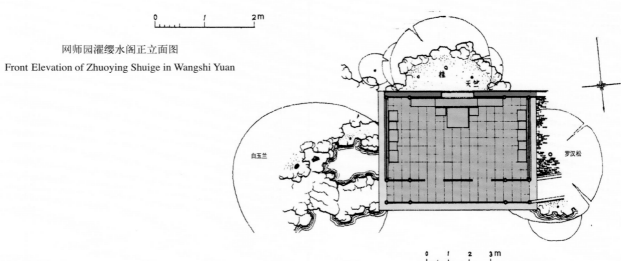

网师园濯缨水阁平面图
Plan of Zhuoying Shuige in Wangshi Yuan

网师园濯缨水阁横剖面图
Cross Section View of Zhuoying Shuige in Wangshi Yuan

（十二）轩

轩式类车，取轩轩欲举之意，宜置高敞，以助胜则称。

(12) Xuan

The form of a Xuan is similar to a carriage in ancient times. In architecture, we take its open and tall properties. It should be built on a higher place, and as long as it can enhance an architectural space's prominence, it would be considered appropriate.

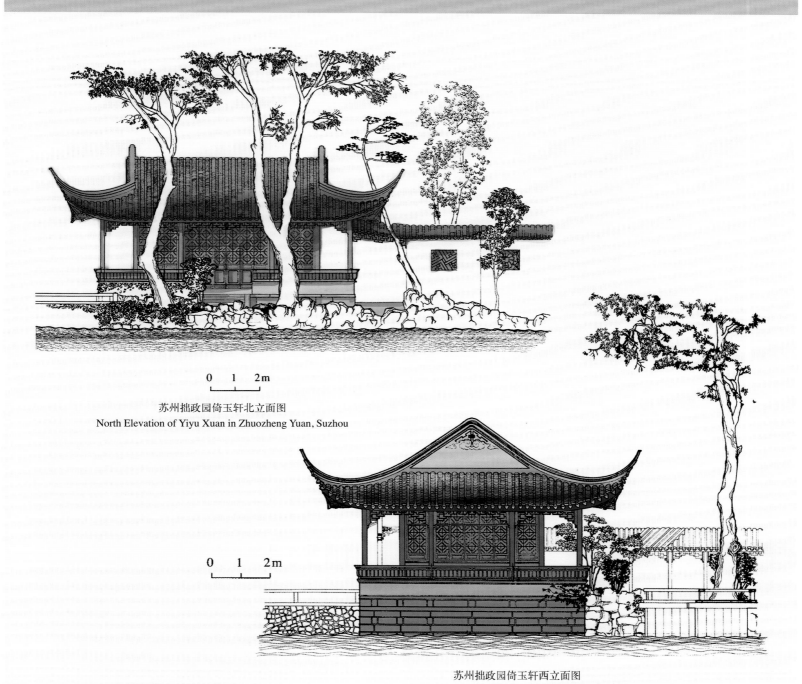

苏州拙政园倚玉轩北立面图
North Elevation of Yiyu Xuan in Zhuozheng Yuan, Suzhou

苏州拙政园倚玉轩西立面图
West Elevation of Yiyu Xuan in Zhuozheng Yuan, Suzhou

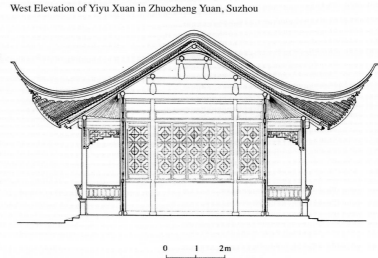

苏州拙政园倚玉轩平面图
Plan of Yiyu Xuan in Zhuozheng Yuan, Suzhou

苏州拙政园倚玉轩剖面图
Section View of Yiyu Xuan in Zhuozheng Yuan, Suzhou

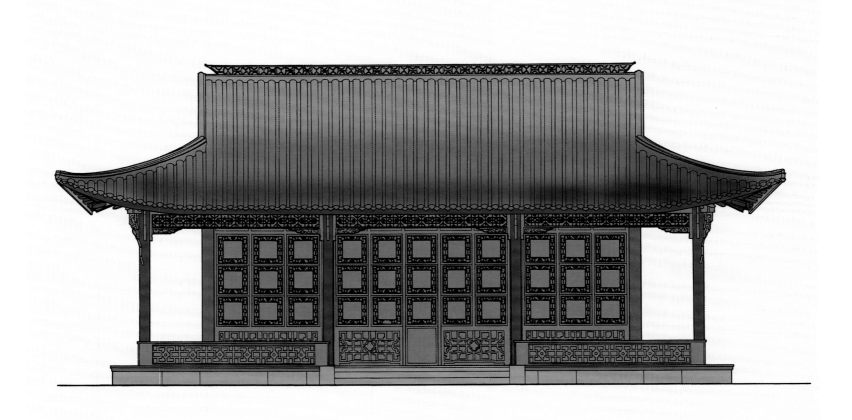

扬州何园静香轩南立面图
South Elevation of Jingxiang Xuan in Heyuan, Yangzhou

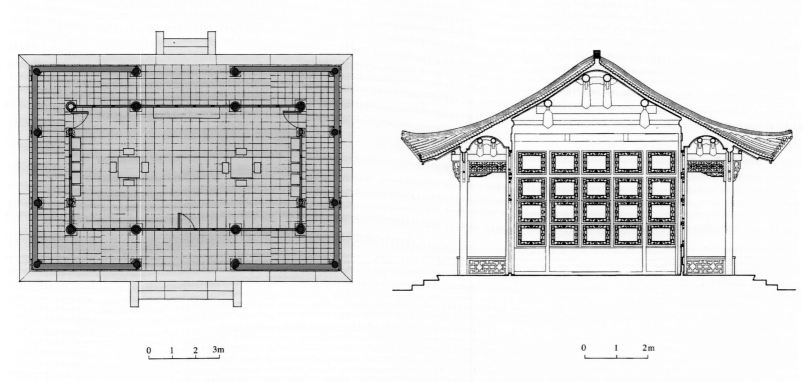

扬州何园静香轩平面图
Plan of Jingxiang Xuan in Heyuan, Yangzhou

扬州何园静香轩剖面图
Cross Section View of Jingxiang Xuan in Heyuan, Yangzhou

苏州虎丘悟石轩西立面图
West Elevation of Wushi Xuan in Huqiu, Suzhou

苏州虎丘悟石轩南立面图
South Elevation of Wushi Xuan in Huqiu, Suzhou

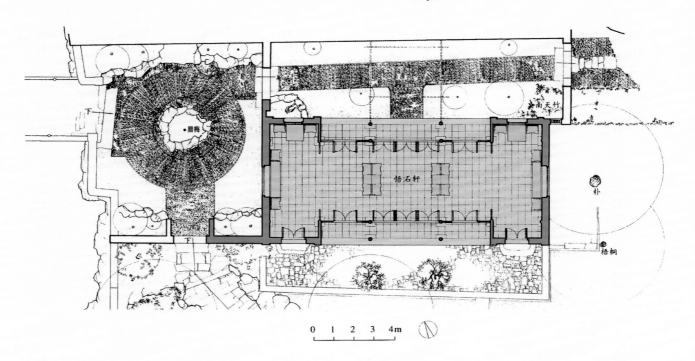

苏州虎丘悟石轩平面图
Plan of Wushi Xuan in Huqiu, Suzhou

（十三）卷

卷者，厅堂前欲宽展，所以添设也。或小室欲异人字，亦为斯式。惟四角亭及轩可并之。

(13) Juan

A Juan is an extension added to the front of a hall-like building and its purpose is to make the front space a little wider. Another reason is to reduce the perception and impact of using an A-shaped roof structure. There are only two forms of Juan, a four cornered pavilion and Xuan, and they can be used together.

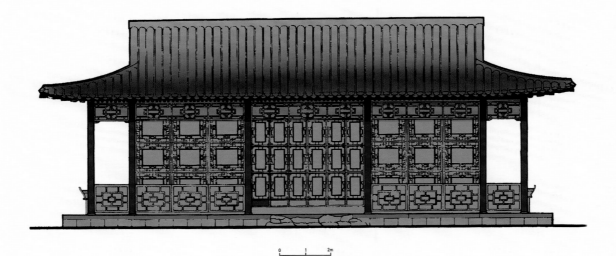

扬州个园宜雨轩南立面图
South Elevation of Yiyu Xuan in Geyuan, Yangzhou

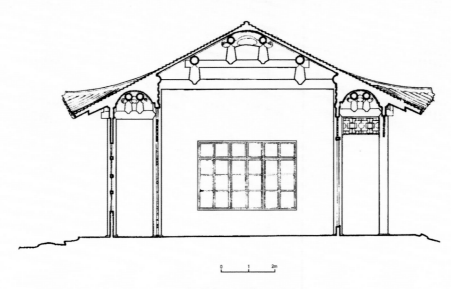

扬州个园宜雨轩剖面图
Cross Section View of Yiyu Xuan in Geyuan, Yangzhou

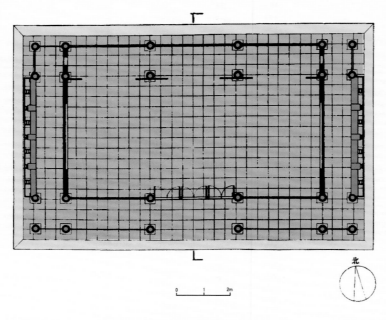

扬州个园宜雨轩平面图
Plan of Yiyu Xuan in Geyuan, Yangzhou

（十四）广

古云：因岩为屋曰"广"，盖借岩成势，不成完屋者为"广"。

(14) Guang

The ancients told us that a building with one side leaning against a mountain or palisade is called "Guang". Because it leans against a mountain with half pitched roof, it is referred to as Guang.

拙政园"别有洞天"正立面图
Front Elevation of Bieyou Dongtian in Zhuozheng Yuan

拙政园"别有洞天"横剖面图
Cross Section of Bieyou Dongtian in Zhuozheng Yuan

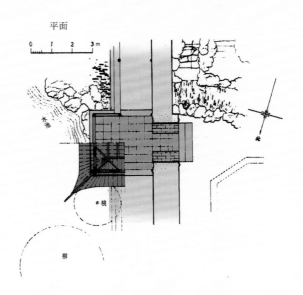

拙政园"别有洞天"仰视平面图
Plan/Ceiling View of Bieyou Dongtian in Zhuozheng Yuan

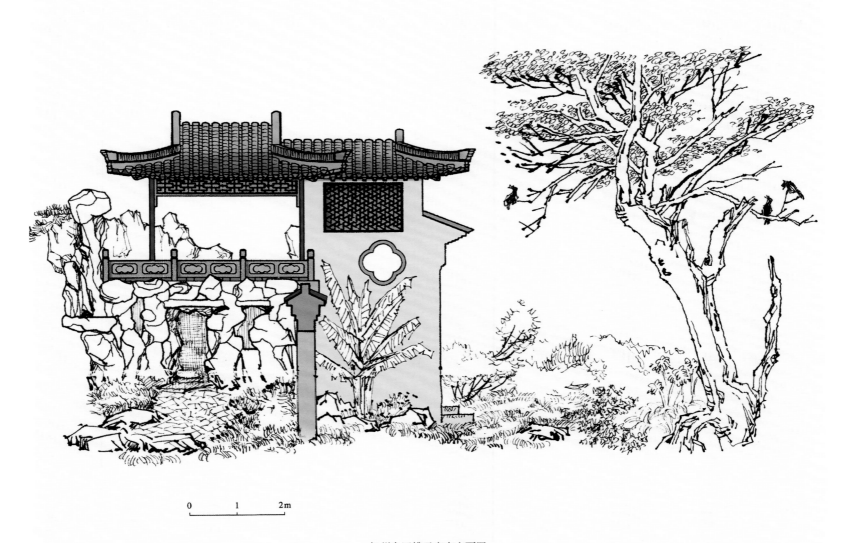

扬州个园拂云亭东立面图
East Elevation of Fuyun Pavilion in Geyuan, Yangzhou

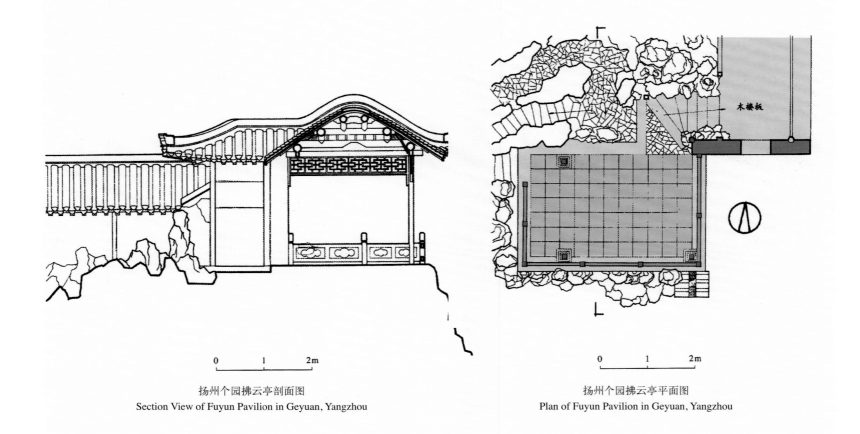

扬州个园拂云亭剖面图
Section View of Fuyun Pavilion in Geyuan, Yangzhou

扬州个园拂云亭平面图
Plan of Fuyun Pavilion in Geyuan, Yangzhou

苏州网师园竹外一枝轩西立面图
West Elevation of Zhuwai Yizhixuan in Wangshi Yuan, Suzhou

苏州网师园竹外一枝轩前亭子剖面图
Section View of Zhuwai Yizhixuan in Wangshi Yuan, Suzhou

苏州网师园竹外一枝轩南立面图
South Elevation of Zhuwai Yizhixuan in Wangshi Yuan, Suzhou

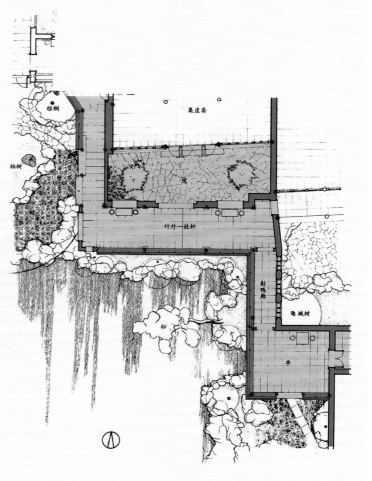

苏州网师园竹外一枝轩平面图
Plan of Zhuwai Yizhixuan in Wangshi Yuan, Suzhou

拙政园倚虹亭正立面图
Front Elevation of Yihong Pavilion in Zhuozheng Yuan, Actual Survey

拙政园倚虹亭外观图
Appearance of Yihong Pavilion in Zhuozheng Yuan

拙政园倚虹亭横剖面图
Section View of Yihong Pavilion in Zhuozheng Yuan

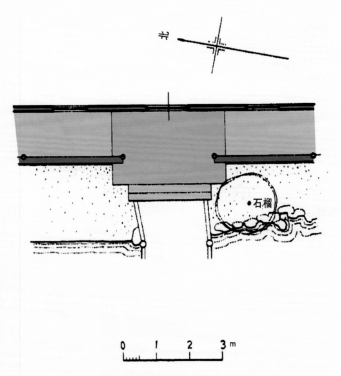

拙政园倚虹亭平面图
Plan of Yihong Pavilion in Zhuozheng Yuan

（十五）廊

廊者，庑出一步也，宜曲宜长则胜。古之曲廊，俱曲尺曲。今予所构曲廊，之字曲者，随形而弯，依势而曲。或蟠山腰，或穷水际，通花渡壑，蜿蜒无尽，斯寤园之"篆云"也。予见润之甘露寺数间高下廊，传说鲁班所造。

(15) Lang

A Lang is a building that extends Wu a little further and it should be curved and long. The ancient corridors are all like carpenter's square. The corridors that I built now were zigzag-shaped and they turn and curve according to the terrain. They have unlimited possibilities to change directions. They would either go around a hill side, meander along a water bank, go through flowers, or cross a brook. In the past, Zhuanyun Lang of Wu Yuan was like this. Also, I saw a few units of Gaoxia Lang at Ganlu Temple in Zhenjiang, which are said to be built by great carpenter master Lu Ban.

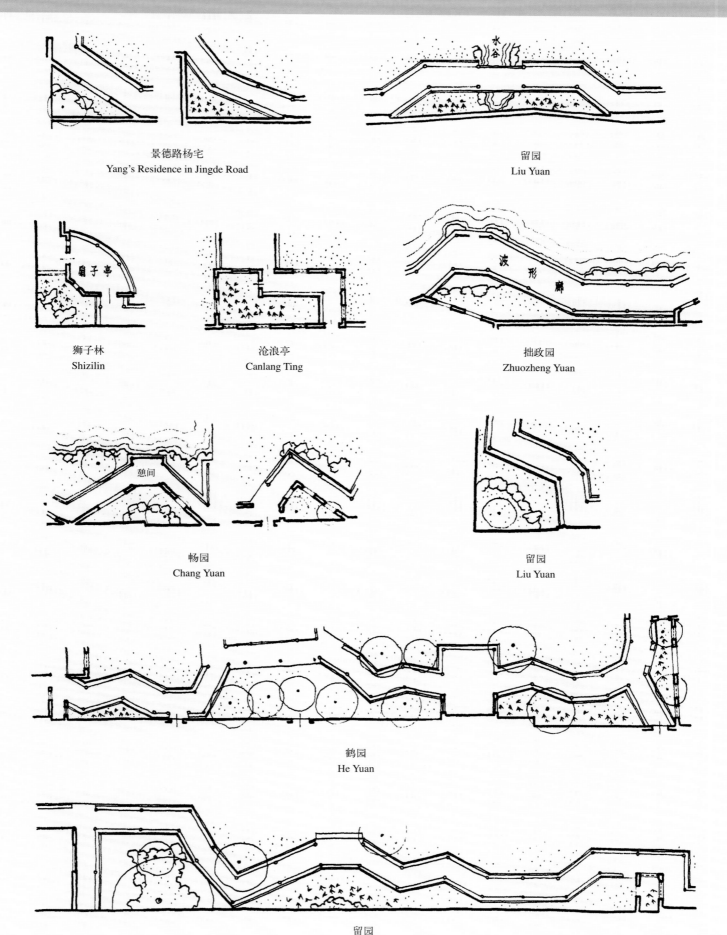

景德路杨宅 Yang's Residence in Jingde Road　　留园 Liu Yuan

狮子林 Shizilin　　沧浪亭 Canlang Ting　　拙政园 Zhuozheng Yuan

畅园 Chang Yuan　　留园 Liu Yuan

鹤园 He Yuan

留园 Liu Yuan

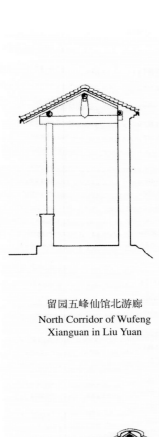
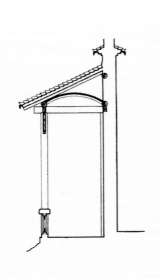
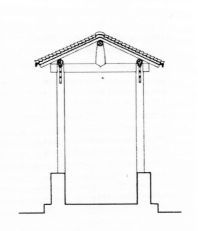

留园五峰仙馆北游廊
North Corridor of Wufeng Xianguan in Liu Yuan

网师园月到风来亭游廊
Corrifor of Yuedao Fenglai Pavilion in Wangshi Yuan

怡园书舫斋南游廊
South Corridor of Shufang Zhai in Yi Yuan

留园远翠阁西空廊
Xikong Lang of Yuancui Ge in Liu Yuan

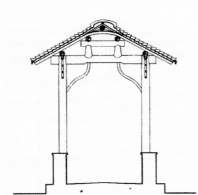
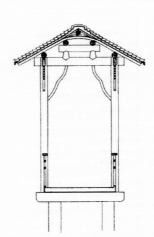
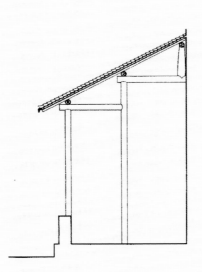

拙政园柳阴路曲空廊
Kong Lang of Liuyin Luqu in Zhuozheng Yuan

拙政园小飞虹水廊
Shui Lang of Xiao Feihong in Zhuozheng Yuan

怡园书舫斋南游廊
South Corridor of Shufang Zhao in Yi Yuan

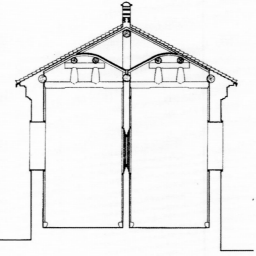

怡园拜石轩西复廊
West Double Lang of Baishi Xuan in Yi Yuan

狮子林立雪堂前复廊
Double Lang in Front of Lixue Tang in Shizilin

狮子林小方厅前复廊
Double Lang in Front of Xiaofang Ting in Shizilin

廊的剖面实测图
Section Views, Actual Survey

0　　1　　2　　3m

（十六）五架梁

五架梁，乃厅堂中过梁也。如前后各添一架，合七架梁列架式。如前添卷，必须草架而轩敞。不然前檐深下，内黑暗，斯故也。如欲宽展，前再添一廊。又小五架梁，亭、榭、书房可构。将后童柱换长柱，可装屏门，有别前后，或添廊亦可。

(16) Five-Frame Girder

Five-frame girder, is the main beam in a hall-like building. If one extra frame were added to its front and back, it would becoma a seven-frame girder structure. If a Cao Juan needs to be added in front of the hall, then only using Cao Jia format can make it open and bright enough; otherwise, the eave in front would be too low and it would be too dim indoors. If you want to make indoor more spacious, then there should add a Lang Xuan in front. Also, there is also a minor five-frame girder format that can be used to build a pavilion, Xie, and study. If the back Tong column is replaced by a regular column that rises from the ground, then a screen door can be installed to separate the front room from the back room. If no screen door is used, then there could also add a corridor.

屋宇图式

五架过梁式（图一－一）

前或添卷，后添架，合成七架列。

Five Frame Girder (Figure 1-1)

Extra unit may be added to the front and back, making it a seven frame girder structure.

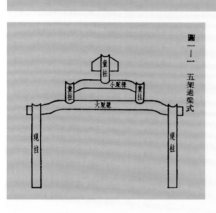

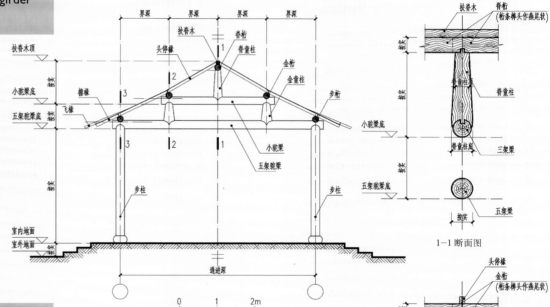

五架过梁式立面图
Section View of a Five Frame Structure

小五架梁式（图一－八）

凡造书房、小斋或亭，此式可分前后。

Small Five Frame Girder (Figure 1-8)

This can be applied to study building, small Zhai or pavilion, and it can have a distinction between front and back.

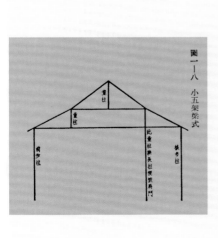

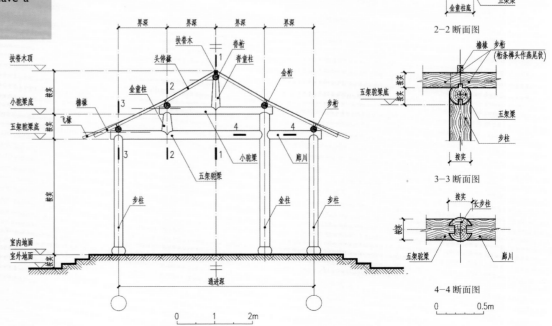

小五架梁式立面图
Section View of a Small Five Frame

（十七）七架梁

七架梁，凡屋之列架也，如厅堂列添卷，亦用草架。前后再添一架，斯九架列之活法。如造楼阁，先算上下檐数，然后取柱料长，许中加替木。

(17) Seven-Frame Girder

A Seven-frame girder is the one used in common buildings. If a Juan is to be added in front of a hall, then a Cao Jia is also needed. If one more frame is added in its front and back, then this becomes a flexible extension of a nine-frame girder. If a Ge needs to be built, one needs to first calculate the heights of the upper and lower rafters, and then estimate the length of the columns. A substitute wood can be added in the middle.

七架列式（图一－三）

凡屋以七架为率。

Seven Frame Illustrations (Figure 1-3)

Any building should follow a seven frame structure.

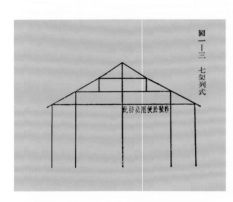

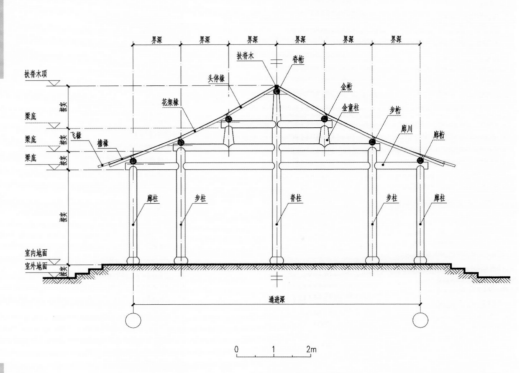

七架列式立面图
Section View of a Seven Frame Structure

七架酱架式（图一－四）

不用脊柱，便于挂画，或朝南北，屋傍可朝东西之法。

Seven Frame Jiangjia Pattern (Figure 1-4)

No middle column is used. This is for the convenience of picture/painting hanging. It may face south or north, with its side facing east or west.

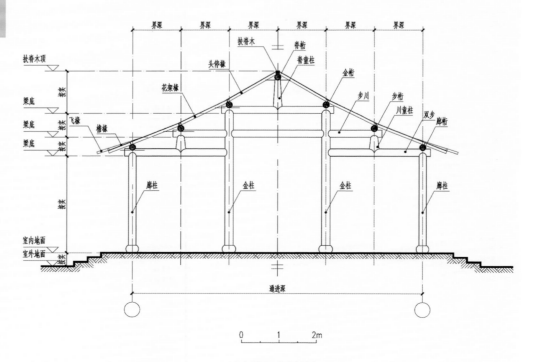

七架酱架式立面图
Section View of a Seven Frame Jiangjia Structure

（十八）九架梁

九架梁屋，巧于装折，连四、五、六间，可以东、西、南、北。或隔三间、二间、一间、半间，前后分为。须用复水重椽，观之不知其所。或嵌楼于上，斯巧妙处不能尽式，只可相机而用，非拘一者。

(18) Nine-Frame Girder

For buildings with nine-frame girders, its fitments must be cleverly designed. For its depth, it can run four, five, or six units. It can also extend to east, west, south, and north all four directions. For separations, it can be divided into three rooms, two rooms, one room, half a room, and the front and back can be divided into relatively independent spaces. When it is built, Fushui double eave structure should be used to hide the structure. Above it, another story could be added. These ingenuities cannot be fully visualized by drawings and the designer should be adaptable and not confined to one fixed form.

九架梁式（图一－五）、（图一－六）、（图一－七）

此屋宜多间，随便隔间，复水或向东、西、南、北之活法。

Nine Frame Girder Illustrations (Figures 1-5, 1-6, 1-7)

It is more appropriate to have multiple rooms with this structure, and the separations can be flexible. Fushui can face east, west, south, or north any direction.

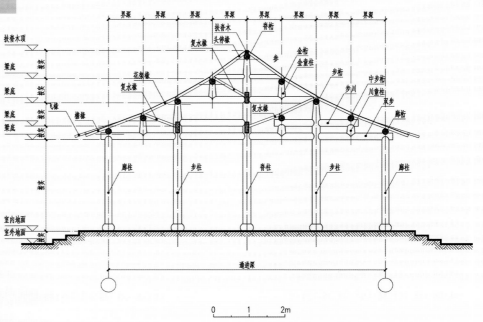

九架梁五柱式立面图
Section View of Nine Frame Girder with Five Columns

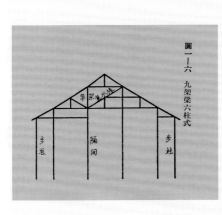

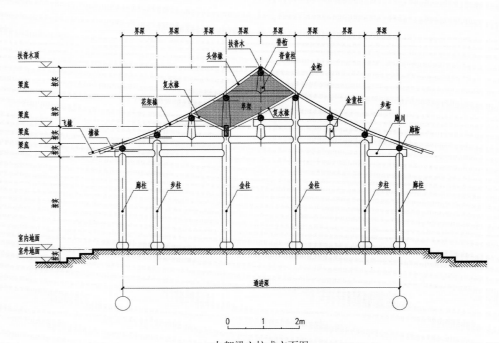

九架梁六柱式立面图
Section View of Nine Frame Girder with Six Columns

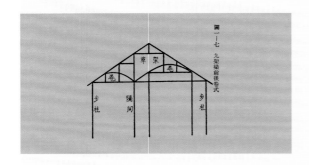

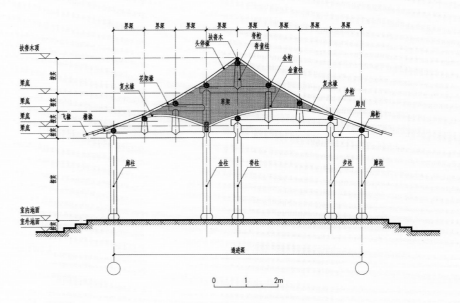

九架梁前后卷式立面图
Section View of Nine Frame Girder Front and Back Juanshi

（十九）草架

草架，乃厅堂之必用者。凡屋添卷，用天沟，且费事不耐久，故以草架表里整齐。向前为厅，向后为楼，斯草架之妙用也，不可不知。

(19) Cao Jia

Cao Jia is a structure that a Ting or Tang must use. If a Juan is added in front of a building, a gutter needs to be used. This is time-consuming and prone to damage. So an alternative is to use Cao Jia to have a consistent decent look outside and inside the building. If the Juan is added in front, it can become a Ting; if it is added at the back, it can become a Ge. This is the beauty of Cao Jia and one cannot afford not knowing about it.

草架式（图一-二）

惟厅堂前添卷，须用草架，前再加之步廊，可以磨角。

Cao Jia Illustrations (Figure 1-2)

If an extra unit is added in front of a Ting or Tang, a Cao Jia is needed. A walking corridor can be added and a Mo Jiao can be applied.

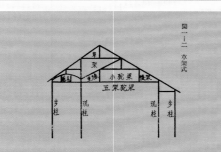

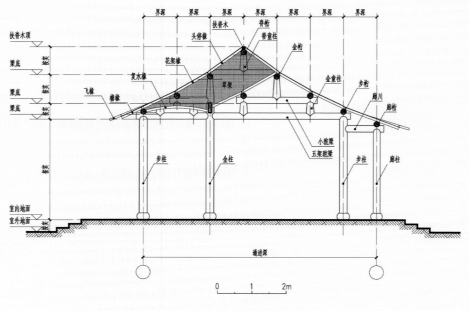

草架式立面图
Section View of a Cao Jia

（二十）重椽

重椽，草架上椽也，乃屋中假屋也。凡屋隔分不仰顶，用重椽复水可观。惟廊构连屋，构倚墙一披而下，断不可少斯。

(20) Chong Chuan (Double Rafter)

Chong Chuan is a rafter used on a Cao Jia. It is actually a fake roof under the real roof. To make a separated room, one does not have to use a ceiling; a fake ceiling with Chong Chuan and Fushui would make the inner space complete and look good. Especially at the connection point between a corridor and a building, or at a wall leaning single pitched building, it is definitely necessary to have this structure.

（二十一）磨角

磨角，如殿阁蹴角也。阁四敞及诸亭决用。如亭之三角至八角，各有磨法，尽不能式，是自得一番机构。如厅堂前添廊，亦可磨角，当量宜。

(21) Mo Jiao

A Mo Jiao is similar to a building's corner, and it is a mandatory building method on an open Ge and pavilion. The format and style of Mo Jiao for different kind of pavilions, from triangular shaped to octagonal shaped, all depend and as such we cannot elaborate on them all. Designers and builders should have some careful design ideas before construction. If there is an addition of a corridor in front of a Ting or Tang, a Mo Jiao can be placed at the corner, but it must be appropriate.

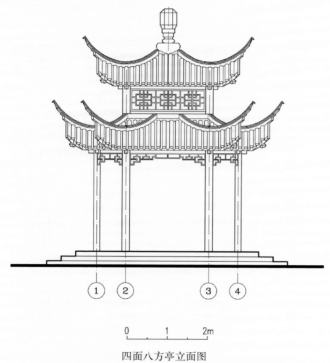

四面八方亭立面图
Elevation of Simian Bafang Pavilion

水戗构造
Structure of a Shui Qiang

1-1 断面图
Section View

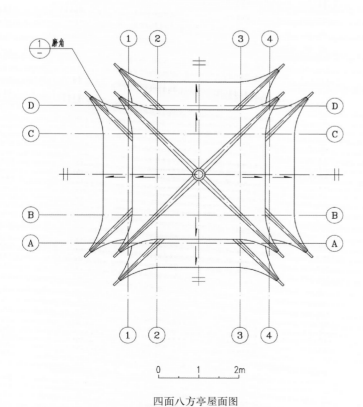

四面八方亭屋面图
Roof View of Simian Bafang Pavilion

磨角仰视图
Mo Jiao Ceiling View

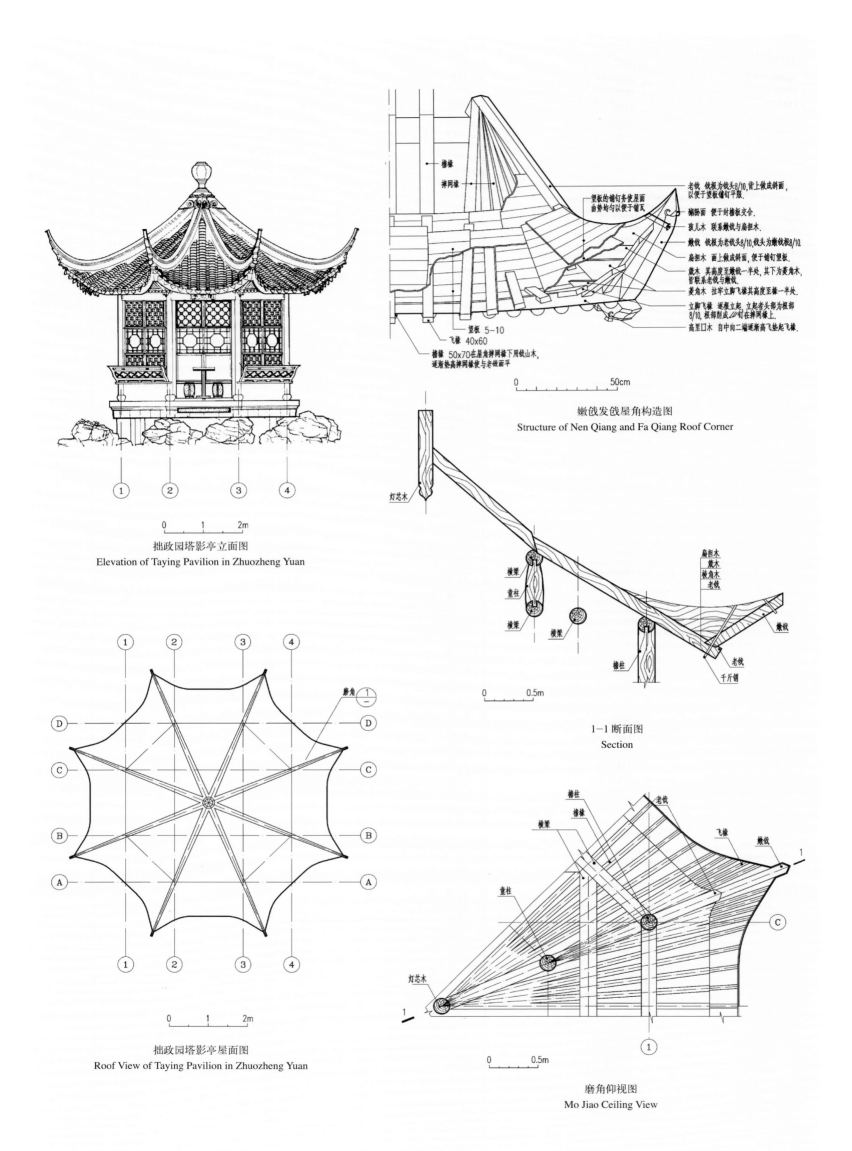

屋宇
BUILDINGS

（二十二）地图

凡匠作，止能式屋列图，式地图者鲜矣。夫地图者，主匠之合见也。假如一宅基，欲造几进，先以地图式之。其进几间，用几柱着地，然后式之，列图如屋。欲造巧妙，先以斯法，以便为也。

(22) Floor Plan

In the old time, when craftsmen constructed a building, they usually only made a building's structure drawing and rarely drew a floor plan. However, a floor plan reflects a consensus between the designer and the craftsmen. Assuming there is a house site and we would like to build a house with a few units of depth, we should first draw it on a floor plan. Based on this we could determine how many units to be built and how many columns are needed. From here, we could further make the plan and structure drawings and display and label the house's units and columns locations and other structural elements. If one wants to build a house skillfully, he must adopt this method, because it facilitates the construction.

地图式（图一 — 九）

凡兴造，必先式斯。偷柱定磉，量基广狭，次式列图。凡厅堂中一间宜大，傍间宜小，不可匀造。

Floor Plan Patterns (Figure 1-9)

Normally when a building is built, a plan must be drawn. To decide how to erect columns, one must inspect the size of a site, and then make a drawing plan. When building a Ting or Tang, the middle one should be large and the side ones should be small; they cannot be made equally.

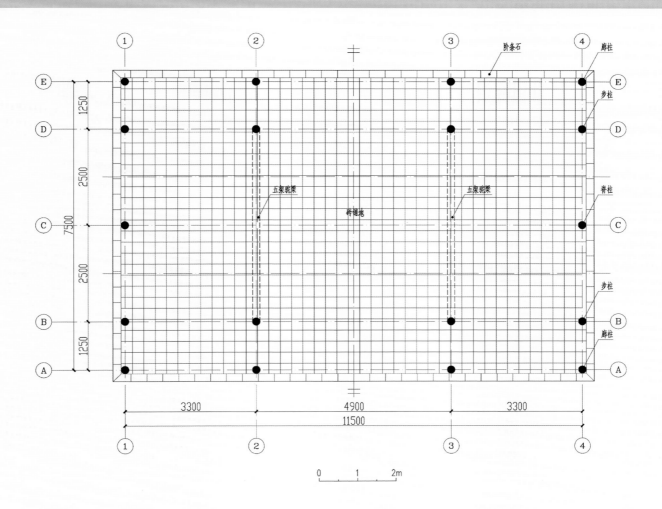

地图式
Floor Plan

梅花亭地图式（图一－十）
先以石砌成梅花基，立柱于瓣，结顶合檐，亦如梅花也。
Floor Plan of Plum Flower Pattern (Figure 1-10)

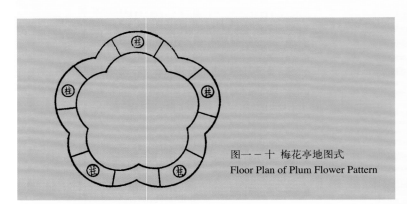

图一－十 梅花亭地图式
Floor Plan of Plum Flower Pattern

注：柱位置不对，疑印刷错误

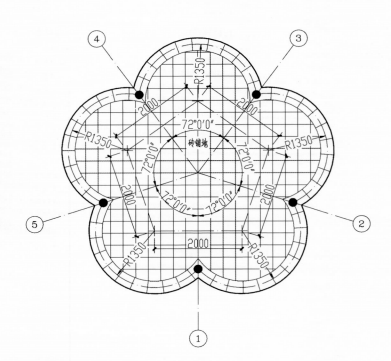

梅花亭地图式
Floor Plan of Plum Flower Pattern

十字亭地图式（图一－十一）
十二柱四分而立，顶结方尖，周檐亦成十字。
Cross Pattern Floor Plan (Figure 1-11)

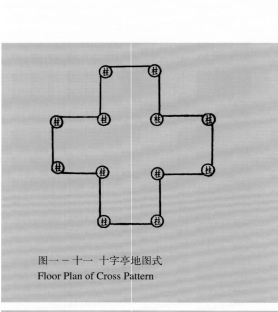

图一－十一 十字亭地图式
Floor Plan of Cross Pattern

诸亭不式，为梅花、十字，自古伪造者，故式之地图，聊识其意可也。斯二亭，只可盖草。

The remaining pavilion styles are not illustrated, except for plum and cross patterns, whose real ones were not built since the ancient times. Their plans are shown here for illustration purposes only, and if they're built, their roofs could only be thatched ones.

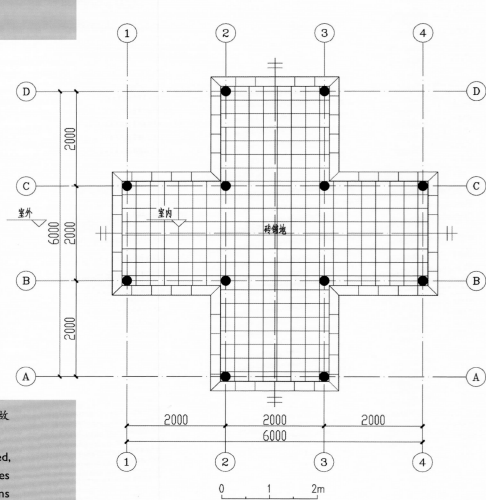

十字亭地图式
Floor Plan of Cross Pattern

园冶图释

Yuan Ye Illustrated
—Classical Chinese Garden Explained

四 装折

4 Decorations

凡造作难于装修，惟园屋异乎家宅，曲折有条，端方非额，如端方中须寻曲折，到曲折处还定端方，相间得宜，错综为妙。装壁应为排比，安门分出来由。假如全房数间，内中隔开可矣。定存后步一架，余外添设何哉？便径他居，复成别馆。砖墙留夹，可通不断之房廊；板壁常空，隐出别壶之天地。亭台影罅，楼阁虚邻。绝处犹开，低方忽上。楼梯仅乎室侧，台级借矣山阿。门扇岂异寻常，窗棂遵时各式。掩宜合线，嵌不窥丝。落步栏杆，长廊犹胜，半墙户槅，是室皆然。古以菱花为巧，今之柳叶生奇。加之明瓦斯坚，外护风窗觉密。半楼半屋，依替木不妨一色天花；藏房藏阁，靠虚檐无碍半弯月牖。借架高檐，须知下卷。出幕若分别院，连墙拟越深斋。构合时宜，式征清赏。

Generally decoration is a difficult part of building construction, especially for garden buildings, which is different than that for residential buildings. In garden buildings, there must be certain orders among various changes and neat and tidy is not an utmost concern. We seek changes in uniformity and once we introduce a variation, we still need to keep order. A division should be proper and scattered design should be appropriate. A placement of a bulkhead should keep balance with the rest and an installation of a door should follow after travel direction is made clear. If a building has several units deep, an indoor separation along the depth line would suffice. Why do we need to keep one frame of Xuan and others are not necessary? That is because from here we could reach other places through small paths or we could extend to build another one. An alley should be added between a gable and a fence wall and this promotes connections. There should open a few hollow windows on a wall, when it is appropriate, to reveal sceneries of other courtyards. A pavilion or terrace should have openings to allow light and shadows and a Ge Lou should lean against some void places. Sometimes it seems to be at a dead end but then again you see a whole new series of scenes. You may reach a low place before you find you would suddenly go up. A stair is suitable to be at a side of a room and outdoor steps can take advantage of a natural slope to be built up. Although there is no big difference in doors and windows used here, we should use newer and more fashionable styles for windows lattices. When the doors and windows are closed, they should be seamless and no apparent cracks should be visible. Railings should be installed by steps and it would be better for them to be connected with a corridor. Window panes should be installed on semi-walls in all rooms. For window pane decorations, in the old time a delicate pattern is that of a water chestnut flower, and nowadays people favor willow leaf patterns. It becomes sturdier if they are covered by tiles and they would be protected well by air windows. For half-single, half multi-stored buildings, a uniform colored ceiling would be preferred. For a deep, quiet and secluded dwelling, a semi-crescent shaped window could be installed by the proxy rafters. For high eave buildings, an open Juan should be added. If there needs a connection between an indoor space and another courtyard, a door opening is necessary in the wall separating the two, to provide a passage. Generally speaking, the decorations should be up to date and the styles should be elegant and tasteful.

（一）屏门

堂中如屏列而平者，古者可一面用，今遵为两面用，斯谓"鼓儿门"也。

(1) Screen Doors

A screen door is like a screen to be arranged neatly in an indoor space. In ancient times, a screen door can be only used at one side but nowadays we use it at both sides. Since it is hollow inside, it is called "drum door".

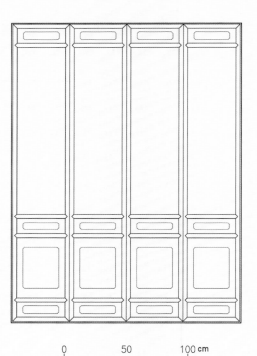

屏门
Screen Door

（二）仰尘

仰尘即古天花版也。多于棋盘方空画禽卉者类俗，一概平仰为佳，或画木纹，或锦，或糊纸，惟楼下不可少。

(2) Yang Chen

Yang Chen, is actually what meant ceilings in ancient times. Most of them used bird and flower patterns in the grid, which is almost philistine. The best way is to use flat surfaces painted with wood textures, or mounted with brocade or silk, or plain paper. It is indispensable for downstairs decorations.

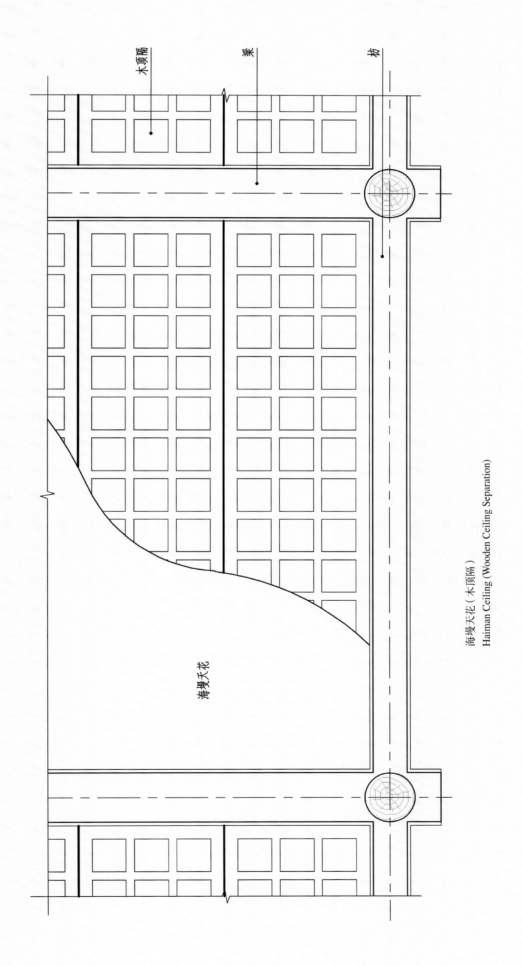

海墁天花（木顶隔）
Haiman Ceiling (Wooden Ceiling Separation)

（三）户槅

古之户槅，多于方眼而菱花者，后人减为柳条槅，俗呼"不了窗"也。兹式从雅，予将斯增减数式，内有花纹各异，亦遵雅致，故不脱柳条式。或有将栏杆竖为户槅，斯一不密，亦无可玩，如棂空仅阔寸许为佳，犹阔类栏杆风窗者去之，故式于后。

(3) Hu Ge

In ancient times, Hu Ge, or house separation panels, used to have square openings with water chestnut pattern decorations. Later generations generalized it into willow leaf patterned ones, commonly known as "Buliao Window" (Note: endless window). The purpose is to be more elegant. I made some modifications by adding and dropping, and derived several patterns. Following the same principle for elegance, the patterns in the grid still use the same basic willow form. Some try to use erected railings for separation panels; however, they are too scattered apart to have pasted paper over them and their look is not that appealing either. The distance between lattices is about one inch and if it is greater than that, it would become railings or protective windows and so I wouldn't use it. Now I illustrate various patterns below.

装折图式
长槅式（图一-十二）

古之户槅棂版．分位定于四、六者，观之不亮。依时制，或棂之七、八，版之二、三之间。谅槅之大小，约桌几之平高，再高四、五寸为最也。

Illustrations
Long Separation (Figure 1-12)

In ancient times, the ratio of the upper lattice section vs. lower dado section is about 4:6, and that makes the indoor space a bit dimmer. The current way is to have a ratio of 7:3 or even 8:2. We would determine its size and make the dado part about the same height as that of a table, or a few inches higher to be the maximum.

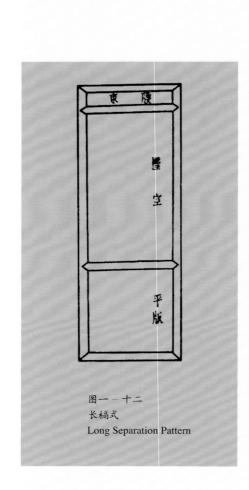

图一-十二
长槅式
Long Separation Pattern

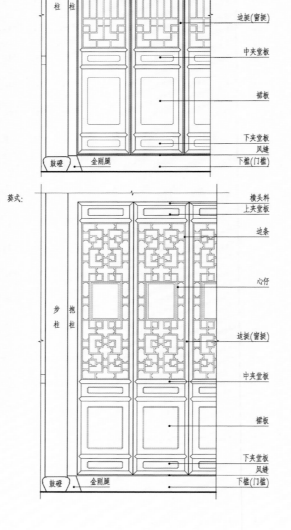

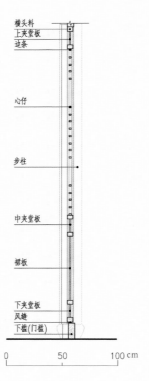

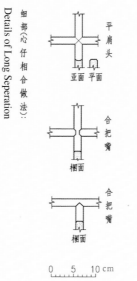

Details of Long Seperation

短楹式（图一-十三）

古之短楹，如长楹分楔版位者，亦更不亮。依时制，上下用束腰，或版或楔可也。

Short Separation (Figure 1 – 13)

The ratio for short separations in ancient times is the same as that for long ones, making the indoor space even darker. The modern way of doing it is what is called girdling at both the top and bottom ends, with the middle section being either lattice or solid dado.

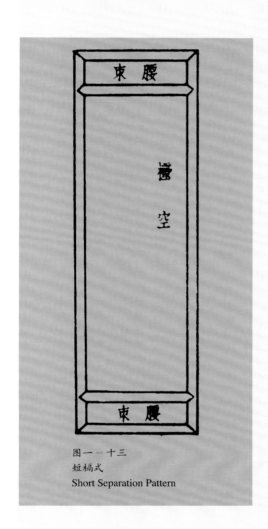

图一-十三
短楹式
Short Separation Pattern

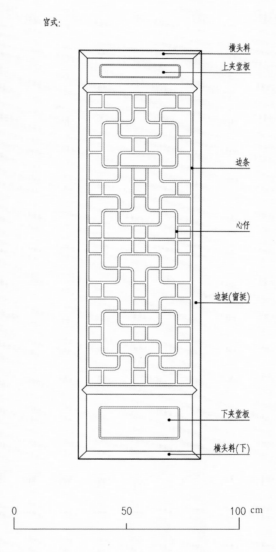

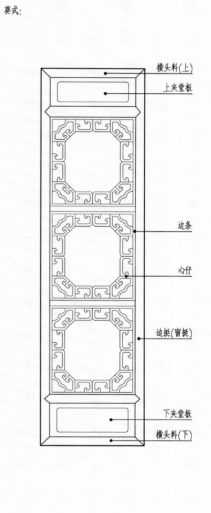

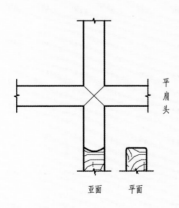

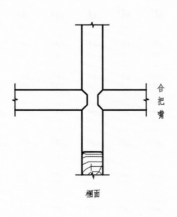

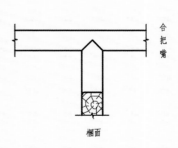

细部（心仔相合做法）
Details of short Seperation

宫式长窗
Gong Style Long Window

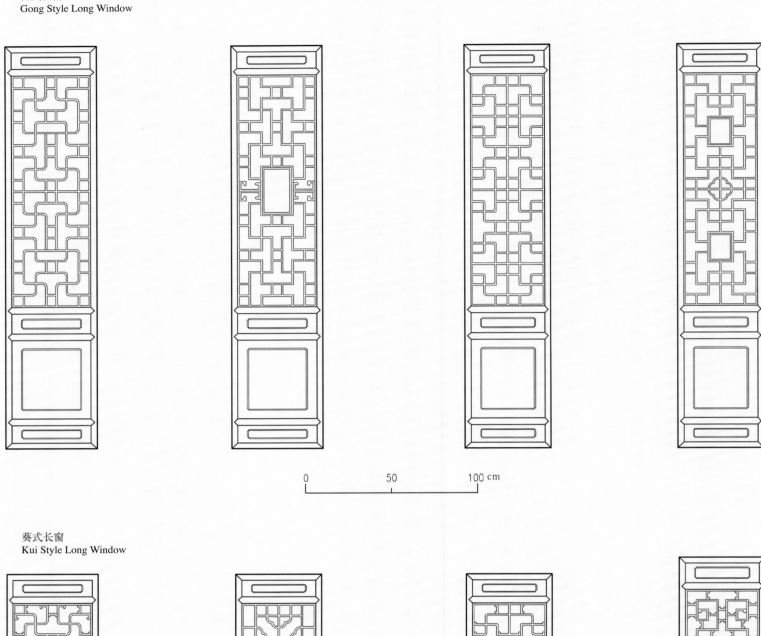

葵式长窗
Kui Style Long Window

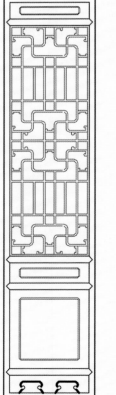
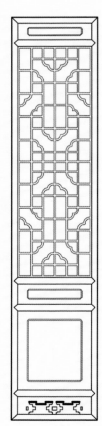
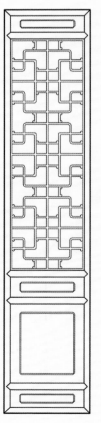
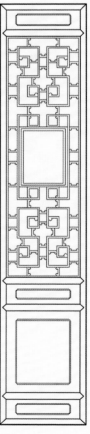

宫式短窗
Gong Style Short Window

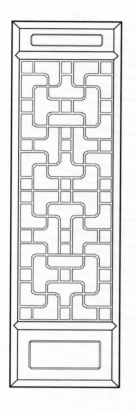
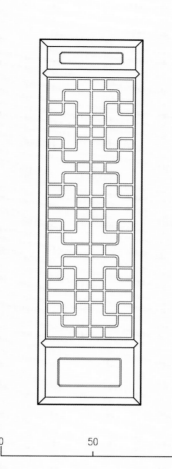
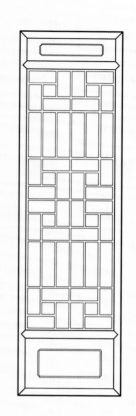

葵式短窗
Kui Style Short Window

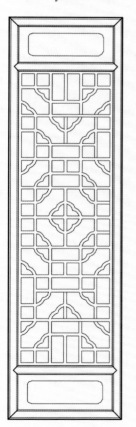
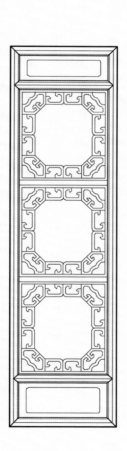

常用长窗装折形式
Commonly Used Long Window Styles

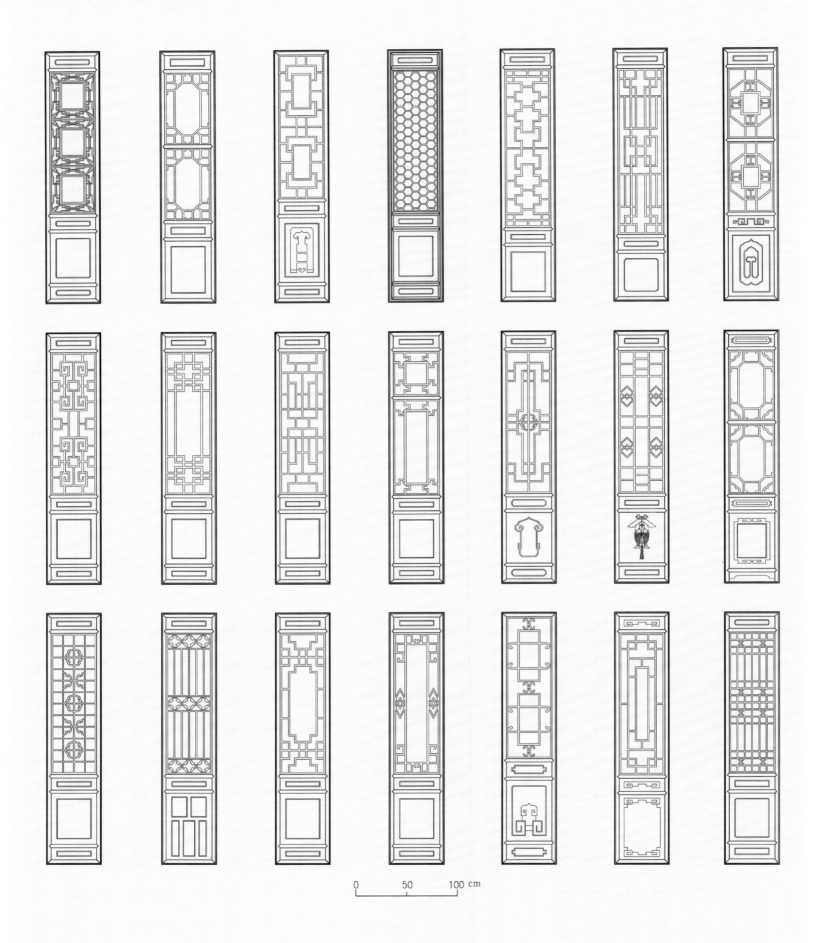

常用短窗装折形式
Commonly Used Short Window Styles

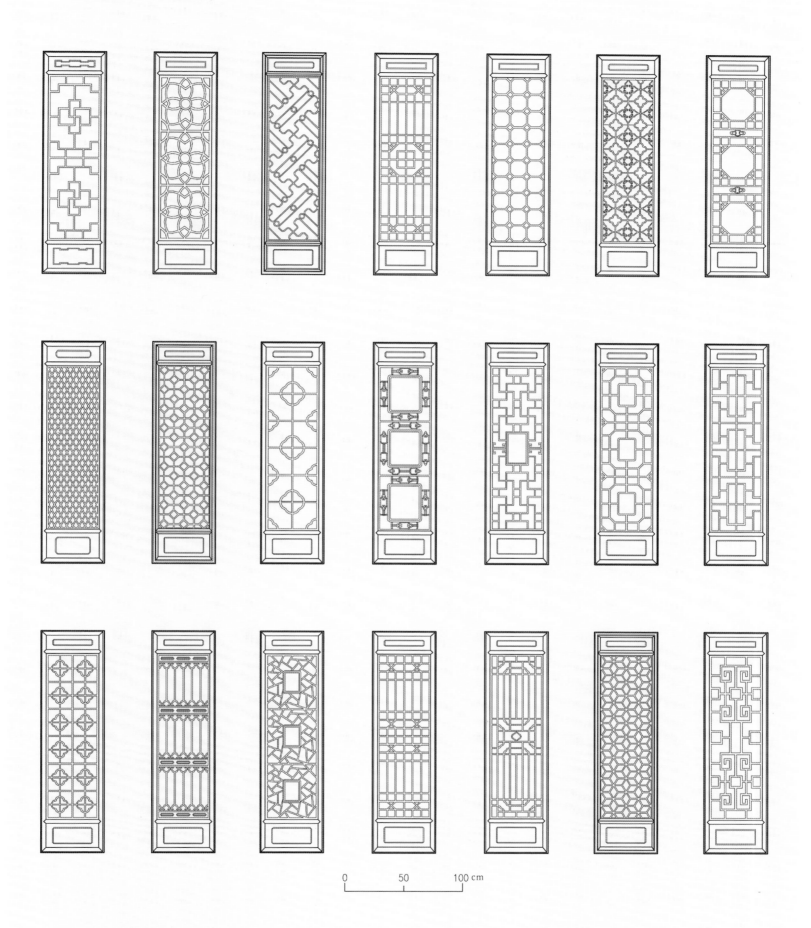

槅棂式
户槅柳条式（图一-十四至五十六）
时遵柳条槅，疏而且减，依式变换，随便摘用。

Separation Lattice Patterns
Hu Ge Wicker Style (Figures 1–14 to 1–56)
Wicker style is popular in modern Hu Ge. It is sparse and not complicated. One can pick anyone of them and improvise for different situations.

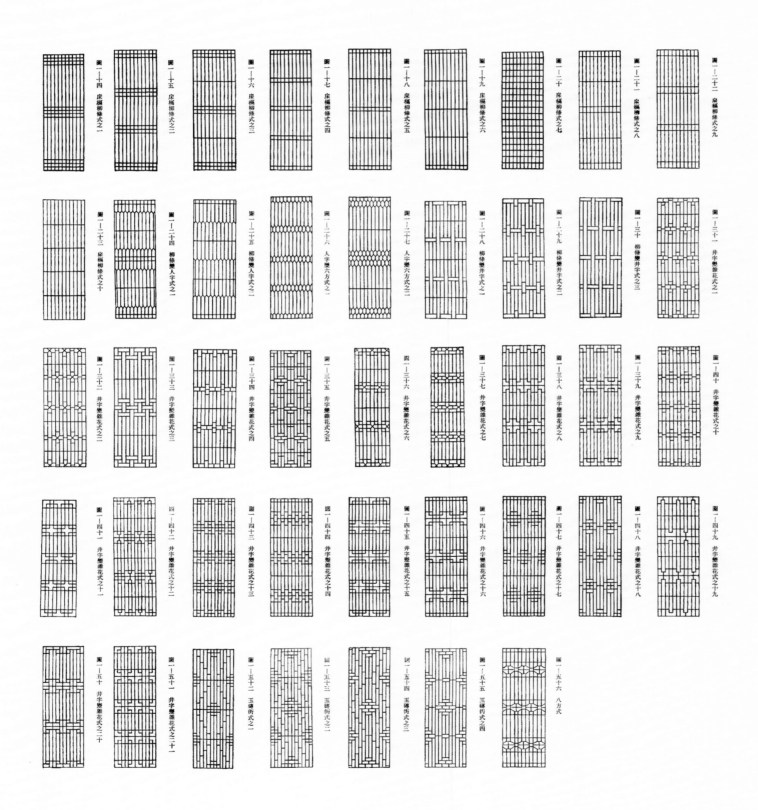

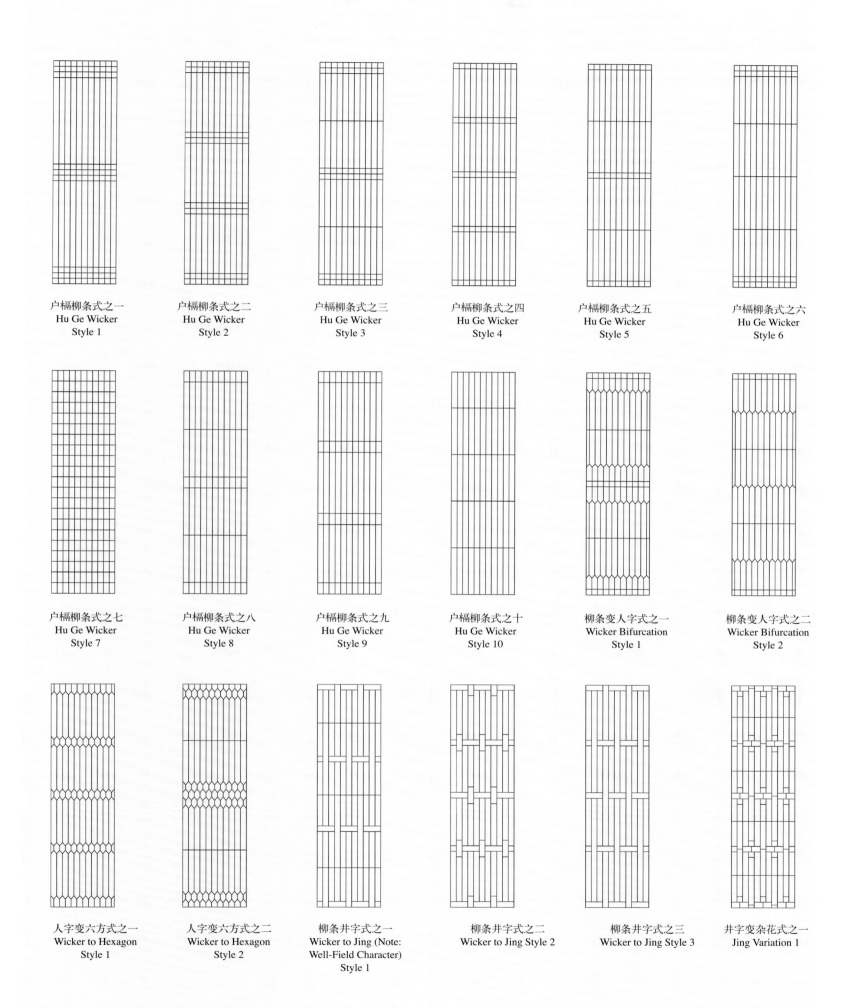

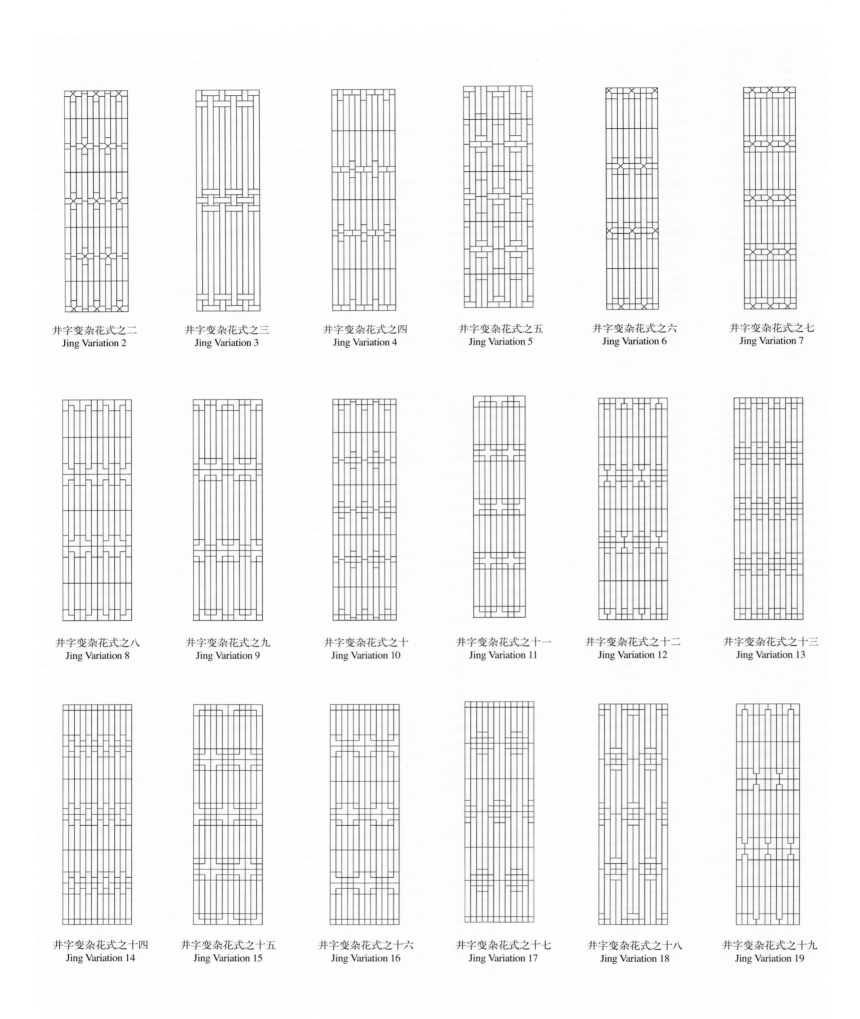

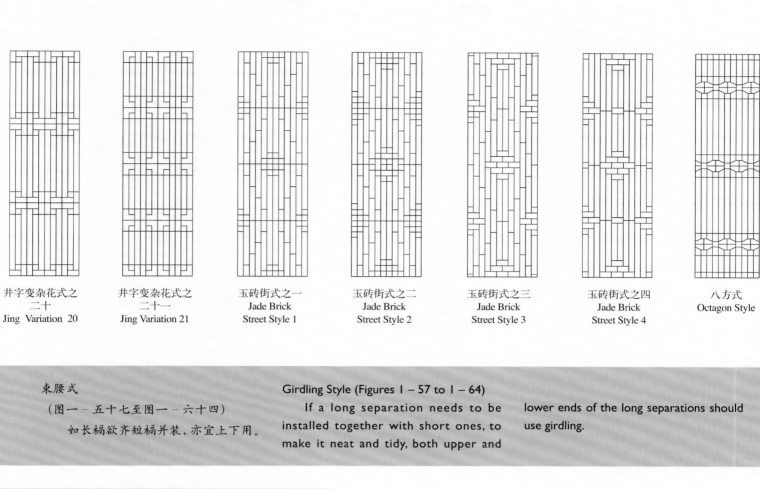

| 井字变杂花式之二十
Jing Variation 20 | 井字变杂花式之二十一
Jing Variation 21 | 玉砖街式之一
Jade Brick Street Style 1 | 玉砖街式之二
Jade Brick Street Style 2 | 玉砖街式之三
Jade Brick Street Style 3 | 玉砖街式之四
Jade Brick Street Style 4 | 八方式
Octagon Style |

束腰式
（图一－五十七至图一－六十四）
如长槅欲齐短槅并装，亦宜上下用。

Girdling Style (Figures 1 – 57 to 1 – 64)
If a long separation needs to be installed together with short ones, to make it neat and tidy, both upper and lower ends of the long separations should use girdling.

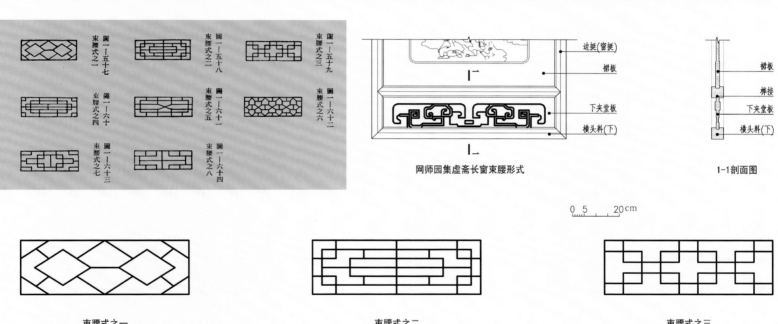

网师园集虚斋长窗束腰形式　　　　1-1 剖面图

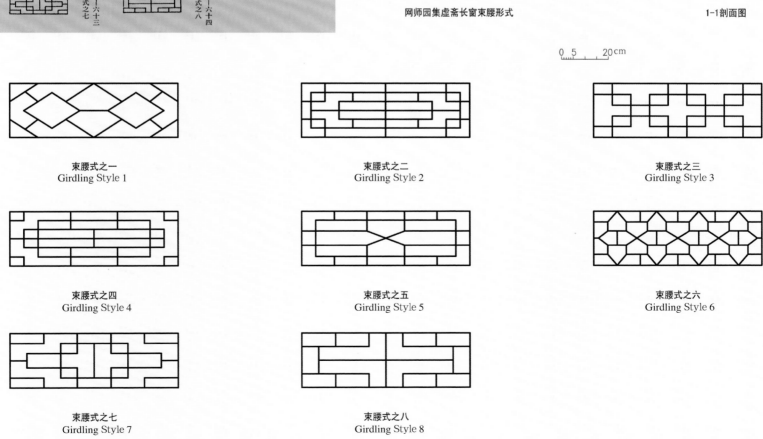

束腰式之一　Girdling Style 1　　束腰式之二　Girdling Style 2　　束腰式之三　Girdling Style 3

束腰式之四　Girdling Style 4　　束腰式之五　Girdling Style 5　　束腰式之六　Girdling Style 6

束腰式之七　Girdling Style 7　　束腰式之八　Girdling Style 8

（四）风窗

风窗，槅棂之外护，宜疏广减文，或横半，或两截推关，兹式如栏杆，减者亦可用也。在馆为"书窗"，在闺为"绣窗"。

(4) Feng Chuang

Feng Chuang (Note: wind window) is the protective windows outside of the window lattices. Its pattern is simple, few and far between another. Some are split into upper and lower parts for convenience of opening and closing. Its form is like a railing, simple and practical. Feng Chuang is called "Shu Chuang" (Note: book window), if it is in a study, and "Xiu Chuang" (Note: embroider window), if it is in a boudoir.

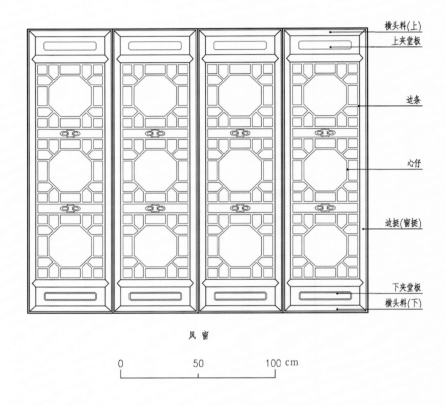

风窗

风窗式
（图一－六十五、图一－六十六）

风窗宜疏，或空框糊纸，或夹纱，或绘，少饰几棂可也。检栏杆式中，有疏而减文，竖用亦可。

Feng Chuang Patterns (Figures 1 - 65, 1 – 66)

A Feng Chuang pattern should be sparse, simple, and appropriate. Pasting paper on empty boxes, or using a thin gauze, probably painted with some drawings. Fewer lattices may be more appropriate. One may select a few from the railing's patterns, especially those with simple and sparse ones, as they could be used as a Feng Chuang when it is upright.

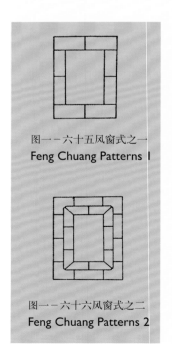

图一－六十五风窗式之一
Feng Chuang Patterns 1

图一－六十六风窗式之二
Feng Chuang Patterns 2

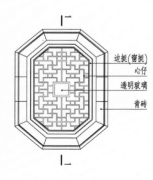

八角万字风窗
Octagonal Wan Character Feng Chuang

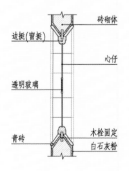

1-1 剖面图
1-1 Section View

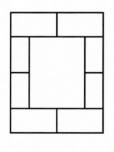

风窗式之一
Feng Chuang Pattern 1

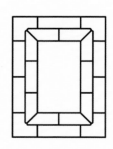

风窗式之二
Feng Chuang Pattern 2

冰裂式（图一-六十七）

冰裂惟风窗之最宜者，其文致减雅，信画如意，可以上疏下密之妙。

Ice Crack Pattern (Figure 1–67)

The ice crack pattern is probably the most appropriate one to use for Feng Chuang because the pattern is simple, delicate and elegant. One can create it at will but it is better to be sparse on top and dense at bottom.

两截式（图一-六十八）

风窗两截者，不拘何式，关合如一为妙。

Two-Part Style (Figure 1–68)

Two-part-styled Feng Chuang, no matter what pattern is used, it is better to form a complete pattern after it is shut.

冰裂式

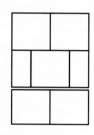
两截式

三截式（图一-六十九）

将中窗挂合上扇，仍撑上扇不碍空处。中连上，宜用铜合扇。

Three-Part Style (Figure 1–69)

For three-part-styled Feng Chuang, the middle pane is connected to the upper one. When the window is open, the middle one holds up the upper one and they do not use much space. The connection part between the middle and upper panes should use copper hinges.

梅花式（图一-七十）

梅花风窗，宜分瓣做。用梅花转心于中，以便开关。

Plum Flower Style (Figure 1–70)

For plum flower type of Feng Chuang, each petal should be made separately. A small plum flower centerpiece handle should be installed in the center of the window for convenience of opening and closing the window.

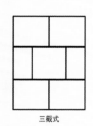
三截式

梅花式

梅花开式（图一-七十一）

连做二瓣，散做三瓣，将梅花转心，钉一瓣于连二之尖，或上一瓣、二瓣、三瓣，将转心向上扣住。

Plum Flower Blooming Style (Figure 1–71)

The lower two petals combine into one, while the upper three petals are separate. The plum flower shaped centerpiece handle should be installed at the top tip of the lower petals. For the three separate upper petals, depending on need, one, or two, or all three can be installed. After installation, twist the center handle upwards and they are all shut close.

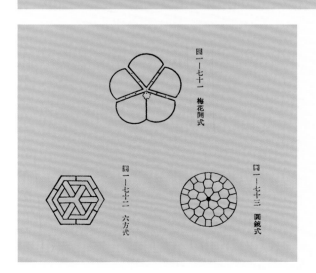

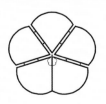
梅花开式

六方式

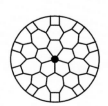
圆镜式

留园远翠阁
Yuancui Ge in Liu Yuan

拙政园海棠春坞
Haitang chunwu in Zhuozheng Yuan

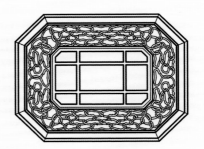
网师园蹈和馆
Daohe Guan in Wangshi Yuan

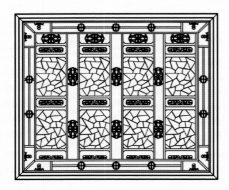
留园鹤所
Hesuo in Liu Yuan

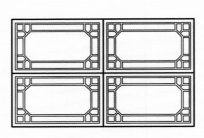
留园冠云楼
Guanyun Lou in Liu Yuan

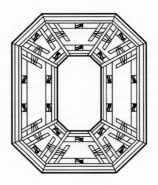
狮子林燕誉堂
Yanyu Tang in Shizilin

狮子林立雪堂
Lixue Tang in Shizilin

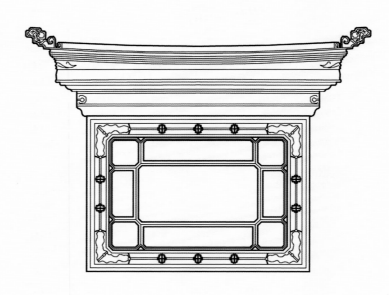
留园还我读书处
Huanwo dushuchu (Note: Return My Study Place) in Liu Yuan

砖框花窗实测图
Brick Frame Flower Window, Actual Survey

园冶图释

Yuan Ye Illustrated
— Classical Chinese Gardens Explained

五 栏杆

5 Railings

栏杆信画而成，减便为雅。古之回文万字，一概屏去，少留凉床佛座之用，园屋间不可制也。予历数年，存式百状，有工而精，有减而文，依次序变幻，式之于左，便为摘用。以笔管式为始，近有将篆字制栏杆者，况理画不匀，意不联络。予斯式中，尚觉未尽，尽可粉饰。

One can draw a railing design at will but the pattern should be simple and easy to build. We should abandon the ancient usage of railing pattern of palindrome or the Chinese character representing ten thousands, and allow only a small portion used on cool beds and Buddhist ornaments. These patterns should find no place in any of the garden buildings. Over the years, I have collected over a hundred design patterns. Some are very delicate while others are simple and elegant. Based on their change order, I list them below for your selection, starting with a Bi Guan style(Note: writing brush shaft style). Recently some people used seal script style for railing patterns; however, since the strokes are not even, it does not look coherent or well balanced, nor does it have any associations with the conceptual image. Among all the patterns I drew, there are still some imperfections and one may modify and polish further.

栏杆图式

笔管式（图二－一至图二－三十七）

栏杆以笔管式为始，以单变双，双则如意。变化以次而成，故有名。无名者恐有遗漏，总次序记之。内有花纹不易制者，亦书做法，以便鸠匠。

Railing Patterns

Bi Guan Style (Figures 2 – 1 to 2 – 37)

The railing patterns start with Bi Guan Style. It evolves from single to double, which can be changed into other styles at will. From the simple ones to complex ones, the cardinal rule is the same, and they are all referred to as "Bi Guan". There may be some missing ones because of their too many changes and difficulty to name, but they are all collected and listed here, with graphic illustrations. Some patterns are not so easy to make but they all come with textual explanations to facilitate craftsmen's work.

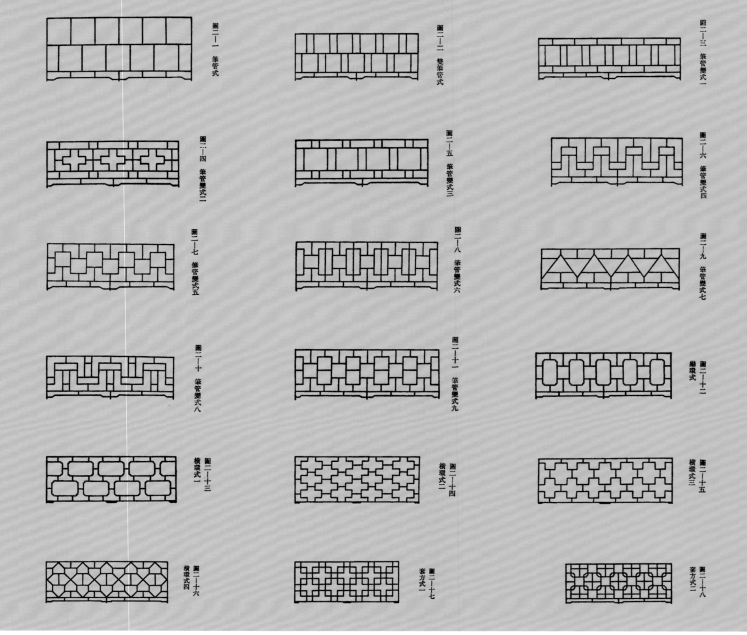

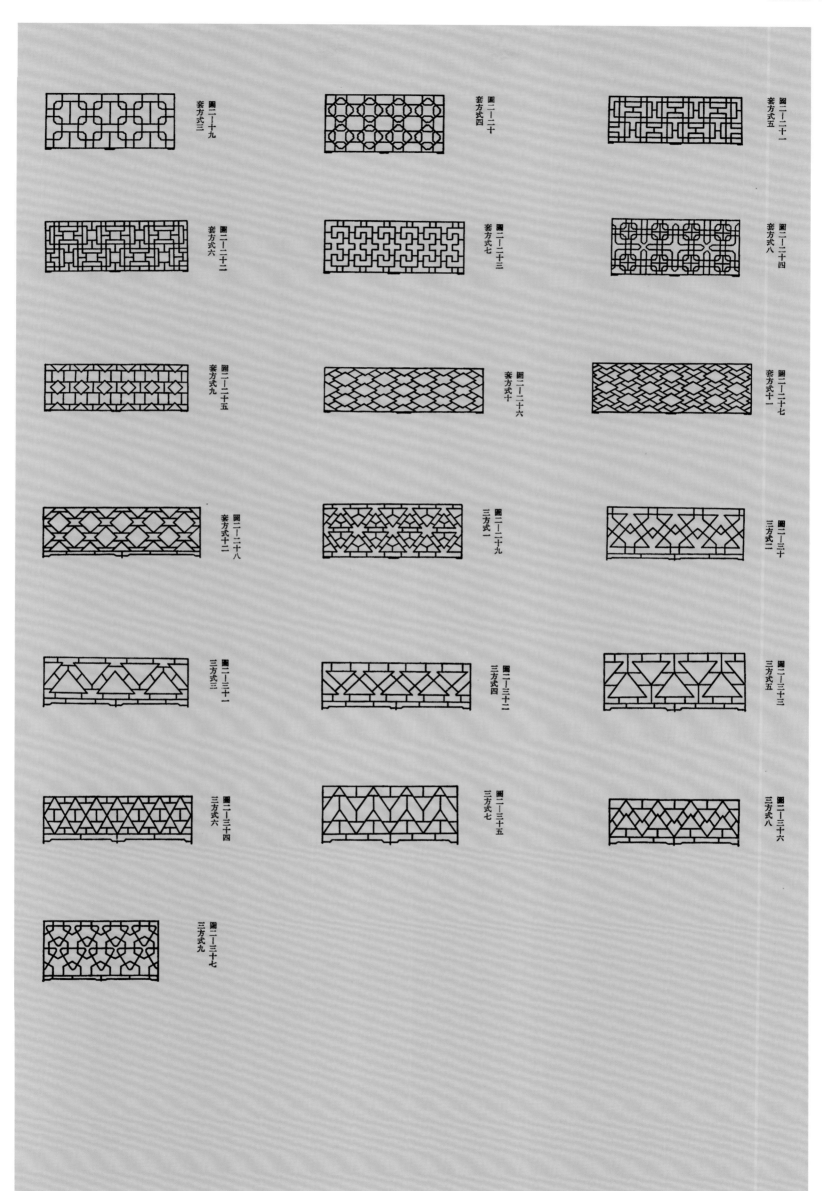

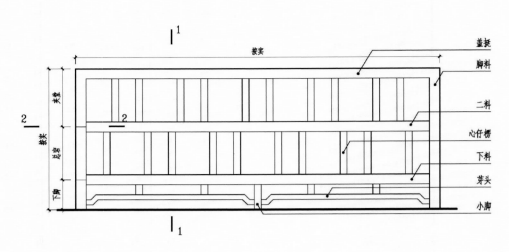
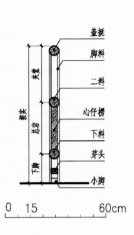

双笔管式立面图
Double Bi Guan Style Facade

1-1 断面图
1-1 Section View

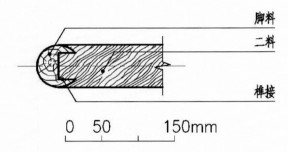

2-2 断面图
2-2 Section View

工程做法：

1. 栏杆之装柱间者，置短抱柱，抱柱宽以较臌瞪面阔出一寸为宜，上置捺槛，而栏杆之面较捺槛进。

2. 栏杆高度以长窗高一丈计，栏杆高度除捺槛厚约三寸外，实高三尺。夹堂高四寸，总宕高一尺八寸下脚为二寸半。脚料及盖挺、二料、下料之看面为一寸八分，深二寸。

Construction Implementations:
1. For parapet installed between columns, there should be a short embracing post, whose width should be one Cun wider than the width of the Gudeng. On top of the post there should install a railing, whose thickness should be greater than the parapet panel.
2. The height of a parapet is one Zhang, based on the long window. The height of a parapet, except for the railing in between whose thickness is three Cuns, should be three Chis. Jiatang is four Cuns tall. Zongdang is one Chi and eight Cuns while Xiajiao is 2.5 Cuns. Jiaoliao, Gaiting, Erliao, and Xialiao are all one Cun and eight Fens with two Cuns of depth.

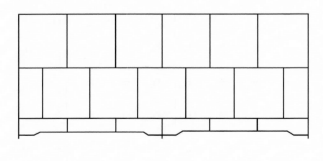

笔管式立面图
Bi Guan Style Facade

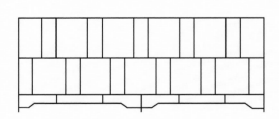

双笔管式立面图
Double Bi Guan Style Facade

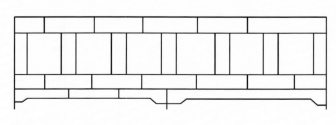

笔管变式一立面图
Bi Guan Variation Style 1 Facade

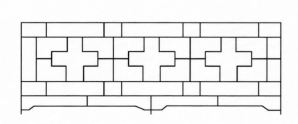

笔管变式二立面图
Bi Guan Variation Style 2 Facade

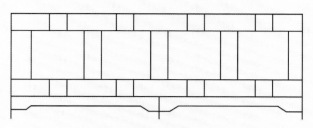
笔管变式三立面图
Bi Guan Variation Style 3 Facade

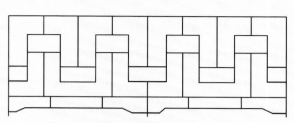
笔管变式四立面图
Bi Guan Variation Style 4 Facade

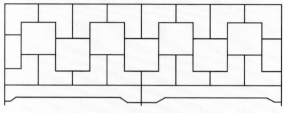
笔管变式五立面图
Bi Guan Variation Style 5 Facade

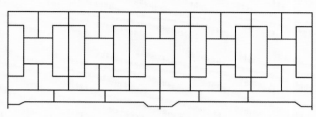
笔管变式六立面图
Bi Guan Variation Style 6 Facade

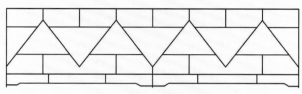
笔管变式七立面图
Bi Guan Variation Style 7 Facade

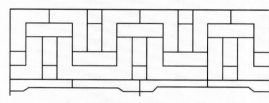
笔管变式八立面图
Bi Guan Variation Style 8 Facade

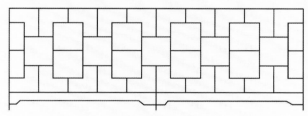
笔管变式九立面图
Bi Guan Variation Style 9 Facade

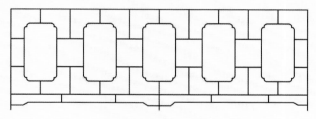
绦环式立面图
Strip Ring Style Facade

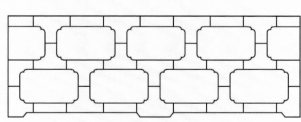
横环式一立面图
Horizontal Ring Style 1 Facade

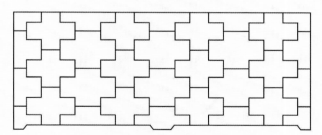
横环式二立面图
Horizontal Ring Style 2 Facade

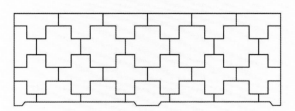
横环式三立面图
Horizontal Ring Style 3 Facade

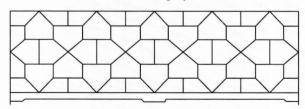
横环式四立面图
Horizontal Ring Style 4 Facade

锦葵式

（图二-三十八至图二-四十五）

先以六料攒心，然后加瓣，如斯做法。斯一料斗瓣。

Mallow Style (Figures 2 – 38 to 2 – 45)

First we need to use six pieces to make the center of the flower and then splice others to the petals. This is how we make mallow flower patterns.

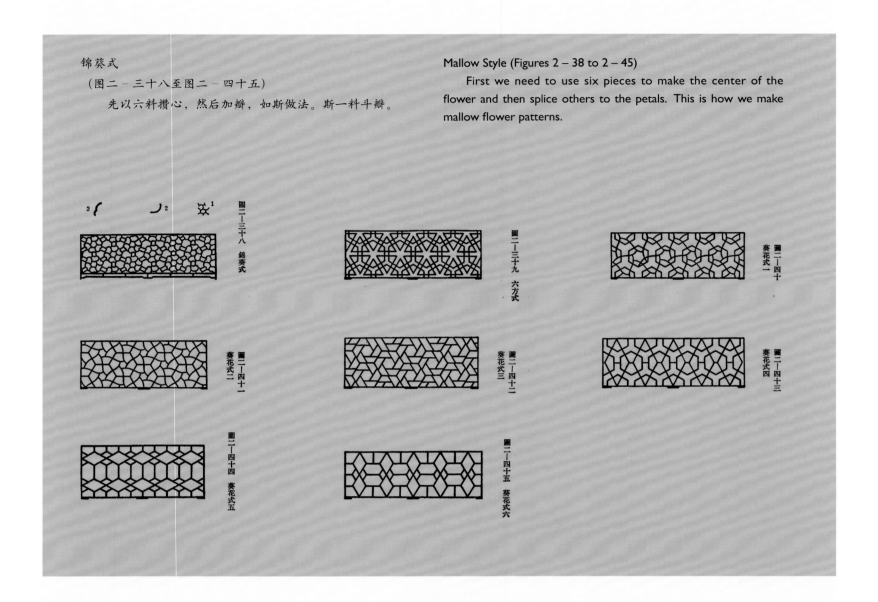

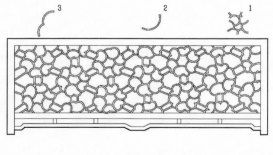

锦葵式立面图
Mallow Style Facade

六方式立面图
Hexagon Style Facade

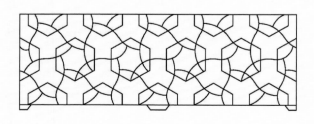

葵花式一立面图
Sunflower Style 1 Facade

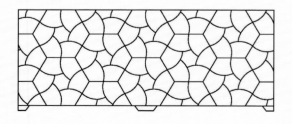

葵花式二立面图
Sunflower Style 2 Facade

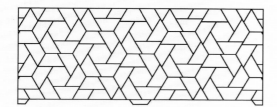

葵花式三立面图
Sunflower Style 3 Facade

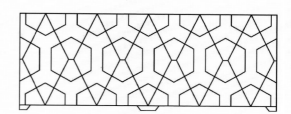

葵花式四立面图
Sunflower Style 4 Facade

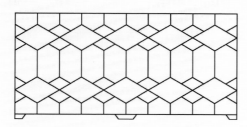

葵花式五立面图

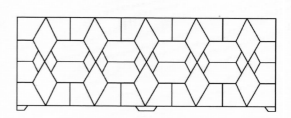

葵花式六立面图
Sunflower Style 6 Facade

波纹式（图二-四十六）
惟斯一料可做。

梅花式（图二-四十七）
用斯一料，斗瓣料直，不攒榫眼。

Ripple Style (Figure 2 - 46)
We can make this with only one material.

Plum Flower Style (Figure 2 – 47)
The petals are made using this material and all others are made using straight ones. There is no need to use mortise on the straight materials.

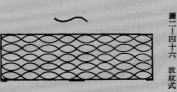

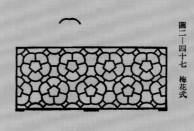

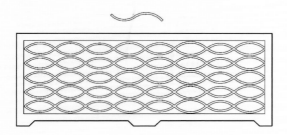

波纹式立面图
Ripple Pattern

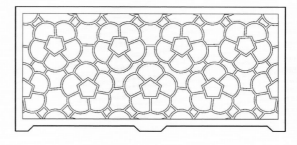

梅花式立面图
Plum Flower Pattern

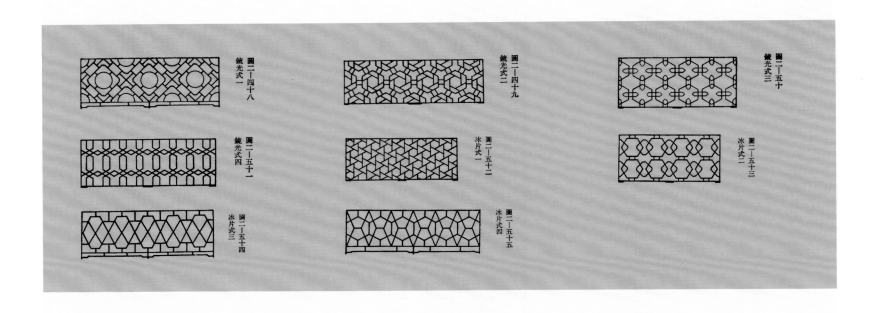

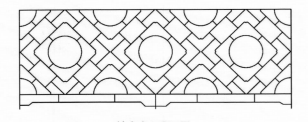

镜光式一立面图
Jingguang Style 1 Facade

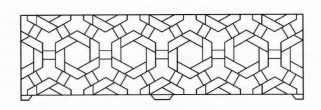

镜光式二立面图
Jingguang Style 2 Facade

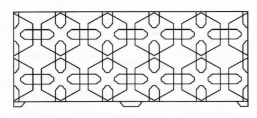

镜光式三立面图
Jingguang Style 3 Facade

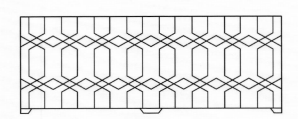

镜光式四立面图
Jingguang Style 4 Facade

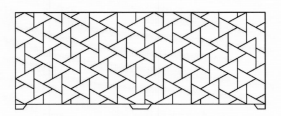

冰片式一立面图
Bingpian Style 1 Facade

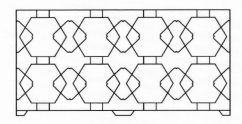

冰片式二立面图
Bingpian Style 2 Facade

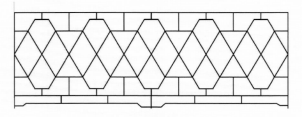

冰片式三立面图
Bingpian Style 3 Facade

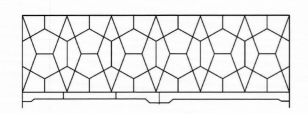

冰片式四立面图
Bingpian Style 4 Facade

联瓣葵花式（图二-五十六至图二-六十）
惟斯一料可做。

Connected Sunflower Style (Figure 2 – 56 to 2 – 60)
Using only one material is sufficient.

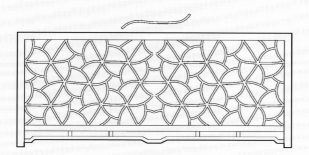
联瓣葵花式一立面图
Joint Petal Sunflower Style 1 Facade

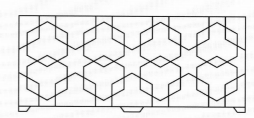
联瓣葵花式二立面图
Joint Petal Sunflower Style 2 Facade

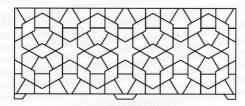
联瓣葵花式三立面图
Joint Petal Sunflower Style 3 Facade

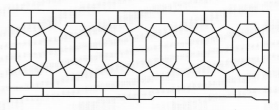
联瓣葵花式四立面图
Joint Petal Sunflower Style 4 Facade

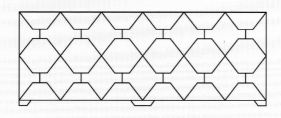
联瓣葵花式五立面图
Joint Petal Sunflower Style 5 Facade

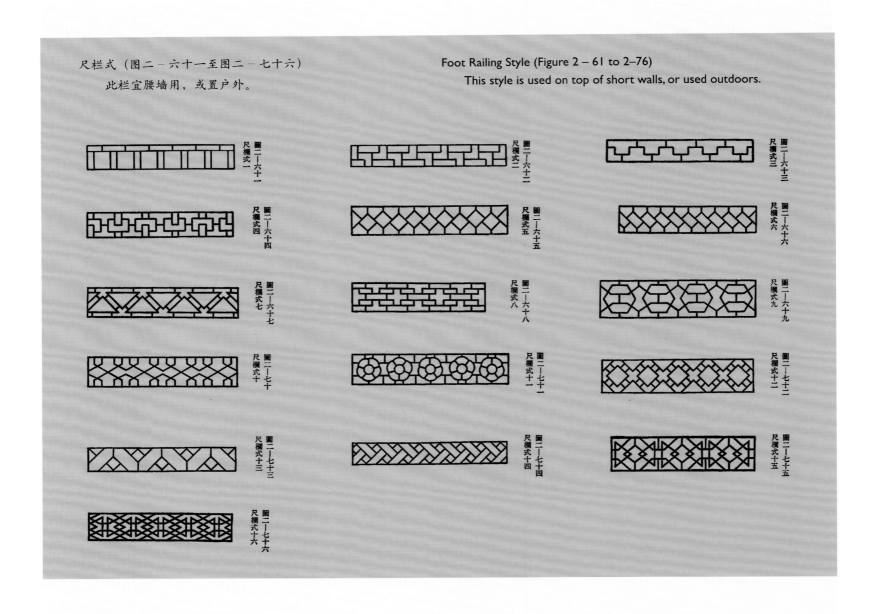

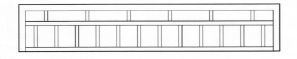

尺栏式一立面图
Foot Railing Style 1 Facade

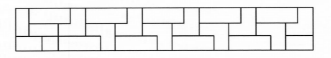

尺栏式二立面图
Foot Railing Style 2 Facade

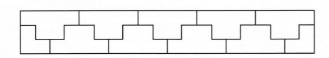

尺栏式三立面图
Foot Railing Style 3 Facade

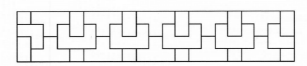

尺栏式四立面图
Foot Railing Style 4 Facade

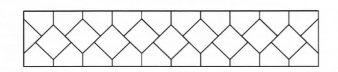

尺栏式五立面图
Foot Railing Style 5 Facade

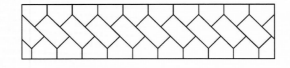

尺栏式六立面图
Foot Railing Style 6 Facade

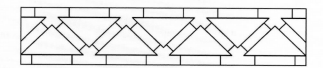

尺栏式七立面图
Foot Railing Style 7 Facade

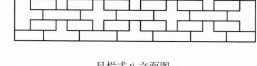

尺栏式八立面图
Foot Railing Style 8 Facade

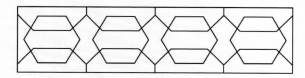

尺栏式九立面图
Foot Railing Style 9 Facade

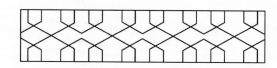

尺栏式十立面图
Foot Railing Style 10 Facade

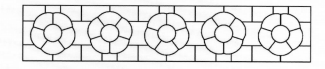

尺栏式十一立面图
Foot Railing Style 11 Facade

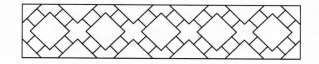

尺栏式十二立面图
Foot Railing Style 12 Façade

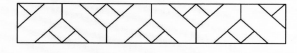

尺栏式十三立面图
Foot Railing Style 13 Facade

尺栏式十四立面图
Foot Railing Style 14 Facade

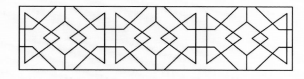

尺栏式十五立面图
Foot Railing Style 15 Facade

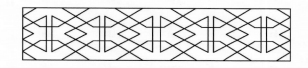

尺栏式十六立面图
Foot Railing Style 16 Facade

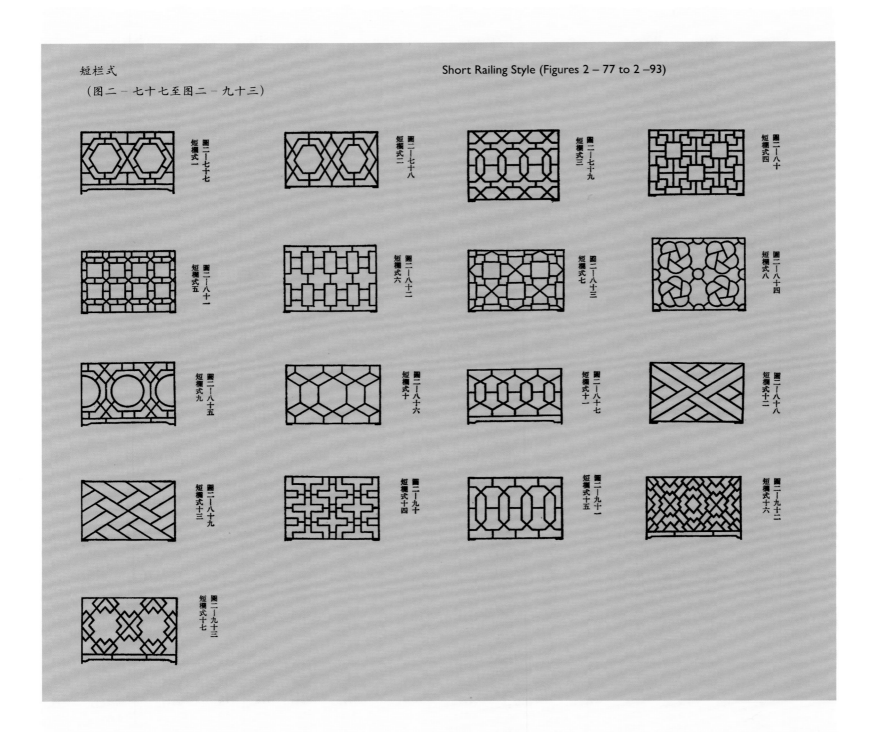

短栏式 Short Railing Style (Figures 2–77 to 2–93)

(图二-七十七至图二-九十三)

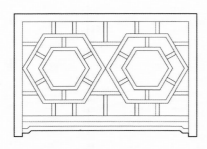

短栏式一立面图
Short Railing Style 1 Facade

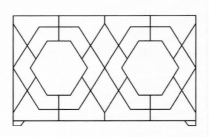

短栏式二立面图
Short Railing Style 2 Facade

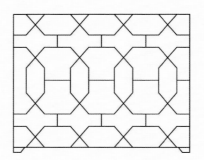

短栏式三立面图
Short Railing Style 3 Facade

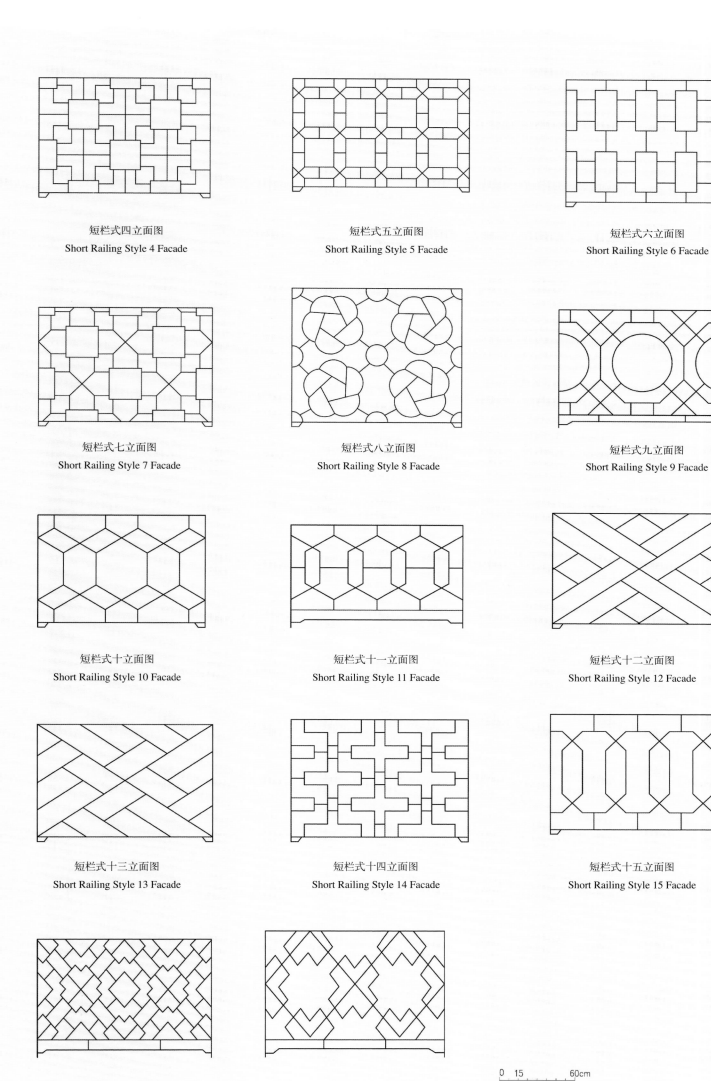

短栏式四立面图
Short Railing Style 4 Facade

短栏式五立面图
Short Railing Style 5 Facade

短栏式六立面图
Short Railing Style 6 Facade

短栏式七立面图
Short Railing Style 7 Facade

短栏式八立面图
Short Railing Style 8 Facade

短栏式九立面图
Short Railing Style 9 Facade

短栏式十立面图
Short Railing Style 10 Facade

短栏式十一立面图
Short Railing Style 11 Facade

短栏式十二立面图
Short Railing Style 12 Facade

短栏式十三立面图
Short Railing Style 13 Facade

短栏式十四立面图
Short Railing Style 14 Facade

短栏式十五立面图
Short Railing Style 15 Facade

短栏式十六立面图
Short Railing Style 16 Facade

短栏式十七立面图
Short Railing Style 17 Facade

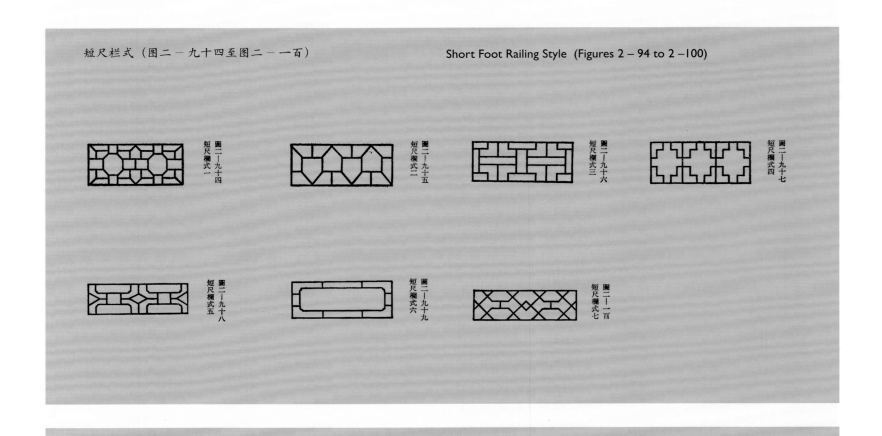

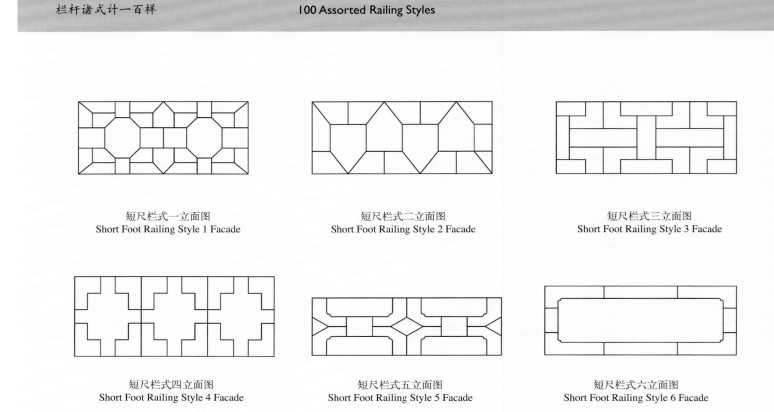

短尺栏式一立面图
Short Foot Railing Style 1 Facade

短尺栏式二立面图
Short Foot Railing Style 2 Facade

短尺栏式三立面图
Short Foot Railing Style 3 Facade

短尺栏式四立面图
Short Foot Railing Style 4 Facade

短尺栏式五立面图
Short Foot Railing Style 5 Facade

短尺栏式六立面图
Short Foot Railing Style 6 Facade

短尺栏式七立面图
Short Foot Railing Style 7 Facade

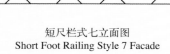

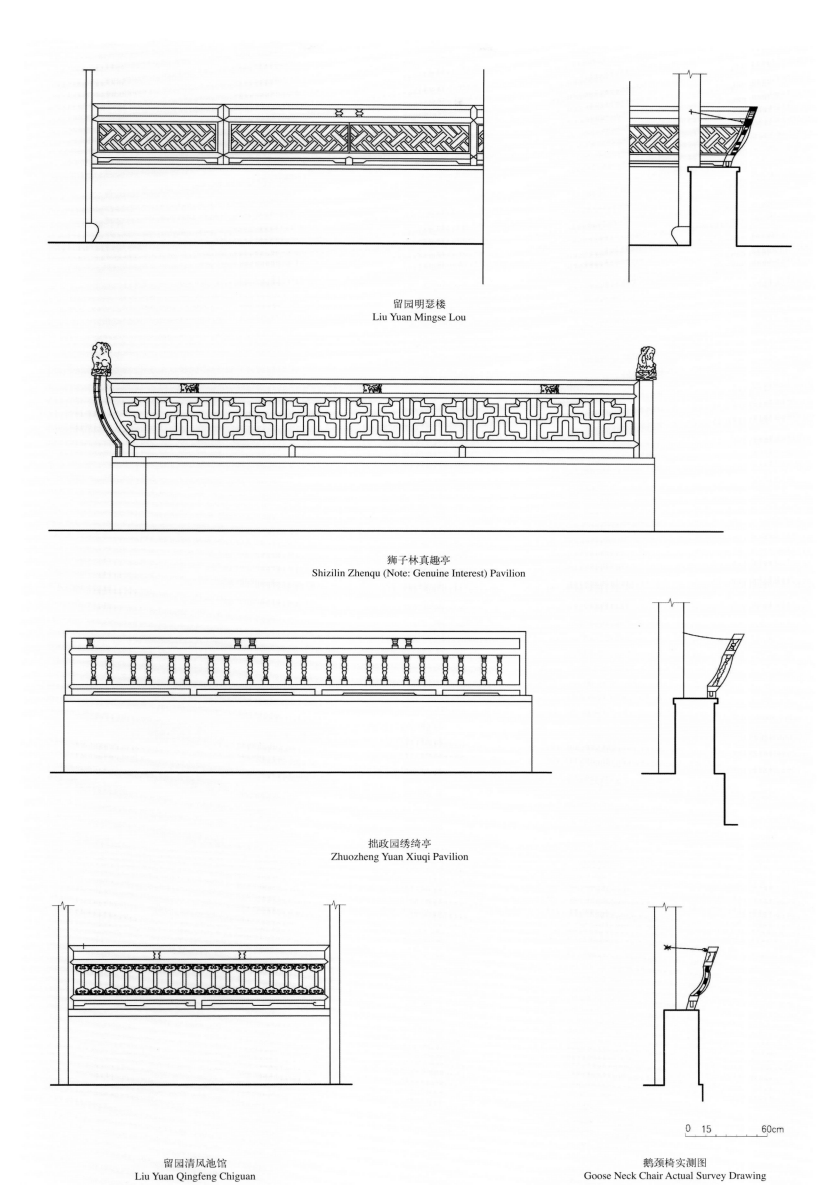

留园明瑟楼
Liu Yuan Mingse Lou

狮子林真趣亭
Shizilin Zhenqu (Note: Genuine Interest) Pavilion

拙政园绣绮亭
Zhuozheng Yuan Xiuqi Pavilion

留园清风池馆
Liu Yuan Qingfeng Chiguan

鹅颈椅实测图
Goose Neck Chair Actual Survey Drawing

栏杆常用样式（1）
Commonly Used Railing Styles (1)

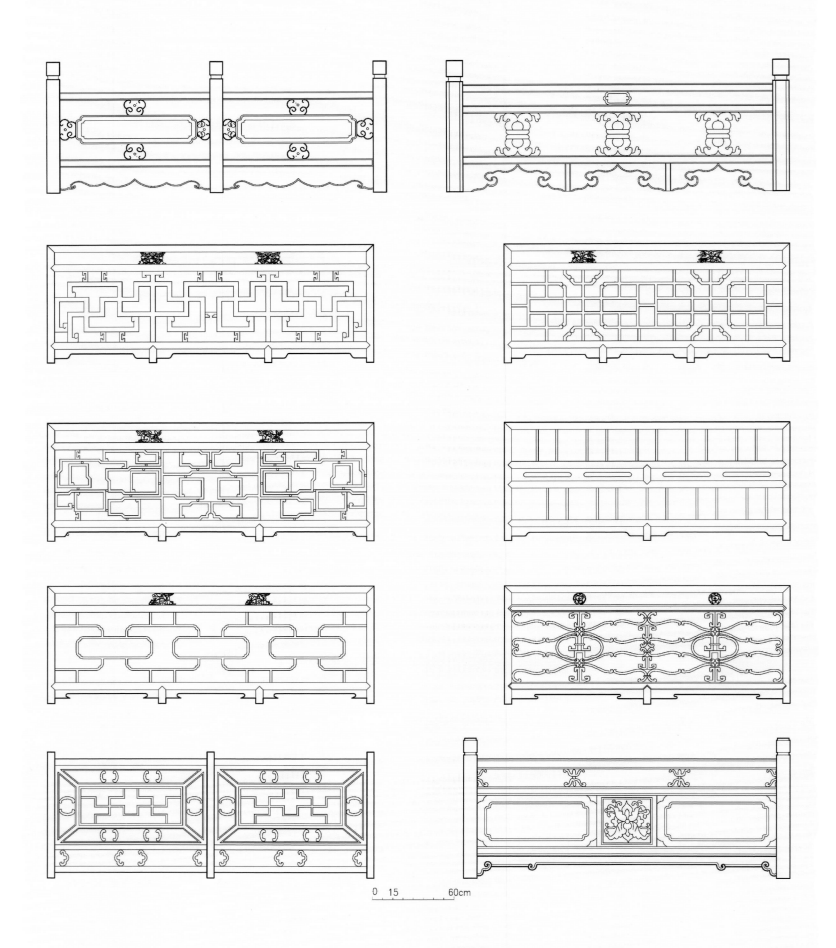

栏杆常用样式（2）
Commonly Used Railing Styles (2)

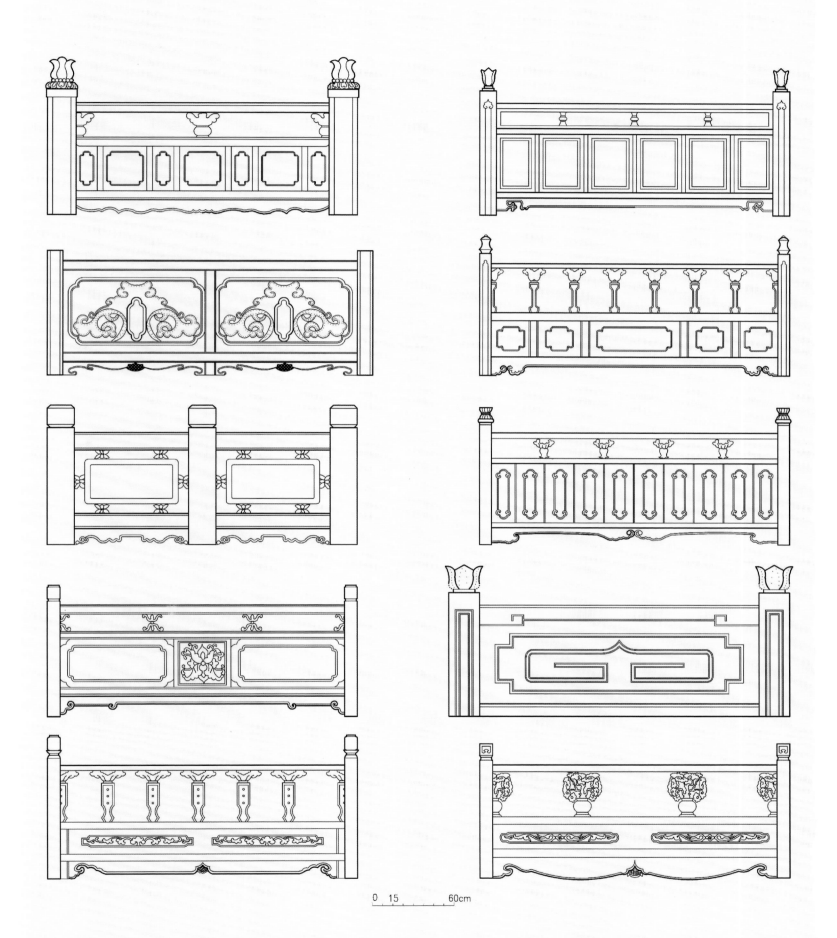

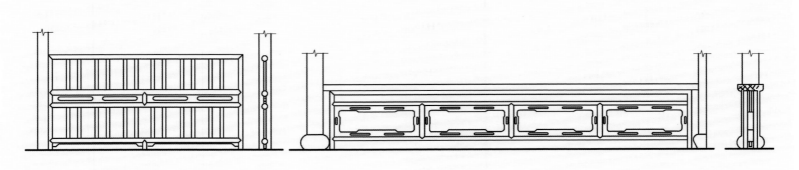

耦园城曲草堂
Ou Yuan (Note: Lotus Garden) Chengqu Caotang

王洗马巷万宅
Mr. Wan's Residence in Wangxima Lane

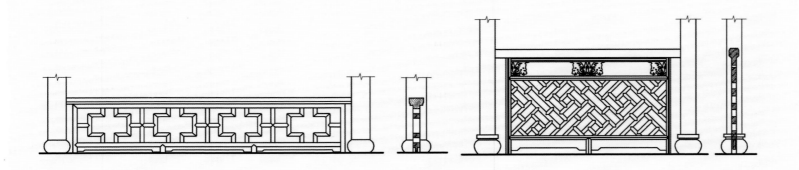

拙政园别有洞天
Zhuozheng Yuan Bieyou Dongtian

狮子林卧云室
Shizilin Woyun Shi (Note: Cloud Perching Room)

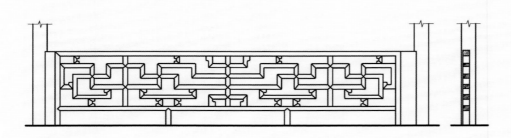

留园冠云楼
Liu Yuan Guanyun Lou (Note: Cloud Topping Building)

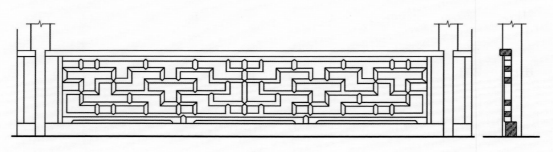

拙政园倒影楼
Zhuozheng Yuan Daoying Lou (Note: Reflection Building)

木栏杆实测图
Wooden Railing Actual Survey Drawing

0　20　　　100cm

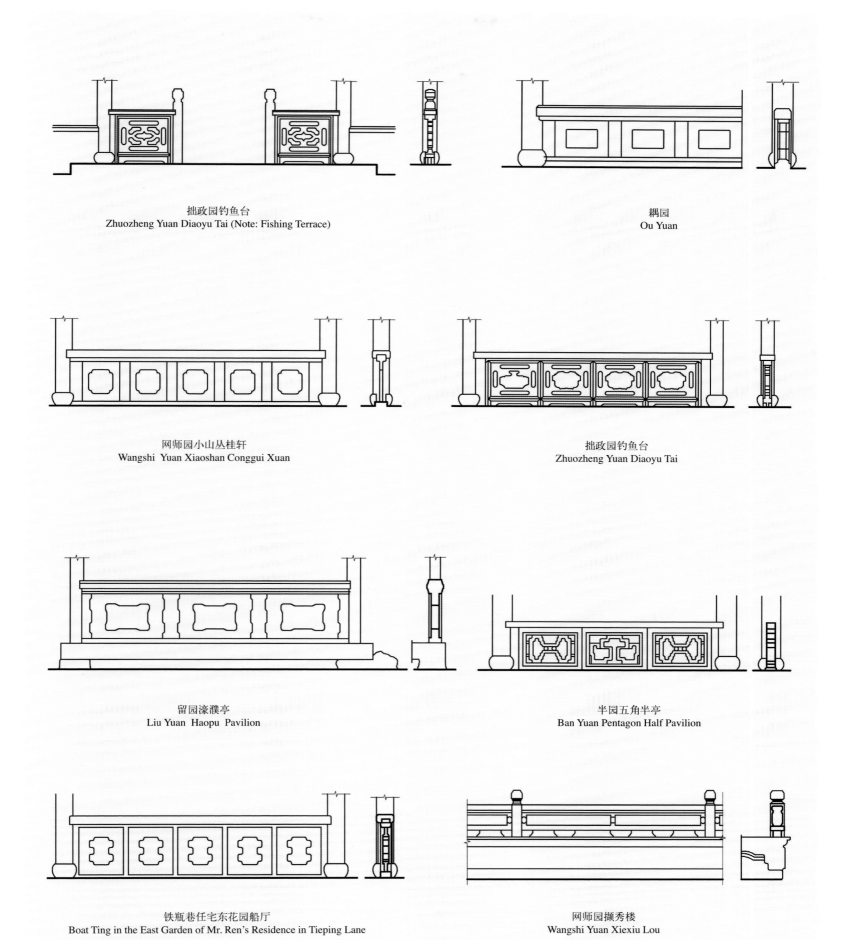

砖栏杆实测图
Brick Railing Actual Survey Drawing

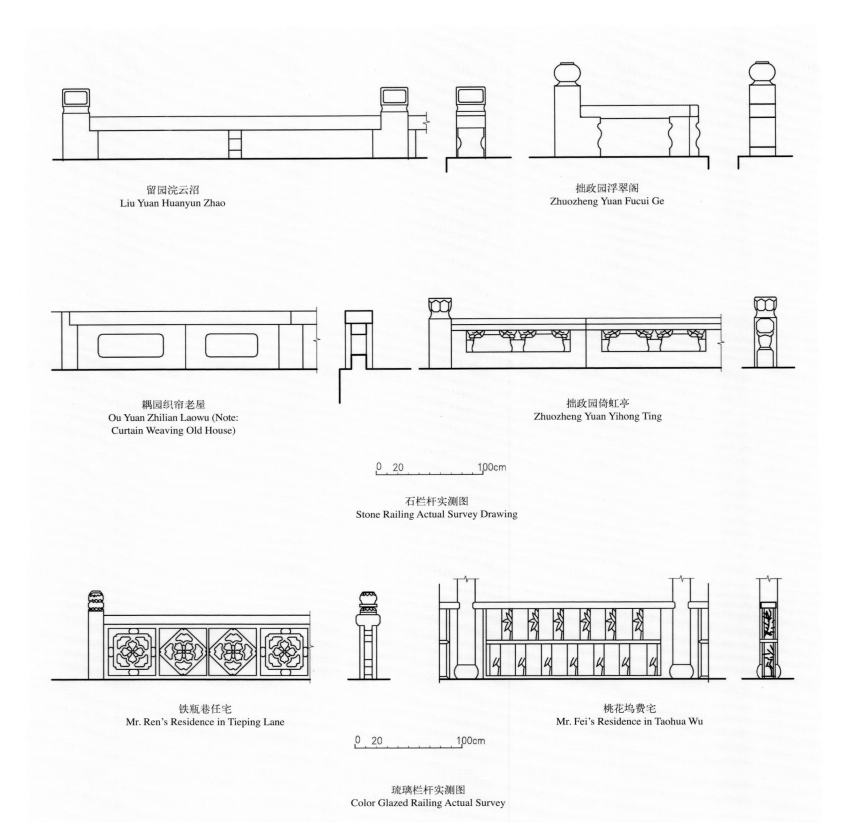

园冶图释

Yuan Ye Illustrated
— Classical Chinese Gardens Explained

图释

六 门窗

6 Doors & Windows

门窗磨空，制式时裁，不惟屋宇翻新，斯谓林园遵雅。工精虽专瓦作，调度犹在得人，触景生奇，含情多致，轻纱环碧，弱柳窥青。伟石迎人，别有一壶天地；修篁弄影，疑来隔水笙簧。佳境宜收，俗尘安到。切忌雕镂门空，应当磨琢窗垣。处处邻虚，方方侧景。非传恐失，故式存余。

For doors and windows without casements, embedded ground brick should be used for ornaments, and current styles should be used. The purpose is not only for novelty but also for elegancy. Although the exquisite workmanship is all dependent on the craftsmen, the original creative concept is still up to the garden designer. The garden scenes can spark wonderful ideas and indefinite exquisite interests. Through a screen window, one can see verdant peaks looming in the distance; via a willow patterned lattice, green mountains and clear water meet the eyes. Tall and rare rocks proudly greet visitors, creating unique landscape. Long and slender bamboos sway gently and cast interesting shadows and it seems there is music passing over the water. All wonderful scenes are obtained from within buildings and there is no room for the mundane and the tacky. It is a no-no to have carvings on door frames but more design thoughts should be stressed on windows. Outside a window there should leave ample space, but to every direction from the window a scene should be close by. If I do not disseminate these design issues, they may get lost, so I drew all the door and window patterns here to keep a record.

门窗图式

Door & Window Illustrations

方门合角式（图三—一）

磨砖方门，凭匠俱做券门，砖土过门石，或过门枋者。今之方门，将磨砖用木栓栓住，合角过门于上，再加之过门枋，雅致可观。

Square Door and Closed Corner Style (Figure 3–1)

To make a ground brick door in the past, it is all at the disposal of the craftsman. To build a door frame, an arch or lintel could be used. To build a contemporary square door, first at each side ground bricks are mounted using wood nails, then splice in the mortises at the upper two corners for connection, and finally a lintel ground brick is inlaid on top. It looks elegant and beautiful.

工程做法：

1. 洞门安装完毕后用油灰嵌缝，并用猪血砖屑灰补转面和线脚上的空洞，待干后再用沙砖打磨平滑。

2. 洞门通常用灰青色方砖镶砌，其上刨成挺秀的线脚，形式多样，与白墙配合成朴素明净的色调。

Construction Details:
1. After the installation of an arched door, putties are used to seal the gap. Ground brick crumbs missed with pig blood are used to fill holes at a corner or architrave, and after it's been dried, an emery brick is used for polishing.
2. A doorway is usually made of gray and cyan colored square bricks, with its upper part planed into a straight architrave with many styles. Together with white walls, it forms a simple, bright and clean color tone.

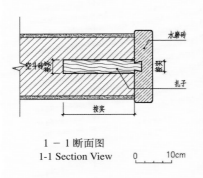

1-1 断面图
1-1 Section View

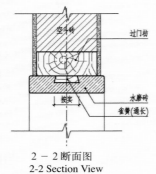

2-2 断面图
2-2 Section View

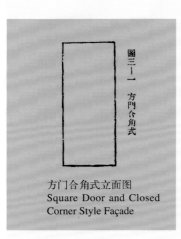

方门合角式立面图
Square Door and Closed Corner Style Façade

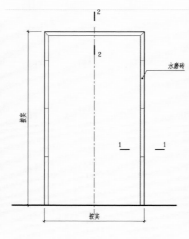

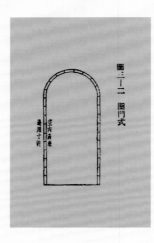

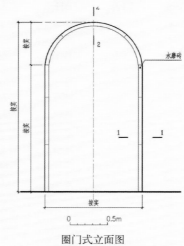

圈门式立面图
Arch Door Style Façade

圈门式（图三—二）

凡磨砖门窗，量墙之厚薄，校砖之大小，内空须用满磨，外边只可寸许，不可就砖，边外或白粉或满磨可也。

Arch Door Style (Figure 3–2)

For ground brick inlaid doors and windows, the thickness of the wall must be measured first to calculate the size of ground bricks. The inner wall inside the door or windows must all use inlaid ground bricks. The outer frame can only be sticked out for about an inch to keep its decent look. If it is too thick, then it needs to be trimmed thinner. The outside area of the door frame can be covered with plaster or ground bricks.

上下圈式（图三-三）
The Upper and Lower Patterns (Figure 3 – 3)

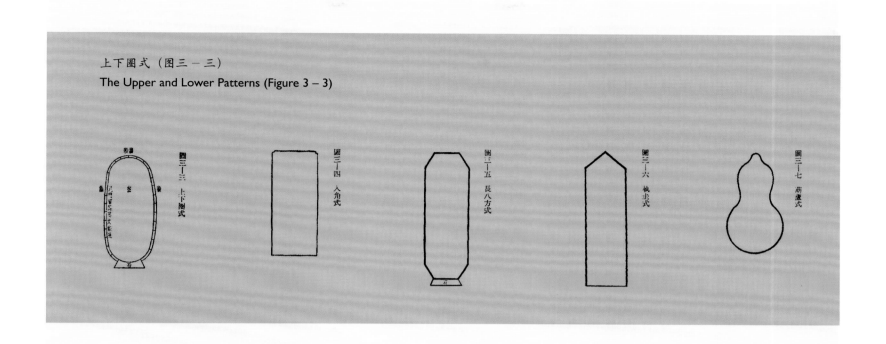

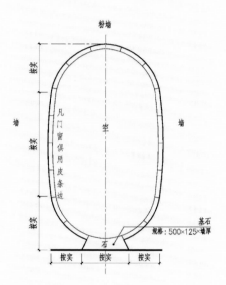

上下圈式立面图
The Upper and Lower Circular Style Facade

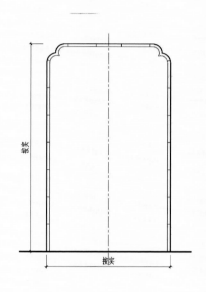

入角式立面图
Corner-in Style Facade

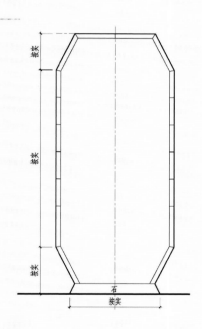

长八方式立面图
Elongated Octagon Style Facade

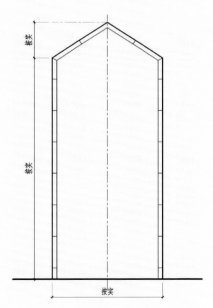

执圭式立面图
Zhigui (Note: an ancient Chinese artifact) Style Facade

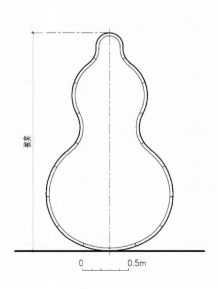

葫芦式立面图
Gourd Style Facade

莲瓣式（图三—八）、如意式（图三—九）、贝叶式（图三—十）
　　莲瓣，如意，贝叶，斯三式宜供佛所用。

Lotus Petal Style (Figure 3 – 8), Ru Yi Style (Figure 3 – 9), Leaf Style (Figure 3 – 10)
　　Lotus Petal, Ruyi, and Pattra Leave styles are all appropriate for Buddha worshipping occasions.

图三—八　莲瓣式
图三—九　如意式
图三—十　贝叶式
图三—十一　剑环式
图三—十二　汉瓶式之一
图三—十三　汉瓶式二
图三—十四　汉瓶式三
图三—十五　汉瓶式四
图三—十六　花觚式
图三—十七　蓍草瓶式
图三—十八　月窗式
图三—十九　片月式
斯亦可为门空　八方式（图三—二十）
图三—二十一　六方式　斯亦可为门空
图三—二十二　菱花式
图三—二十三　如意式
图三—二十四　梅花式
图三—二十五　葵花式
图三—二十六　海棠式
图三—二十七　鞋子式
图三—二十八　贝叶式
图三—二十九　六方嵌桃子式
图三—三十　桃子花式
图三—三十一　藤式

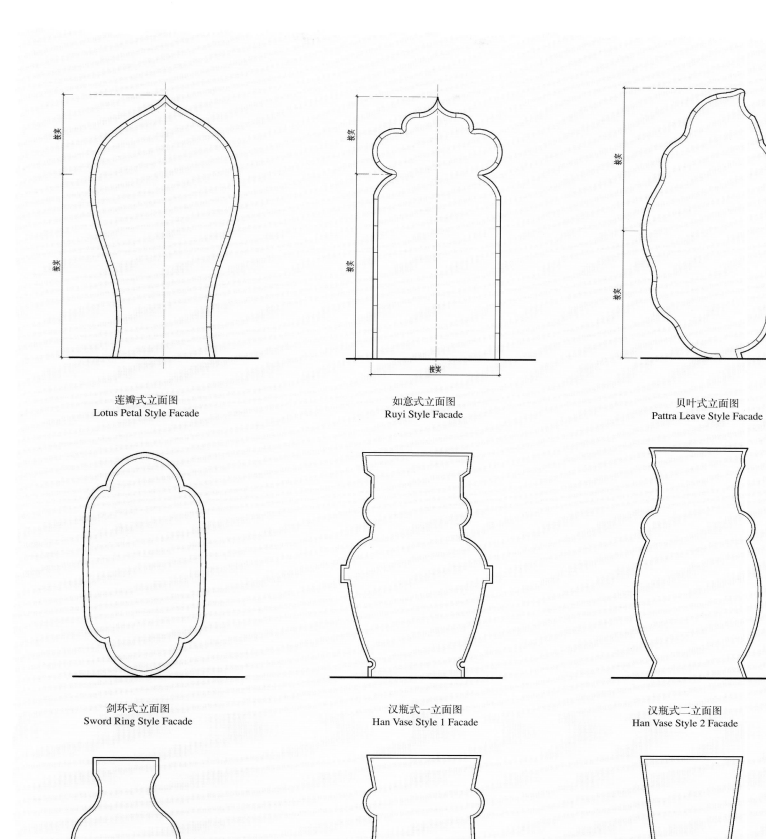

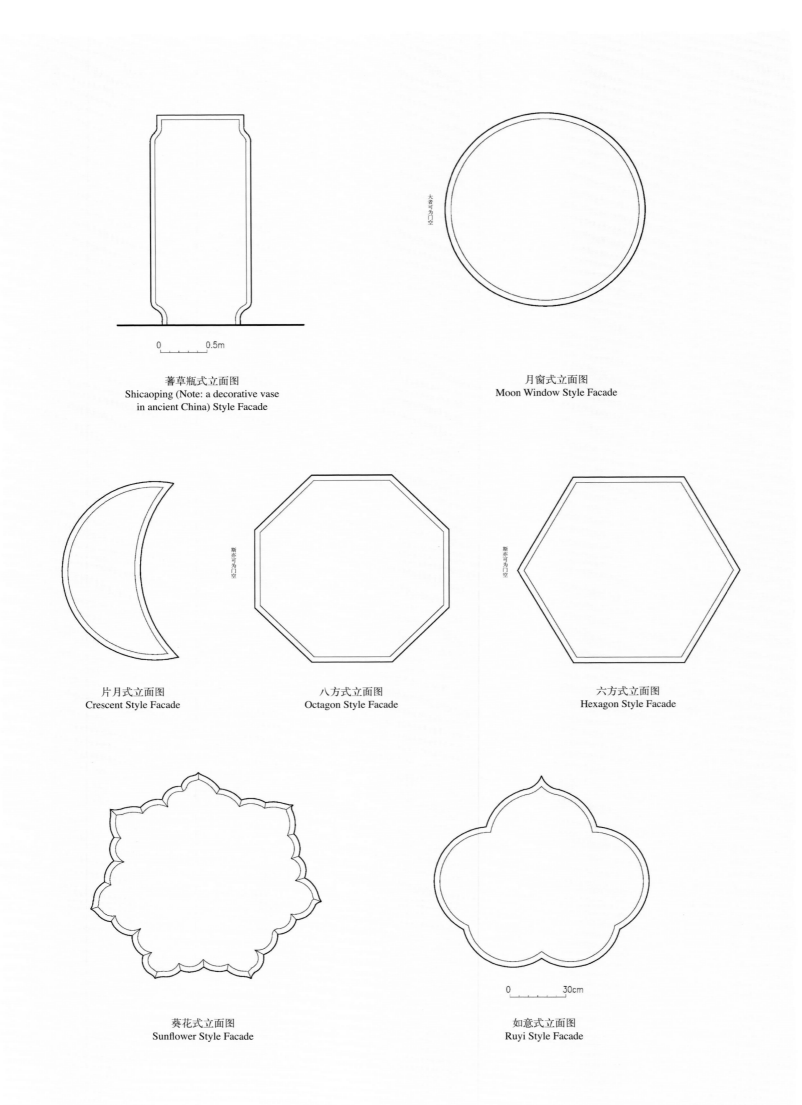

蓍草瓶式立面图
Shicaoping (Note: a decorative vase in ancient China) Style Facade

月窗式立面图
Moon Window Style Facade

片月式立面图
Crescent Style Facade

八方式立面图
Octagon Style Facade

六方式立面图
Hexagon Style Facade

葵花式立面图
Sunflower Style Facade

如意式立面图
Ruyi Style Facade

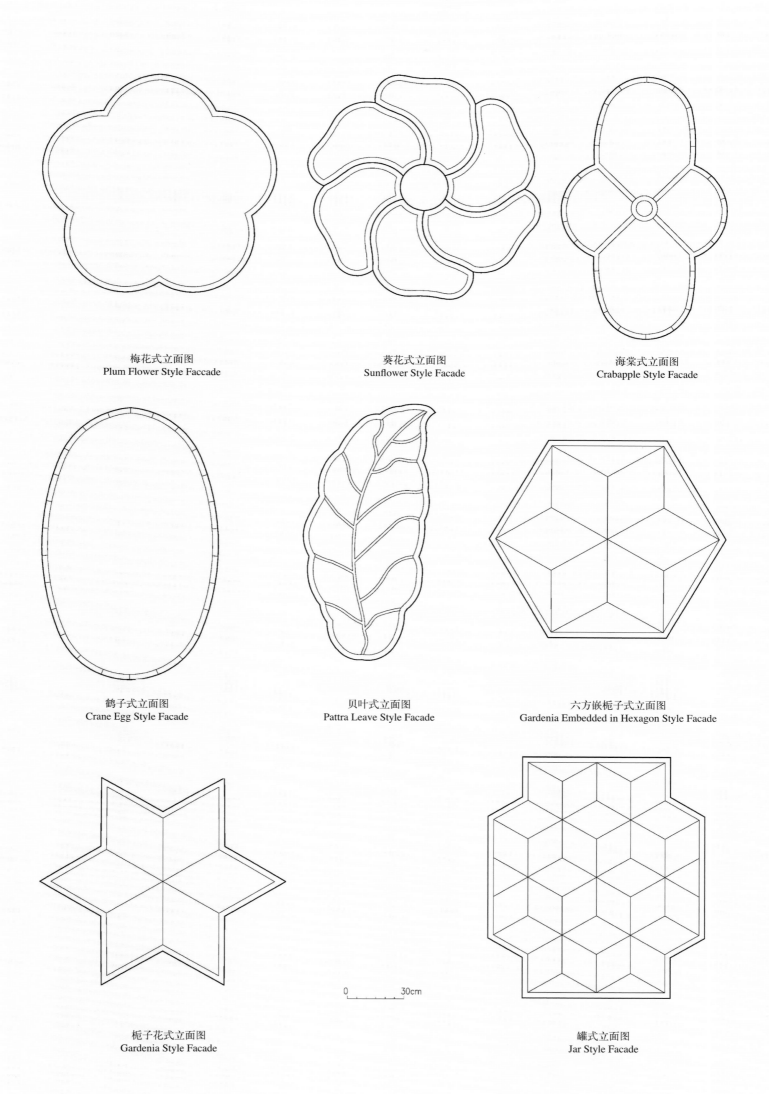

梅花式立面图
Plum Flower Style Faccade

葵花式立面图
Sunflower Style Facade

海棠式立面图
Crabapple Style Facade

鹤子式立面图
Crane Egg Style Facade

贝叶式立面图
Pattra Leave Style Facade

六方嵌栀子式立面图
Gardenia Embedded in Hexagon Style Facade

栀子花式立面图
Gardenia Style Facade

罐式立面图
Jar Style Facade

园冶图释

Yuan Ye Illustrated
—— Classical Chinese Gardens Explained

七 墙垣

7 Walls

凡园之围墙，多于版筑，或于石砌，或编篱棘。夫编篱斯胜花屏，似多野致，深得山林趣味。如内花端、水次、夹径、环山之垣，或宜石宜砖，宜漏宜磨，各有所制。从雅遵时，令人欣赏，园林之佳境也。历来墙垣，凭匠作雕琢花鸟仙兽，以为巧制，不第林园之不佳，而宅堂前之何可也。雀巢可憎，积草如萝，祛之不尽，扣之则废。无可奈何者。市俗村愚之所为也，高明而慎之。世人兴造，因基之偏侧，任而造之。何不以墙取头阔头狭，就屋之端正，斯匠主之莫知也。

Generally a garden wall is either made of dirt, or stone, or made of thorny bushes as hedgerows. A hedgerow is better than a flower wall, because it appears more natural and has a taste of wild landscape. In a garden, if a wall is built by a flower patch, a water bank, a path or a foothill, it could be made by stones, bricks, flowers, or ground bricks. The materials and construction methods may vary but the common goal is to make it most beautiful, elegant and enjoyable. However, some walls made by craftsmen were carved into different shapes such as flower plants, birds, beasts and supernatural beings. They might have been made very skillfully and looked quaint but they are very inappropriate to be used in a garden, or even in front of a building. These kinds of walls often attract sparrows to build nests and make pestering noise; they harbor uncontrollable wild weeds that are hard to be eradicated. If you hit them with force then you would damage the wall. You just can't do anything about it. These are all works of the laity, and the wise people should pay more attention. Generally people build walls based and limited by the shape and orientation of a site. One may try to build a wall that is narrow at one end and broad on the other to ensure that the main building is in a regular shape; however this is hard to comprehend for the normal craftsmen and construction lead.

（一）白粉墙

历来粉墙，用纸筋石灰，有好事取其光腻，用白蜡磨打者。今用江湖中黄沙，并上好石灰少许打底，再加少许石灰盖面，以麻帚轻擦，自然明亮鉴人。倘有污渍，遂可洗去，斯名"镜面墙"也。

(1) White Plaster Wall

A white plaster wall has been made with paper strip mixed lime mortar. However, some finical people strive for its smoothness and apply white wax first and then polish it. Now we use river sand mixed with quality lime for base and then spread an outer layer of lime. Finally we use hemp broom to lightly brush on it and make it naturally bright. Even if it gets dirty, we could clean it. This kind of wall is called "mirror wall".

（二）磨砖墙

如隐门照墙、厅堂面墙，皆可用磨成方砖吊角，或方砖裁成八角嵌小方；或砖一块间半块，破花砌如锦样。封顶用磨挂方飞檐砖几层，雕镂花、鸟、仙、兽不可用，入画意者少。

(2) Ground Brick Wall

A screen wall for hiding the main entrance or the wall facing a main hall can both use a ground brick wall. The brick can be obliquely patterned. One pattern is to cut the square brick into an octagonal shape and fill the gaps between the bricks with smaller square bricks. One sized brick interweaves with a half sized brick and that makes Jin Yun pattern (Note: a decorative cloud shape). On top of the wall, use a few layers of square shaped ground brick eaves. Carving flowers, birds, gods, beasts on the wall must be avoided, as this philistine practice rarely creates artistic image.

（三）漏砖墙

凡有观眺处筑斯，似避外隐内之义。古之瓦砌连钱、叠锭、鱼鳞等类，一概屏之，聊式几于左。

(3) Lattice Brick Wall

Wherever there is a distant scene, this kind of wall can be used, because this wall could serve the purposes of blocking outside peek and hiding internal scenes. In ancient times, people used tiles to form round-outer-square-inner coin like pattern, or mimic layered ancient money, or fish scale. We should not use those. Below I list several modern and elegant lattice brick patterns.

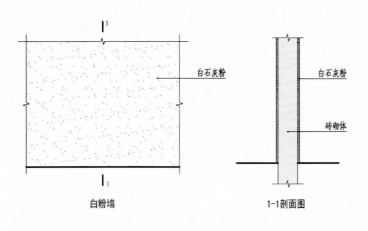

白粉墙　　　　　　1-1剖面图

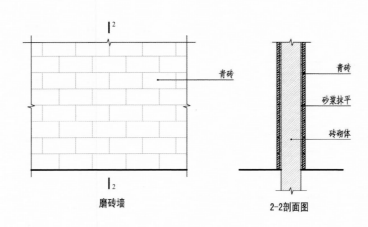

磨砖墙　　　　　　2-2剖面图

墙垣 WALLS

漏砖墙图式（图三—三十二至图三—四十七） Lattice Brick Wall Style (Figures 3-32 to 3-47)

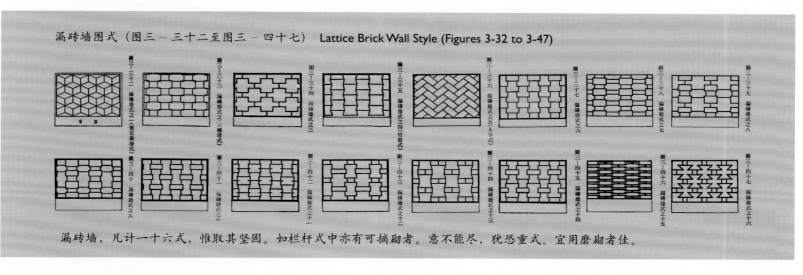

漏砖墙，凡计一十六式，惟取其坚固。如栏杆式中亦有可摘砌者。意不能尽，犹恐重式。宜用磨砌者佳。

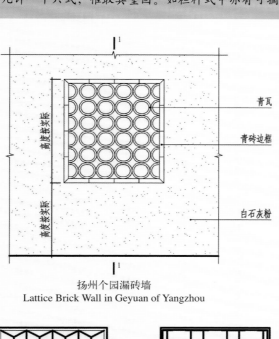

扬州个园漏砖墙
Lattice Brick Wall in Geyuan of Yangzhou

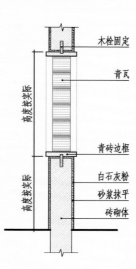

1-1 剖面图
1-1 Section View

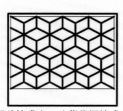

漏砖墙式之一（菱花漏墙式）
Lattice Brick Wall Style 1
(Diamond Flower Style)

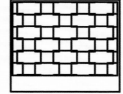

漏砖墙式之二（绦环式）
Lattice Brick Wall Style 1
(X-Ring Style)

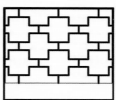

漏砖墙式之三
Lattice Brick Wall Style 3

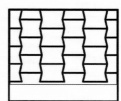

漏砖墙式之四（竹节式）
Lattice Brick Wall Style 4
(Bamboo Joint Style)

漏砖墙式之五（人字式）
Lattice Brick Wall Style 5
(Ren Character Style)

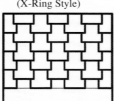

漏砖墙式之六
Lattice Brick Wall Style 6

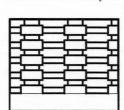

漏砖墙式之七
Lattice Brick Wall Style 7

漏砖墙式之八
Lattice Brick Wall Style 8

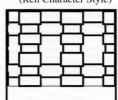

漏砖墙式之九
Lattice Brick Wall Style 9

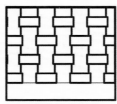

漏砖墙式之十
Lattice Brick Wall Style 10

漏砖墙式之十一
Lattice Brick Wall Style 11

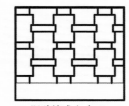

漏砖墙式之十二
Lattice Brick Wall Style 12

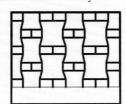

漏砖墙式之十三
Lattice Brick Wall Style 13

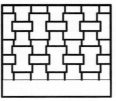

漏砖墙式之十四
Lattice Brick Wall Style 14

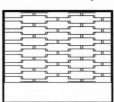

漏砖墙式之十五
Lattice Brick Wall Style 15

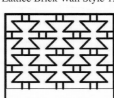

漏砖墙式之十六
Lattice Brick Wall Style 16

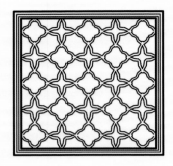

留园古木交柯前走廊之一
Corridor in front of Gumu Jiaoke at Liuyuan 1

留园古木交柯前走廊之二
Corridor in front of Gumu Jiaoke at Liuyuan 2

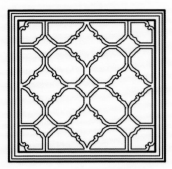

留园古木交柯前走廊之三
Corridor in front of Gumu Jiaoke at Liuyuan 3

狮子林燕誉堂北院走廊之一
Corridor in the yard north of Yanyu Tang in Shizilin 1

狮子林燕誉堂北院走廊之二
Corridor in the yard north of Yanyu Tang in Shizilin 2

狮子林燕誉堂北院走廊之三
Corridor in the yard north of Yanyu Tang in Shizilin 3

狮子林燕誉堂北院走廊之四
Corridor in the yard north of Yanyu Tang in Shizilin 4

狮子林燕誉堂北廊东端
The east end of the north corridor of Yanyu Tang in Shizilin

狮子林小方厅北廊东端
The small square room in the east end of the north corridor in Shizilin

沧浪亭瑶华境界东走廊
The east corridor of Yaohua Jingjie in Canglang Ting

沧浪亭瑶华境界西走廊
The west corridor of Yaohua Jingjie in Canglang Ting

漏砖墙实测图之一
Part 1 of the Actual Survey Drawing of Lattice Brick Wall

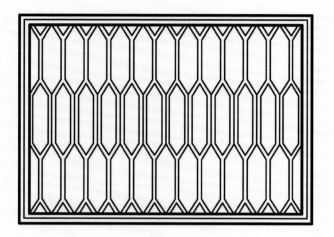

怡园拜石轩南院走廊
Corridor in the south yard of Baishi Xuan (Note: Rock Worship Building) of Yiyuan

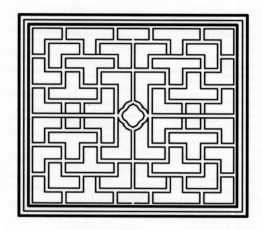

留园古木交柯前走廊之一
The front corridor of Gumu Jiaoke in Liu Yuan 1

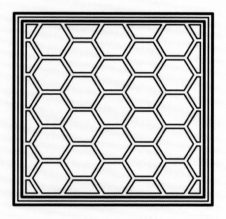

留园古木交柯前走廊之二
The front corridor of Gumu Jiaoke in Liu Yuan 2

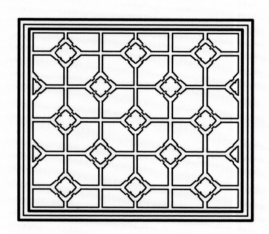

留园古木交柯前走廊之三
The front corridor of Gumu Jiaoke in Liu Yuan 3

留园古木交柯前走廊之四
The front corridor of Gumu Jiaoke in Liu Yuan 4

留园古木交柯前走廊之五
The front corridor of Gumu Jiaoke in Liu Yuan 5

漏砖墙实测图之二
Part 2 of the Actual Survey Drawing of Lattice Brick Wall

沧浪亭假山北游廊之一
The corridor north of the rockery in Canglang Ting 1

沧浪亭假山北游廊之二
The corridor north of the rockery in Canglang Ting 2

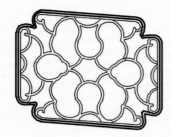
沧浪亭假山北游廊之三
The corridor north of the rockery in Canglang Ting 3

沧浪亭假山北游廊之四
The corridor north of the rockery in Canglang Ting 4

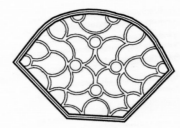
沧浪亭假山北游廊之五
The corridor north of the rockery in Canglang Ting 5

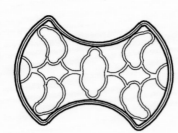
沧浪亭假山北游廊之六
The corridor north of the rockery in Canglang Ting 6

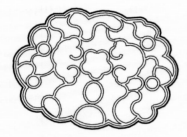
沧浪亭假山北游廊之七
The corridor north of the rockery in Canglang Ting 7

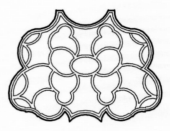
沧浪亭假山北游廊之八
The corridor north of the rockery in Canglang Ting 8

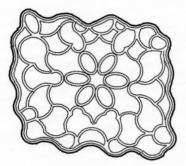
沧浪亭假山北游廊之九
The corridor north of the rockery in Canglang Ting 9

狮子林指柏轩东游廊
The corridor east of Zhibai Xuan in Shizilin

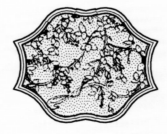
狮子林指柏轩西游廊
The corridor west of Zhibai Xuan in Shizilin

狮子林问梅阁后游廊
The corridor behind of Wenmei Ge in Shizilin

漏砖墙实测图之三
Part 3 of the Actual Survey Drawing of Lattice Brick Wall

（四）乱石墙
是乱石皆可砌，惟黄石者佳。大小相间，宜杂假山之间，乱青石版用油灰抿缝，斯名"冰裂"也。

(4) Random Masonry Wall
Any assorted stones can be used, and it is best to use yellow stone. Large and small piece should be interspersed and it is particularly appropriate to use it between rockeries. If blue flagstone was used, then lime powder with China wood oil stucco should be used to seal the cracks, and thus it is callde "ice crack wall".

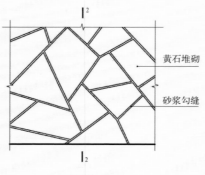
乱石墙
Random Masonry Wall

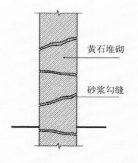
2-2 剖面图
2-2 Section View

园 冶 图 释

Yuan Ye Illustrated
— Classical Chinese Gardens Explained

八 铺地

8 Pavings

大凡砌地铺街，小异花园住宅。惟厅堂广厦中铺一概磨砖；如路径盘蹊，长砌多般乱石。中庭或宜叠胜，近砌亦可回文。八角嵌方，选鹅子铺成蜀锦；层楼出步，就花梢琢拟秦台。锦线瓦条，台全石版，吟花席地，醉月铺毡。废瓦片也有行时，当湖石削铺，波纹汹涌；破方砖可留大用，绕梅花磨斗，冰裂纷纭。路径寻常，阶除脱俗。莲生袜底，步出个中来；翠拾林深，春从何处是。花环窄路偏宜石，堂迥空庭须用砖。各式方圆，随宜铺砌，磨归瓦作，杂用钩儿。

Paving a road or ground is a bit different than that in a garden or house. For large buildings or halls, watermill square bricks should be used for all. If it is a winding small path then assorted stone paving should be used, and the path is long. A garden yard can use Die Sheng pattern (Note: a Chinese decoration pattern shaped like two interlocking diamond) and for areas close to stairs a palindrome pattern paving can be used. In inlaid octagonal pavements, cobblestones can be used to make Shu Jin pattern (Note: another Chinese decoration). At the front of a multi-storied building, a small terrace can be built bordering flowers. Tiles can be used for the outlines and slates for top surface. One could use the pavement as a mat when doing poetry recitation among flowers; others could use the stone pavers as a carpet when having a drink by the moon light. Leftover tiles can have their uses as they could be laid vertically by the lake rockery to form a pattern simulating rough waves. Brocken square bricks also have their uses as they could be used in a yard with plum flowers to form diverse and rich ice crack patterns. Although paving a road or path seems like a mundane construction job, when done well it could really make the ground decorations more refined and transcending. With such a wonderful paving, lotus flowers seem to bloom under your feet, and you feel as if walking on or from flower bouquets. You greet green everywhere in the forest and it is spring everywhere in the yard. The narrow paths among flowers are better paved with rocks while the empty areas around a building are more appropriate to have brick pavers. Patterns vary and they need to be adaptable depending on the context. The paving jobs are done by masons and other miscellaneous construction works are covered by unskilled laborers.

（一）乱石路

园林砌路，堆小乱石砌如榴子者，坚固而雅致，曲折高卑，从山摄壑，惟斯如一。有用鹅子石间花纹砌路，尚且不坚易俗。

(1) Random Masonry Path

To pave garden paths, only using small assorted rocks to form pomegranate seed like patterns can ensure its sturdiness and elegancy. No matter how it changes in direction and elevation, no matter whether it goes on a hill or along a gully, this method can all be applied. Some use cobblestones with intermittent other patterns; however, this is neither sturdy nor elegant.

（二）鹅子地

鹅子石，宜铺于不常走处，大小间砌者佳；恐匠之不能也。或砖或瓦，嵌成诸锦犹可。如嵌鹤、鹿、狮球，犹类狗者可笑。

(2) Cobblestone Path

Cobblestone paver is only suitable for those rarely used areas. It is better to interchangeably use both large and small cobblestones but this is hard to do for common craftsmen. One may also use bricks or tiles to form brocade patterns with cobblestones. If certain patterns such as cranes, deers, or lions playing with balls were attempted, then it would look pretty ridiculous.

（三）冰裂地

乱青版石，斗冰裂纹，宜于山堂、水坡、台端、亭际，见前风窗式，意随人活，砌法似无拘格，破方砖磨铺犹佳。

(3) Ice Crack Path

Using assorted quartzite slate as pavers to form an ice cracking pattern is suitable for ground on a hill, slope by water, terrace of a building, and empty area near a pavilion. For its patterns one can refer to those for afore mentioned Feng Chuang. As to the size and density, one can improvise according to actual situation. There is not fixed rule for paving, and a better application is to use broken square bricks with polished sides.

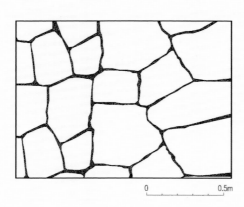

乱石路
Random Masonry Path

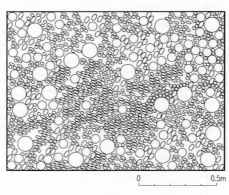

鹅子地
Cobblestone Path

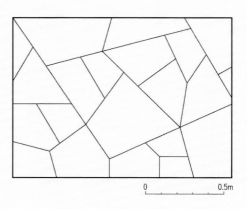

冰裂地
Ice Crack Path

(四)诸砖地

诸砖铺地,室内或磨扁铺,庭下宜仄砌。方胜、叠胜、步步胜者,古之常套也。今之人字、席纹、斗纹,量砖长短合宜可也。有式。

砖铺地图式
    砖铺地图式
(图三—四十八至图三—五十九)

(4) Brick Path

Various bricks can be used for pavers. Ground bricks can be laid flatly indoors. In a courtyard, however, it is more appropriate to lay ground bricks vertically. Patterns like Fang Sheng, Die Sheng, and Bubu Sheng were commonly used in the ancient times. Nowadays people like to use Ren character pattern, reed mat pattern, and cube pattern. It would work if the scale and size of the bricks are appropriate. The following shows some illustrations.

Brick Paving Patterns
Brick Paving Patterns (Figures 3 – 48 to 3 – 59)

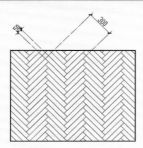  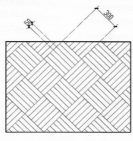 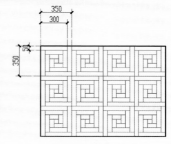

人字式	席纹式	间方式	斗纹式
Ren Character Style	Mat Pattern Style	Square Interlocking Style	Dou (Note: an ancient Chinese measurement device) Style

以上四式用砖仄砌
The above four patterns use brick side laying method.

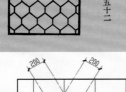 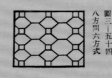

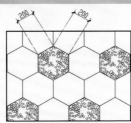 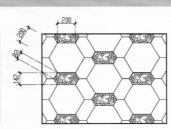 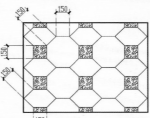 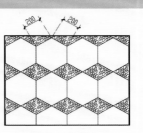

六方式	攒六方式	八方间六方式	套六方式
Hexagon Style	Stacked Hexagon Style	Octagon and Hexagon Interlocking Style	Cased Hexagon Style

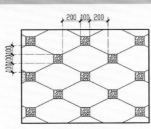 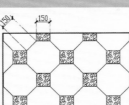 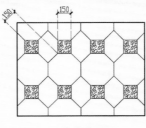 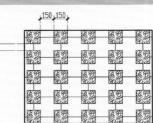

长八方式	八方式	海棠式	四方间十字式
Elongated Octagon Style	Octagon Style	Crabapple Style	Square and Cross Interlocking Style

以上八式用砖嵌鹅子砌
The above eight patterns use bricks inlaid with cobblestones.

香草边式（图三—六十）

用砖边、瓦砌。香草中或铺砖，或铺鹅子。

Vanilla Border Pattern (Figure 3 – 60)

The vanilla border style is to use bricks at the border and use tiles to lay in vanilla pattern. In the center area, either bricks or cobblestones can be used.

圖三—六十　香草邊式

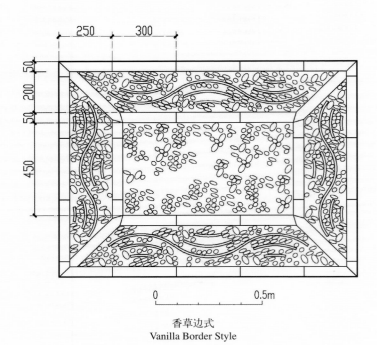

香草边式
Vanilla Border Style

球门式（图三—六十一）

鹅子嵌瓦，只此一式可用。

Ball Goal Pattern (Figure 3 – 61)

Inlaying cobblestones inside tile made ball goals. This is the only style.

圖三—六十一　毬門式

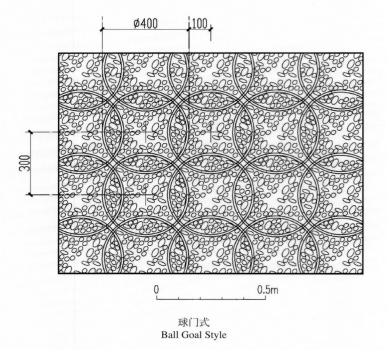

球门式
Ball Goal Style

波纹式（图三—六十二）

用废瓦捡厚薄砌，波头宜厚，波傍宜薄。

Ripple Pattern (Figure 3 – 62)

This is to use broken and leftover tiles of varying thickness. Use the thick ones at the wave start and the thinner one at its sides.

圖三—六十二　波紋式

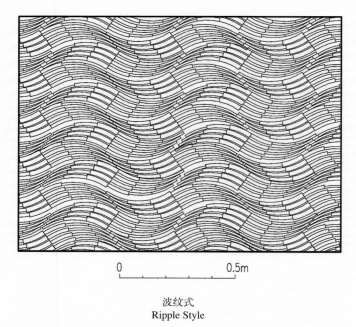

波纹式
Ripple Style

狮子林修竹阁
Xiuzhu Ge in Shizilin

狮子林燕誉堂
Yanyu Tang in Shizilin

狮子林古五松园
Ancient Five Pine Garden in Shizilin

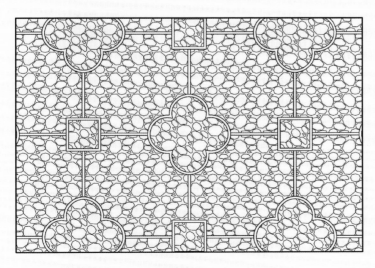
狮子林指柏轩
Zhibai Xuan in Shizilin

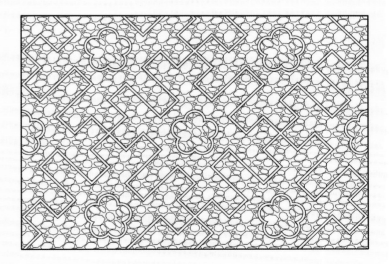
狮子林荷花厅
Hehua Ting (Note: Lotus Hall) in Shizilin

铺地图案实测图之一
Part 1 of Actual Pavement Survey Drawing

留园东园一角
A Corner of East Garden in Liuyuan

狮子林小方厅
Little Square Hall in Shizilin

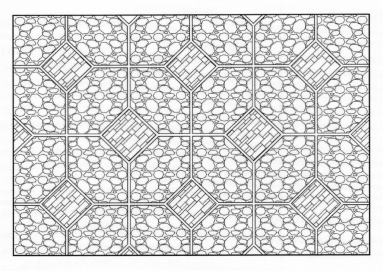
留园东园一角
A Corner of East Garden in Liuyuan

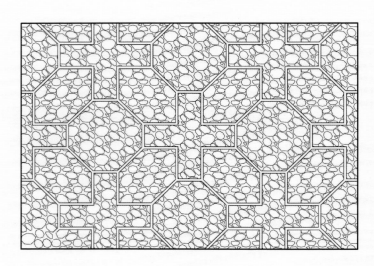
狮子林燕誉堂
Yanyu Tang of Shizilin

拙政园枇杷园
Pipa Yuan (Note:Loquat Garden) in Zhuozheng Yuan

铺地图案实测图之二
Part 2 of Actual Pavement Survey Drawing

铺地工程做法，地基夯实，其上垫 50mm 细土，砖、瓦、坐灰浆排砌图案，内嵌鹅卵石。
Pavement Construction Details: After the base is compacted, lay 50 mm deep fine dirt, and use bricks, tiles, and mortars to lay out design patterns, with cobblestones embedded.

园冶

Yuan Ye Illustrated
— Classical Chinese Gardens Explained
III

吴肇钊　陈　艳　吴迪　著
By Wu Zhaozhao, Chen Yan and Wu Di
马劲武　译（英文）
English Translation by Jinwu Ma

图释

【下卷】

中国建筑工业出版社

目 录

序一——孟兆祯 6
序二——陈晓丽 9
序三——王绍增 11
序四——刘秀晨 13
序五——甘伟林 王泽民 15
序六——王金满 18
自序一 20
自序二 22
《园冶图释》缘起 24
《园冶》冶叙——阮大铖 26
《园冶》题词——郑元勋 26
《园冶》自序——计成 28
兴造论 001
园说 013

一 相地 025
（一）山林地 045
（二）城市地 055
（三）村庄地 081
（四）郊野地 084
（五）傍宅地 095
（六）江湖地 109

二 立基 121

三 屋宇 141
（一）门楼 152
（二）堂 153
（三）斋 154
（四）室 157
（五）房 158
（六）馆 159
（七）楼 161
（八）台 169
（九）阁 174

（十）亭 176
（十一）榭 190
（十二）轩 191
（十三）卷 194
（十四）广 195
（十五）廊 199
（十六）五架梁 201
（十七）七架梁 202
（十八）九架梁 203
（十九）草架 204
（二十）重椽 204
（二十一）磨角 205
（二十二）地图 207

四 装折 209
（一）屏门 210
（二）仰尘 211
（三）户槅 212
（四）风窗 222

五 栏杆 225
栏杆图式 226

六 门窗 245
门窗图式 246

七 墙垣 253
（一）白粉墙 254
（二）磨砖墙 254
（三）漏砖墙 254
（四）乱石墙 258

八 铺地 259
（一）乱石路 260
（二）鹅子地 260

（三）冰裂地 260
（四）诸砖地 261
砖铺地图式 261

九 掇山 265
（一）园山 272
（二）厅山 276
（三）楼山 279
（四）阁山 281
（五）书房山 282
（六）池山 283
（七）内室山 284
（八）峭壁山 285
（九）山石池 286
（十）金鱼缸 286
（十一）峰 289
（十二）峦 292
（十三）岩 293
（十四）洞 294
（十五）涧 295
（十六）曲水 296
（十七）瀑布 297

十 选石 301

十一 借景 311
《园冶》自识 329
海外交流 331
主要参考文献 365
跋一 366
跋二 369
译后记 370
鸣谢 374

CONTENTS

Preface 1, by Meng Zhaozhen 7
Preface 2, by Chen Xiaoli 10
Preface 3, by Wang Shaozeng 12
Preface 4, by Liu Xiuchen 14
Preface 5, by Gan Weilin,
Wang Zemin 16
Preface 6, by Wang Jinman 19
Preface by Author 1, by Wu Zhaozhao
.... 21
Preface by Author 2, by Chen Yan,
Wu Di 23
The Origin of *Yuan Ye Illustrated* 25
Preface of *Yuan Ye*, by Ruan Dacheng ... 26
Inscription of *Yuan Ye*,
by Zheng Yuanxun 27
Preface of *Yuan Ye*, by Ji Cheng 28
On Design and Construction 001
On Garden Design 013

1. Siting 025
(1) Hilly Forest Land 045
(2) Urban Land 055
(3) Village Land 081
(4) Suburban Land 084
(5) Residence Side Land 095
(6) River and Lake Land 109

2. Base 121

3. Buildings 141
(1) Men Lou 152
(2) Tang 153
(3) Zhai 154
(4) Shi 157
(5) Fang 158
(6) Guan 159

(7) Lou 161
(8) Tai 169
(9) Ge 174
(10) Ting 176
(11) Xie 190
(12) Xuan 191
(13) Juan 194
(14) Guang 195
(15) Lang 199
(16) Five-Frame Girder 201
(17) Seven-Frame Girder 202
(18) Nine-Frame Girder 203
(19) Cao Jia 204
(20) Chong Chuan (Double Rafter) 204
(21) Mo Jiao 205
(22) Floor Plan 207

4. Decorations 209
(1) Screen Doors 210
(2) Yang Chen 211
(3) Hu Ge 212
(4) Feng Chuang 222

5. Railings 225
Railing Patterns 226

6. Doors & Windows 245
Door & Window Illustrations 246

7. Walls 253
(1) White Plaster Wall 254
(2) Ground Brick Wall 254
(3) Lattice Brick Wall 254
(4) Random Masonry Wall 258

8. Pavings 259

(1) Random Masonry Path 260
(2) Cobblestone Path 260
(3) Ice Crack Path 260
(4) Brick Path 261
Brick Paving Patterns 261

9. Rockery 265
(1) Yuan Rockery 272
(2) Ting Rockery 276
(3) Lou Rockery 279
(4) Ge Rockery 281
(5) Study Rockery 282
(6) Chi Rockery 283
(7) Indoor Rockery 284
(8) Cliff Rockery 285
(9) Rockery Pond 286
(10) Gold Fish Pond 286
(11) Peaks 289
(12) Luan 292
(13) Yan 293
(14) Caves 294
(15) Jian 295
(16) Meandering Streams 296
(17) Waterfalls 297

10. Rock Selection 301

11. Scene Leveraging 311
Postscript of *Yuan Ye* 329
Overseas Exchange Activities 331
Major Reference 365
Postscript 1 368
Postscript 2 369
Postscript by
the English Translator 372
Acknowledgement 374

目 次（日本語）

序一　孟兆禎
序二　陳暁麗
序三　王紹増
序四　劉秀晨
序五　甘偉林　王沢民
序六　王金満
自序一
自序二
「園冶図釈」縁起
冶叙――阮大鋮
題詞――鄭元勛
自序――計成
興造論
園説

一　相地
（一）山林地
（二）城市地
（三）村荘地
（四）郊野地
（五）傍宅地
（六）江湖地

二　立基

三　屋宇
（一）門楼
（二）堂
（三）斎
（四）室
（五）房
（六）館
（七）楼
（八）台
（九）閣

（十）亭
（十一）榭
（十二）軒
（十三）巻
（十四）広
（十五）廊
（十六）五架梁
（十七）七架梁
（十八）九架梁
（十九）草架
（二十）重椽
（二十一）磨角
（二十二）地図

四　装折
（一）屏門
（二）仰塵
（三）戸槅
（四）風窓

五　欄杆
欄杆図式

六　門窓
門窓図式

七　墻垣
（一）白粉墻
（二）磨磚墻
（三）漏磚墻
（四）乱石墻

八　舗地
（一）乱石路
（二）鵝子路
（三）氷裂路
（四）諸磚地

磚舗地図式

九　掇山
（一）園山
（二）廳山
（三）楼山
（四）閣山
（五）書房山
（六）池山
（七）内室山
（八）峭壁山
（九）山石池
（十）金魚缸
（十一）峰
（十二）巒
（十三）岩
（十四）洞
（十五）澗
（十六）曲水
（十七）瀑布

十　選石

十一　借景

自識
海外交流
参考文献
跋一
跋二
後書き
謝辞

园冶图释

Yuan Ye Illustrated
——Classical Chinese Garden Explained

九 掇山

9 Rockery

摄山之始，桩木为先，较其短长，察乎虚实。随势挖其麻柱，谅高挂以称竿。绳索坚牢，扛台稳重。立根铺以粗石，大块满盖桩头；堑里扫于查灰，着潮尽钻山骨。方堆顽夯而起，渐以敛文而加；瘦漏生奇，玲珑安巧。峭壁贵于直立，悬崖使其后坚。岩、峦、洞、穴之莫穷，涧、壑、坡、矶之俨是。信足疑无别境，举头自有深情。蹊径盘且长，峰峦秀而古。多方景胜，咫尺山林，妙在得乎一人，雅从兼于半土。假如一块中竖而为主石，两条傍插而呼劈峰，独立端严，次相辅弼，势如排列，状若趋承。主石虽忌于居中，宜中者也可；劈峰总较于不用，岂用乎断然。排如炉烛花瓶，列似刀山剑树，峰虚五老，池凿四方，下洞上台，东亭西榭。㙷堪窥管中之豹，路类张孩戏之猫。小藉金鱼之缸，大若酆都之境。时宜得致，古式何裁？深意画图，余情丘壑。未山先麓，自然地势之嶙嶒；构土成冈，不在石形之巧拙。宜台宜榭，邀月招云；成径成蹊，寻花问柳。临池驳以石块，粗夯用之有方；结岭挑之土堆，高低观之多致。欲知堆土之奥妙，还拟理石之精微。山林意味深求，花木情缘易逗。有真为假，做假成真；稍动天机，全叨人力；探奇投好，同志须知。

When a rockery construction starts, pile wood should be set in the ground. The depth of the piles is determined by the degree of solidity of the dirt ground. Based on the actual situation, construction workers dig dirt to erect the piles, and set up the hoisting structure. The ropes for lifting the rocks should be sturdy enough and the tying of the rope should be firm and secure. The lifting and dropping of the rocks should be steady. Rubbles need to be laid first and larger rocks need to cover the pile heads. The base pit needs to be filled with smashed limes. If it is too wet rocks should replace smashed limes. Rockery construction should start with tamping stones and then gradually built up with real textured rocks according to artistic painting theory. In this way, the intriguing "Shou" (thin) and "Lou" (hollow) properties of the rocks may show naturally. Its exquisite appearance lies in clever application of the rocks textures. If one wants to make a cliff, the rock needs to be upright. To build a precipice, its back end must be rooted securely. Rocks and caves should be deep and unfathomable. Gullies and hillocks should appear natural and real. You may feel that you are at a dead end at times, but looking around you may discover a complete new world. Paths should be winding and long; rock peaks should be beautiful and unique. Careful arrange the landscapes around and so you have easy access to them. Although the wonders of the garden are mostly accredited to one person (the designer), a lot should also go to the use of "semi-dirt" (i.e. rocks built on top of dirt). If we erect one big rock in the middle and two smaller ones at the side, then the priority order shows, just the same relationship as a monarch and his subjects. Even though the main rock generally is not supposed to be placed in the middle, if situation requires, it is also acceptable. In this situation, it is better not to use the subordinate rocks. Who said we must use them? A rigid and awkward arrangement of rocks is just like candles and vases placed in a shrine, or aligned weapon stands at an execution ground. Mountains would be like a repetitive array element and water pond would be in a dull and monotonous square shape. Similarly, cave under and terrace upper, a pavilion at east and a Xie at west. The holes on the rockeries are so tiny that not much can be seen through. The paths are so tortuous that visitors go astray and get lost. The small mound is built like toy rock in a gold fish tank, and large mounds are like those in hell. Some contemporary folks think this is graceful and elegant. Where on earth can we find gardens like this in ancient times? Therefore I argue that to build a rockery one should have a careful overall pre-conception of the natural landscape and master the characteristics of the appearance and texture. Before building it up, first set the base well and the momentum grows naturally. Then add dirt to form a hillock not to worry about the look of the rocks. Build a Tai to enjoy a drink over the moon or Xie to invite the clouds for a dance, wherever it is appropriate. Build a path based on the terrain and plant flowers depending on situations. Hard rocks should be laid to form water banks and more dirt should be added to create interests at high and low grounds. To learn the secret of making a dirt mound one should first learn the essence of mixed use of dirt and rocks. It is worth of a deeper study about the natural landscape and it is relatively easy to learn about the characteristics of plants. If you get the essence of real mountains then when you build an imitated one it would look real. To build an artificial rockery, ingenuity is needed for its conception but unending effort is a must for its completion. This is something that we as garden design enthusiasts must understand.

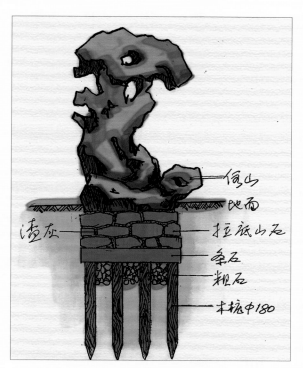

假山基础图（丛置山石基础则拉底山石扫以渣灰即可）
Rockery Base

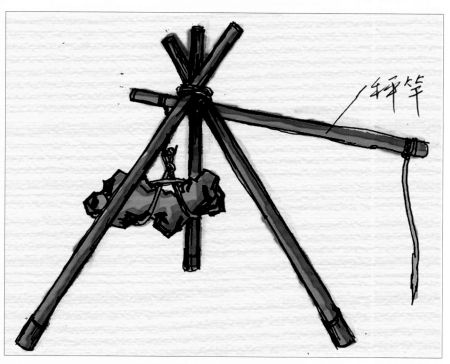

吊石秤架
Hoisting Structure for Hanging Rocks

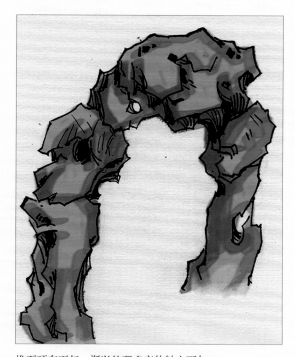

堆石顽夯而起，渐以纹理多变的皴文而加。
Rockery construction should start with tamping stones and then be gradually built up with real textured rocks according to artistic painting theory.

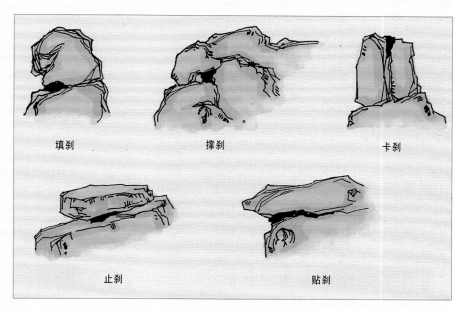

刹是指石之间的石垫块，起到稳定作用
Sha means a rock wedge between rocks; its purpose is to stabilize.

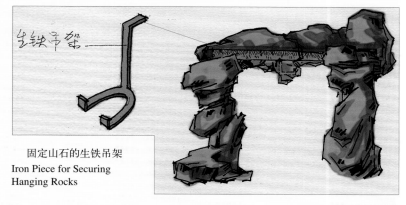

固定山石的生铁吊架
Iron Piece for Securing Hanging Rocks

用石梁架洞法
Using a Rock Beam to Form a Cave

扣石绳结法
Method for Tying Ropes for Hanging Rocks

移动大石的方法
Method to Move Large Rocks

吾师孟兆祯院士是国内研究掇山叠石的权威,他将北方、南方民间传统山石结体总结归纳为十种基本形式,即挑、垂、悬、撑、卡、斗、挎、连、接、拼。下面图文皆孟师指导完成。

挑:又称"出挑",即上石借下石为支承而挑伸于外,再以数倍于上石重量之山石平衡之。

垂:从一山石上面之企口处,用另一山石之企口从上面相接而垂挂其下端称"垂"。由于山岫的衬托,轮廓分外明显。

悬:在下层山石共同组成的竖向洞口上,放进一块上大下小的山石,从竖向洞口悬于当空的做法。

撑:即用斜撑的力量来加固于山石,撑石要掌握合宜的支承点,加撑以后在外观上犹如自然的整体。

卡:下面用两块或两组山石组成。二石各以斜面相对而不相连,形成一个上大下小的缺口,再用一块上大下小的山石从缺口中徐徐放下,这块山石因卡住而稳定,外观十分自然。

斗:置石成拱状,腾空而起,两端搭架于二石之间。

挎:如石之某侧平板,可旁挎一石以全其美,挎石常用铁件加固。

连:山石之间水平搭接称为"连"。要使其方向进出多变,高低错落,间距不等方可生巧。

接:竖向叠接以构成一大块完整的山石形象。

拼:由于空间大小的差别,当石材太小,可以将数块甚至数十块山石拼掇成一整体山石的外观。

My mentor, Academician Meng Zhaozhen is an authority on rockery building in China. He summarized the northern and southern Chinese rockery building practices and generalized ten basic categories: Tiao, Chui, Xuan, Cheng, Ka, Dou, Kua, Lian, Jie, and Pin. The following graphics and texts were completer under Professor Meng's supervision.

Tiao, also referred to as "Chu Tiao", means the upper rock hangs over the lower rock, from which it gets the support at the base. A balance is achieved by a counterd weight that weighs several times as that of the upper rock.

Chui, from a tongued and grooved connection point of one rock connects to another rock's tongued and grooved connection and its drooping part is called "Chui". Set off by the main peak, the drooped rock's silhouette becomes very distinctive.

Xuan, at the vertical hole formed jointly by lower rocks, place a rock, whose upper part is larger than its lower part, through the hole and dangles in the air.

Cheng, is to reinforce the rockery by buttress. The key is to find proper supporting points and after connection, it should appear natural as a whole.

Ka, is made of two rocks at the base and one on top. The two bottom rocks have their bevel faces up forming a gap. From the top, the third rock, larger upper and smaller lower, is lowered down slowly and finally stuck in the gap steadily, with a very natural appearance.

Dou, refers to an arch formation of rocks. It rises up in the air, with both feet landing on another pair of rocks.

Kua, is at a side of a rock, which spans another one for a complete balanced composition. The spanned rock usually needs some iron piece, for security.

Lian, is a horizontal connection between rocks. The key is to have variant orientations, elevations, and distances to make it intriguing and perfect.

Jie, is the connection by stacking rocks vertically to form a larger and complete rockery image.

Pin, relates to a method of splicing several or many smaller rocks to form a larger and more complete rockery. This is done mostly because of space differences.

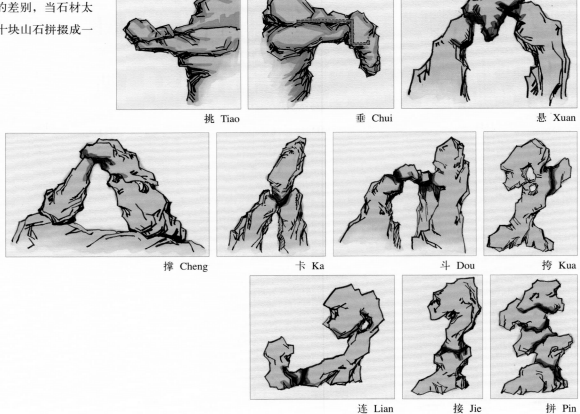

挑 Tiao　　垂 Chui　　悬 Xuan

撑 Cheng　　卡 Ka　　斗 Dou　　挎 Kua

连 Lian　　接 Jie　　拼 Pin

岩、峦、洞、穴（扬州片石山房）
Yan, Luan, Dong, Xue (Pianshi Shanfang, Yangzhou)

高远景观
High and Distant View

涧、壑、坡、矶（南京瞻园）
Jian, He, Po, Ji (Zhan Yuan, Nanjing)

深远景观
Deep and Distant View

自然界山体急徐、曲深等变化
The Abrupt and Relaxed, Curved and Deep Changes in Natural Mountains

平远景观
Flat and Distant View

中国山水画追求"高远、深远、平远"三远境界，园林掇山山意境亦立足三远造诣。

The traditional Chinese landscape paintings seek three distant views – high and distant, deep and distant, flat and distant – and the mound and rockery constructions in a garden should also have the same pursuits.

两石以上的组合需主次分明、顾盼生情。

Combinations of more than two pieces of rocks should have priority levels and corresponding relationships.

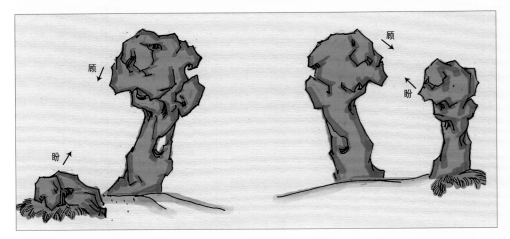

二石的组合（顾盼生情）
Combination of two rocks (reacting to each other to generate 'affections')

相背的呼应
Pay attention to the directions that a rock faces to

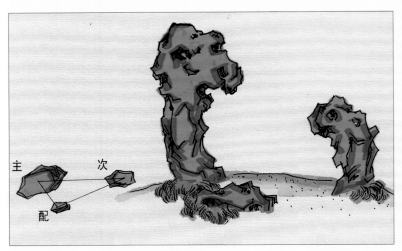

三石的配合
Combinations of three rocks

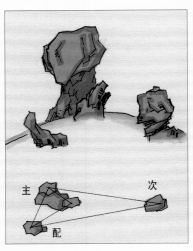

不等边三角形类型（一）
Scalene triangle type 1

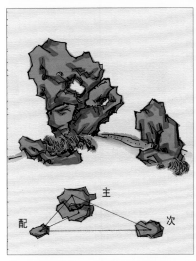
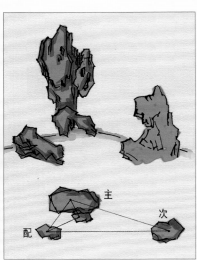

不等边三角形类型（二）
Scalene triangle type 2

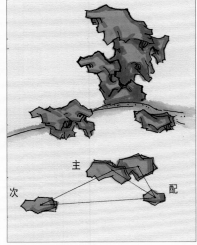
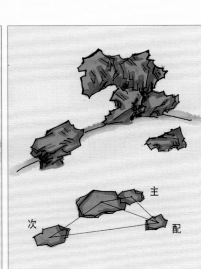

不等边四边形类型
Trapezoid type

掇山败笔大体归纳为以下几种，见图示：
Failed examples of rockery building are listed as below:

堆山像刀山剑树
Rockery looks like a weapons' stand

堆山过于追求象形，俗不可耐
Rockery seeks to resemble something, a poor taste

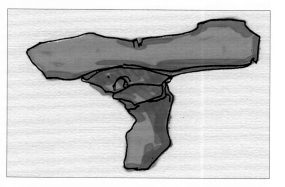

竖峰像蜻蜓
The vertical rock looks like a dragonfly

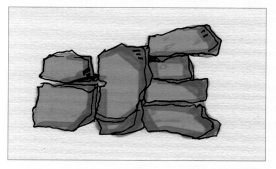

叠石像垒墙
Rockery built like a retaining wall

山石组成犬牙交错
Rockery built like interlocking teeth

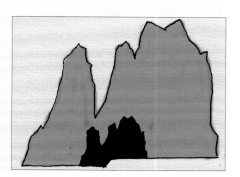

堆山像模型
Rockery made like a model

临池驳以石块，粗夯用之有方
Hard rocks should be laid to form water banks

叠石嶙峋，成径成蹊
No matter how a rockery is built, a path should be made according to terrain change

未山先麓，自然地势之嶙峋
Build the foot of a hill before the hill proper and proceed to higher part

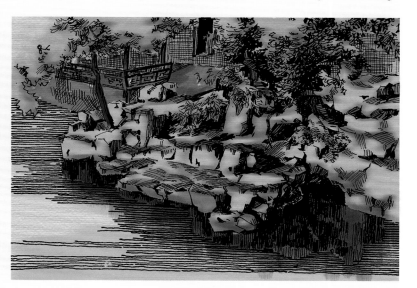

亭台前山石要求成径成蹊
Proper paths should be built in front of a terrace or pavilion

(一) 园山

园中掇山，非士大夫好事者不为也。为者殊有识鉴。缘世无合志，不尽欣赏，而就厅前三峰，楼面一壁而已。是以散漫理之，可得佳境也。

(1) Yuan Rockery

To build a rockery inside a garden is really something that can only be accomplished by a garden connoisseur, who has the knowledge, the gust and appreciation for such works. Because there are very few like-minded people who possess this taste and caliber, a garden rockery is often reduced to three peaks in front of a hall, or a rock cliff facing a building. However, if we follow the nature's rule to make a rockery varying in their heights with a balanced composition, we could create a beautiful and elegant work of art.

计成一生活动之地江苏，其宅园均系士大夫退居林下寄情山水的园林，故园林的主体是山水，小中见大的山水景观既是以名人山水画为蓝本，同时又体现出造园家"有真为假，做假成真"的技艺。现存的园林山水诣高者有二，其一，以春、夏、秋、冬四季为特色创意独特的四季假山园"扬州个园"；其二，戈裕良手笔的"苏州环秀山庄"，远看势、近看质均达到以假乱真的效。苏州环秀山庄如下图。

The main activity area of Ji Cheng in his life time is Jiangsu, and his residential gardens all fall into the category of secluded landscape empathy gardens of retired or exiled officials. Therefore, the main body of a garden is mountain and water; the "big from small" garden scenes are based on famous landscape paintings and reflect the designer's mastery skills to "use real as fake and make fake as real". There are two existing well accomplished landscape gardens around. One is Geyuan in Yangzhou, which is a unique and creative four-season characterized rockery garden. The other is Huanxiu Shanzhuang of Suzhou, which has the signatures of Ge Yuliang. Seeing it from afar at its momentum or enjoying it up close at its texture, we could be almost unable to tell it from a real landscape. Illustrations of HuanXiu Shangzhuang are as follows.

未山先麓的坡岩十分逼真，其冈峦又向主峰有所奔趋，体现"山欲动而势长"，有奔腾跃动气派，章法非凡

Mountain foot goes before the mountain, and they appear to move towards the main peak. This is a realization of "a seemingly moving mountain with a moving momentum". The dynamic air is extraordinary.

叠石外挑，巉岩壁立，"以近求高"手法奏效，悬崖峭壁突兀惊人，幽涧深隐，左峙池水，水动山摇，令人深悟跃入深山大泽

Piled rocks hang over and the side cliff is in perilous position. It works "to use close range to seek effect of high elevation". The steep cliff appears lofty, and the secluded stream is hidden deep. As the water in the left pond flows and the rock mound seems to shiver. This makes one feel that he lives in a real deep mountain.

环秀山庄假山仅一亩之地，能岩峦耸翠，池水映天，气势磅礴，巧夺天工。"咫尺山林"如真山，入画意，堪称"艺术精品"

The size of Huanxiu Shanzhuang is only one Mu, yet its verdant mound and water are great and momentous, and excel what nature can offer. The "miniature landscape" is like a real one and very picturesque; it deserves to be regarded as a masterpiece.

石壁占地甚微，却有洞、壑、涧谷、悬崖、玲珑有致，自成一体，而又融入主山之中，设意奇特，顿置婉转入洞

Although the rockery covers very small area of land, it has got a cave, a gully, a stream, a cliff etc. Exquisite and orderly, it's self-contained and blended into the main mountain with characteristic design idea. A grotto cave is readily available.

扬州个园四季假山

个园内假山创作传为大师石涛手笔，由于画师具备遍游名山大川的经历和高超的艺术造诣，故能根据扬州山石皆从外地运来，品种杂体量小的特点，独出心裁创作出分峰用石的写意山水境域"四季假山"。

春：修竹迎面，石笋参差而立，构成以粉墙为纸，竹石为绘的画面。入园门则是"十二生肖"象形山石，象征春天到了，动物活动频繁，这是一种"移情"的手法。

夏："夏云多奇峰"，夏山外形为"叠石停云"，多层次叠石使之"步移景异，变幻无穷"。与之相映成趣的涧溪、曲桥、深邃洞府令人置身期间，一身顿爽，两腋生风。

秋：着笔气势磅礴，用石泼辣，虽拔地数仞，然峻峭依云，山路崎岖，时洞时天，时壁时崖，时涧时谷，引人入胜，上下盘旋，造意极险，其突兀惊人景观达到"以假乱真"的艺术效果。

冬：位于园墙北面宣石配置的冬山是寓意积雪未化。当人们叹赏"瑞雪"之时，偶尔阵风吹拂，哨声隐约，原是墙面四排圆洞，每排六个，在外面高墙窄巷的负压作用下产生风动效应。更妙的是西边围墙上两个圆窗，远远招来修竹"春笋"，冬去春来，似无止境，可谓匠心独运。

The Four-Season Rockery of Ge Yuan, Yangzhou

The rockery work in Ge Yuan is said to be that of the great master Shi Tao. Since the painter artist traveled widely and visited all the famous mountains and rivers and with great artistic attainments, he could create exotic "Four Season Rockery" with ingenuity on an impressionistic landscape style, using mixed and small rocks imported from elsewhere.

Spring: Slender bamboos greet visitors. Stalagmite-like rocks stand at varying heights. The white paint wall serves as the background and bamboos and rocks are the contents of the live painting. Once entering the garden entrance, you see the "12 Chinese Zodiac Animal Symbol" pictographic rocks, representing the arrival of spring, when animals become activated. This is a method of "empathy".

Summer: "Summer clouds gather at numerous rare peaks." The conformation of the summer rockery is intended to "stack rocks to attract clouds". Multi-layered stacked rocks make it possible to "change views at every step with infinite possibilities". Co-existing with the rocks are streams and gullies, zigzag bridges, and deep caves where once you are there, you feel chilled and refreshing with cool breeze around you.

Autumn: With magnificent and forceful strokes with rocks, even though it is only a few Rens (Note: An ancient Chinese measurement unit, which is about seven or eight feet long), it feels as if the peaks can reach the clouds. The rock paths are rugged, and walking on it you sometimes see the sky and other times only a hole on top to peek through. Passing by gullies and valleys and trekking on winding paths, you are attracted by its seemingly perilous views. Its unexpected and towering scenes can seriously make you doubt that this is a man-made rockery.

Winter: The quartzite rockery north of the garden wall is intended to imitate mountains with accumulated snow. When people marvel at the "snow" they would hear the wuthering sound on and off. This is actually caused by four rows of holes, each containing six, on the wall. Due to the negative pressure in the narrow alley outside, it makes such noise. A more wonderful design is the two round holes on the west, with "spring bamboo" scenes peeking through, hinting the arrival of spring after winter and continuous seasonal cycle, an ingenious idea indeed.

春——春笋"春山淡冶而如笑"

Spring – The spring bamboo shoots growing on "spring mountains that seem to smile"

夏——夏云"夏山苍翠而如滴"

Summer – The summer cloud on "summer mountain that's among luxuriant green"

秋——秋爽"秋山明净而如妆"
Autumn – The cool autumn shown on "autumn mountain that's clear and seemingly adorned"

冬——积雪"冬山惨淡而如睡"
Winter – The accumulated snow on "winter mountain make one feel gloomy and dormant"

（二）厅山

人皆厅前掇山，环堵中耸起高高三峰排列于前，殊为可笑。加之以亭，及登，一无可望，置之何益？更亦可笑。以予见或有嘉树，稍点玲珑石块；不然，墙中嵌理壁岩，或顶植卉木垂萝，似有深境也。

(2) Ting Rockery

People often build rockeries in front of a Ting (Note: hall). It is ridiculous to erect a three-peak rockery too neatly aligned, surrounded in an enclosed yard with walls at four sides. It is even more ridiculous to build a pavilion on top of the rockery since there is no view to enjoy from the pavilion. In my opinion, if there are beautiful trees one might want to place a few exquisite rocks around. Otherwise, embed a cliff on a wall and grow some vines on top to let their branches hang. This may create some natural landscape look scenes.

江南园林的厅均与庭院组合，鉴于面积不大，故多采用山石小品类手法造景，配植卉木垂萝，塑造出饱含深境的画幅。

此部分均选用江南园林现存不同类型的实例佳作，用画作形式表现出来，力求体现出计成的创作宗旨。

A Ting in Southern Chinese gardens always integrates with a yard. Since it is not large, using small garden pieces such as small rocks, coupled with flowers and vines, can create pretty and meaningful scenes.

This part shows existing exemplar of Southern Chinese gardens in painting format, striving to reflect Ji Cheng's creative ideas.

芜湖"翠明园"主建筑"映壑"厅外眺的景观为山水景"鸣珮"，突出该园延山引泉的造园手法

Outside the main building "Yinghe" (Note: Gully Reflection) Hall of "Cuiming Yuan" in Wuhu, the main scene is a mound-water combo "Ming Pei", manifesting the garden building techniques of extending mountains and channeling waters.

狮子林庭园叠石小品"牛吃蟹"似显幽默
The yard rockery opusculum "ox eating crab" shows some humor.

留园"华步小筑"院景为稍点玲珑石块，顶植卉木垂萝，恰似一幅国画横幅
In "Buhua xiaozhu" yard scene of Liu Yuan, a few exquisite rockeries with top grown plants and vines resemble a horizontal Chinese traditional painting.

网师园"梯云室"庭院系墙中嵌理岩壁的佳作
The yard of Tiyun Shi (Note:Building Cloud Room) of Wangshi Yuan shows a great example of a rockery embedded wall.

留园鸳鸯厅北以冠云峰为主题
North of Yuanyang Ting (Note: Madarin Duck Hall) of Liu Yuan, Guanyun Feng (Note: Cloud Crowning Peak) is the main theme.

ROCKERY 277

峰峦有若狮形，妙在似与不似之间
Jiushi feng (Note: Nine Lions Peak) is located north of Xiaofang Ting (Note: Little Square Hall). It has a collection of cleverly arranged Tai Hu rockeries, with their peaks resembling lions. The wonder lies in their quasi resemblance to lions.

狮子林古五松园庭院湖石峰小品
Lake rockery opusculum in the yard of Gu wusong Yuan (Note: Ancient Five Pine Garden) of Shizilin

芜湖"翠明园"内山水庭院"探碧"
"Tan Bi" (Note: Exploring Green), a mound-water yard of "Cuiming Yuan" in Wuhu

九狮峰在小方厅北，是一组堆叠巧妙的太湖石，峰峦有若狮形，妙在似与不似之间
Jiushi feng (Note: Nine Lions Peak) is located north of Xiaofang Ting (Note: Little Square Hall). It has a collection of cleverly arranged Tai Hu rockeries, with their peaks resembling lions. The wonder lies in their quasi resemblance to lions.

怡园岁寒草庐南庭院山石小品

Rockery opusculum in the south yard of Suihan Caolu of Yiyuan

扬州"静香书屋"引瘦西湖水入院,并与假山建筑融为一体

Yangzhou "Jingxiang Shuwu" introduced Shouxi Lake's water into the yard, and it integrates with the rockeries and buildings.

扬州卷石洞天"旱景水意"峭壁山石景,塑造石隙漫流的自然景观,配植芦苇加强水意

The rockery scene "Hanjing Shuiyi" (Note: Dry Scene Water Feeling) of Juanshi Dongtian in Yangzhou. It simulates a rock crack water scene landscape, with reeds for the water scene enhancement.

上海古猗园五老峰石散置逸野堂右侧的五座石峰,因状似聆听山水清音的老人而得名

The five rockeries located to the right of Yiye Tang is referred to as Wulao Fengshi (Note: Five Old Men Peak). They're from Guyi Yuan in Shanghai. It got its name because the five rocks resemble old men listening to the sound of nature.

（三）楼山

楼面掇山，宜最高，才入妙，高者恐逼于前，不若远之，更有深意。

(3) Lou Rockery

To build a rockery facing a multi-storied building, the taller it is the more attractive it can be. However, if the rockery is too high, it may create certain pressure towards the building. So it is better to leave some distance between them, as it also creates a sense of layered depth.

楼山的做法大体可分三类，其一，假山做成山石巉崖的基座，楼阁建在其上，然较平整的顶部亦有台的功能；其二，假山在楼侧，与楼组合一体，亦可通过假山登阶上楼；其三，假山在楼前后，作为对景，产生山林野趣的效果。

There are basically three types of Lou (Note: Building) rockeries. First, a rockery can be built as a base, on which a building can be built. If the top of the rockery is relatively flat, it can also serve as a terrace. Second, place the rockery at the side of the building, and integrate it with the building, so that people can get onto the building via the rockery. Third, set the rockery in front of or behind the building and make it a scene facing the building and create a wild mountain landscape scene.

扬州瘦西湖"卷石洞天"岧崖洞壑与楼阁、庭院结为一体，相得益彰，山顶"四面八方亭"供观览坡下瘦西湖景色
Yangzhou Shouxi Lake's "Juanshi Dongtian" integrates rockeries, caves, buildings and yards, with each part promoting mutually. The mound top "Simian Bafang Ting" (Note: All Direction All Side Pavilion) can be used for enjoying the sceneries of Shouxi Lake down slope.

瘦西湖熙春台侧假山不仅可以登阶上楼，同时体现"叠石停云"的意境，使楼有腾云驾雾的艺术魅力
The side rockery by Xichun Tai of Shouxi Lake has a dual purpose – it can be used as a stair to get onto the upper floor and at the same time realize the idea of "Dieshi Tingyun" (Note: Rock Piling Cloud Stalling), making the building appear loftier and people on top feel giddy.

上海豫园双层楼阁屹立于池畔湖石假山之上，上层称快楼，下层名燕爽阁，眺望园景如诗如画
The double storied building of Shanghai Yu Yuan was erected on top of rockeries by the lake. The upper floor is named Kuai Lou (Note: Fast Building) and the lower one Yanshuang Ge. It gets beautiful scenes overlooking the garden.

扬州个园悬崖峻壁黄石山既有深境，亦为进楼登阶

The cliff rockery in Yangzhou Ge Yuan has a sense of depth while serving as a stair to climb on top.

御风楼由于高居假山之顶，是昔日杭州城南最高建筑

Because Yufeng Lou (Note: Royal Influence Building) is perched high on top of a rockery, it used to be the highest building in the south part of Hangzhou city.

苏州古镇木渎羡园环山草庐楼前侧掇山，使楼与爬山廊皆融入山水环境，诗情画意油然而生

The rockery in front of Caolu Lou (Note: Thatched Building) of Xian Yuan in Mudu town of Suzhou makes the building and the climbing stair corridor well integrated into the recreated natural scene, generating rich artistic and poetic interests.

（四）阁山

阁皆四敞也，宜于山侧，坦而可上，便以登眺，何必梯之。

阁功能为登眺观景，在园林中多为高者，故阁山主要功能是提升高度，由于山堆得高峻，所以山体多采用中空，形式为石屋、洞府；若在高地建阁，则阁山利用山石磴阶造景，深意图画。

(4) Ge Rockery

All four sides of a Ge are open, and so it is appropriate to build a Ge by a mountain side. It can leverage the mountain slope to get onto it for viewing distant scenes. Why bother to build stairs there?

The function of a Ge is to view distant scenes from a higher place and so they mostly are built at a higher ground. So a Ge rockery's function is to elevate to a higher place. Since the rockery is high, its body should mostly be hollow in the forms of rock room or grotto. If it is built on a higher place, then the Ge rockery can use the rock stairs to create scenes, adding more meaning and contents to the landscape.

沧浪亭看山楼建于印心石屋之上，前轩后楼，造型优美，昔日登轩可望苏州西南诸山
Kanshan Lou (Note: Mountain View Building) of Canglang Ting was built on top of Yinxin Shiwu (Note: Seal Heart Stone House). The Xuan is in front and the Lou is at the back. The architectural style is beautiful. In the old time, one could see mountains in the southwest of Suzhou from there.

狮子林屹立在池畔假山上的修竹阁
Xiuzhu Ge (Note: Slender Bamboo Building) on top of pond side rockery in Shizilin.

苏州拥翠山庄月驾轩充分利用地形，假山结合磴阶并彰显兀立险峻
In Yongcui Shanzhuang of Suzhou, Yuejia Xuan takes full advantage of the terrain by integrating the rockery with stairs, making the rockery appear even more precipitous.

上海秋霞圃霞阁建在湖石堆叠的仙人洞上，北洞口有磴道至冈上，入阁从南洞口可步至晚香居
In Shanghai's Qiuxia Pu, Jixia Ge was built on top of lake rockeries that has a Xianren Dong (Note: Immortals Cave). There's a northern cave entrance that leads to a stairway to the top of the rockery, and a southern cave is an entry way to wanxiang Ju.(Note: Evening Fragrance Dwelling)

（五）书房山

凡掇小山，或依嘉树卉木，聚散而理。或悬岩峻壁，各有别致。书房中最宜者，更以山石为池，俯于窗下，似得濠濮间想。

(5) Study Rockery

To build small rockeries in a study courtyard, one way is to rely on beautiful flowers and trees scattered around, diversified in sizes but with an order. Another way is to build a cliff to create a different scene and interest. The most appropriate scene is to use rocks to form a pond and one can enjoy the scene of water against the rails, sojourning mentally in the landscape.

书房是园主人书画场所，故环境注重诗情画意，达到寄情山水的意境。有的直接将画幅营造出立体的山水，有的寓情于季节或某种植物。

A study is where a garden owner writes and paints; therefore its environment should stress on poetic and artistic ambient atmosphere creation, to facilitate the landscape sojourning artistic concept. Some create real artistic landscapes directly and others have empathetic associations upon seasons and certain plants.

《细雨虬松图》主峰（石涛作品）
《游华阳山图轴》主峰（石涛作品）

The main peak in "Xiyu Qiusong Tu" (Note: Drizzle and Twisting Pine Drawing), by Shi Tao
The main peak in "You Huayangshan Tuzhou" (Note: Mount Huayang Sightseeing Drawing), by Shi Tao

片石山房原为明代大画师石涛手笔，故掇山酷似其画作，山高水长的造景恰似一幅立体的国画长卷，在书房里眺望画意盎然，更有濠濮间想
Pianshi Shanfang is the work of the great master Shi Tao of the Ming dynasty. So the rockery mimics his work. The landscape scene is just like a real painting scroll. Looking from a study at the scene makes one wander in nature

拙政园听雨轩前玲珑馆东水池畔掇黄石山岗，种植芭蕉，寄情"雨打芭蕉"的曲韵
Basho trees are grown at the east pond by the yellow stone rockery. The location is Linglong Guan in front of Tingyu Xuan of Zhuozhen Yuan. It hints the melody of "Yuda bajiao". (Note: a classic Chinese music)

鹤园馆房前假山景观
Rockery scene in front of Guan Fang of HeYuan

留园揖峰轩外，峰石林立，故名石林小院，花坛中栽植牡丹，与石头配合，依聚散而理
Outside of Liu Yuan's Yifeng Xuan, there are numerous rockeries; therefore it's named as Shilin Xiaoyuan (Note: Rock Grove Small Yard). Peonies were planted in flower bed and since they're combined with rockeries, they show various clustering groups.

（六）池山

池上理山，园中第一胜也。若大若小，更有妙境。就水点其步石，从巅架以飞梁；洞穴潜藏，穿岩径水；峰峦飘渺，漏月招云；莫言世上无仙，斯住世之瀛壶也。

(6) Chi Rockery

Chi Rockery (Note: pond rockery) is rockeries over a pond. It is the most enjoyable scene in a garden. There are both large and small rocks representing mountains, and this makes it even more attractive. There could be stepping stones in water and rainbow like flying bridges on top. Caves are hidden in the mountain, sometimes passing through the rockery and sometimes crossing over water. Moonlight can pass through holes and crevices at the top of the rockery, and deep in the cavern, fogs can gather. Who said that there is no immortal or celestial being in the world? This is our fairyland and wonderland.

华夏园林自秦以来，形成一池三岛（方丈、瀛洲、蓬莱三仙山）的模式。在江南古典园林中池山多为两种形式：其一，池山位于水中，水面形式前湖后溪；其二，池山位于池畔，崖峦洞穴、桥岛坡矶等元素融入其中，达到以真为假、做假成真的艺术效果。

Since the Qin dynasty, there has formed a one pond and three islets – Fangzhang, Yingzhou, and Penglai – template for Chinese gardens. In southern Chinese gardens, Chi rockery is shown in one of the two forms. One is that it is in the middle of water, with a lake in front and stream at back format. The other is that the rockery is at a side of the pond, with cliffs, caves, bridges, islets and piers etc. well integrated, blurring the boundaries between the real and the fake.

南京瞻园南假山，临池屹立，有绝壁悬崖、峰峦、洞龛、山谷、水洞、瀑布、石径、磴阶、汀步等，为假山中的精品
In Zhan Yuan of Nanjing, the south rockery borders the pond. It has precipitous cliffs, peaks and ranges, caves, valleys, water grottos, water falls, stone steps, stairs, and stepping stones etc. It is a great work among all rockeries.

镇江金山利用塔影湖东北云根岛上山石险峻建风月亭登临观景
In Jinshan of Zhenjiang, the precipitous rocks on Yungen Islet in Taying Hu (Note: Pagoda Reflection Lake) are leveraged to build Fengyue Ting (Note: Wind Moon Pavilion) with great views.

苏州艺圃水池西岸湖石池山，山水相依，花木掩映
The pond rockery is at west of the pond in Suzhou's Yipu. The rockery mound and water rely on each other among vegetations.

南京莫愁山堆叠于池畔，池中山影似幻化倩影
Nanjing's Mount Mochou (Note: Sans Souci) was constructed by lake side, creating its illusive yet beautiful reflections in water.

（七）内室山

内室中掇山，宜坚宜峻，壁立岩悬，令人不可攀。宜坚固者，恐孩戏之预防也。

(7) Indoor Rockery

To build a rockery indoors, one must make it secure, tall, and steep. A steep cliff can be created along a wall surface to have an unattainable feeling. A reason to make it stable and firm is to prevent accidents that may be caused by children playing around it.

江南园林中内室山多为室内竖立石峰，古人喻石为"云根"，峰石使文人产生"满屋烟云"的遐想。清代"扬州以园亭胜"，争奇斗艳的园林创作使内室山亦有匠心独运的作品。

In most Southern Chinese gardens, indoor rockeries are those erected rock peaks, which ancients referred to as "Yun Gen" (Note: cloud root), since those rocks made the literati fancy "room filled with fogs and clouds". In the Qing dynasty, "Yangzhou is known for its gardens and pavilions." The competing garden designs produced creative and unique work.

瘦西湖卷石洞天景区群玉山房内室角隅点缀湖石洞龛，毛竹引泉，使室内顿觉生机勃勃，"山水有清音"诗句油然而生

In Juanshi Dongtian scenic area in Shouxi Lake, Qunyu Shanfang (Note: Group Jade Mountain House) is adorned with lake rocks, caves, and bamboo spring duct, immediately making a lively indoor scene. The poem piece "Shanshui You Qingyin" (Note: nature's sound excels that of musical instrument) is most appropriate here.

扬州二十四桥景区朝西的望春楼，利用建筑外廊营造内室叠石，开南门引入瘦西湖景色，得到楼内外景观融为一体的效果

Yangzhou's Twenty-Four Bridge Scene Area's west facing Wangchun Lou (Note: Spring Watching Building) used building's exterior corridor to make indoor rockeries. It opens its south entrance to have introduced scenes of Shouxi Lake, making the inner and outer scenes an integrated whole.

上海豫园跨水游廊中方亭，竖有立峰，名"美人腰"，是进入大假山的前奏

There erected a rockery named "Meiren Yao" (Note: Beauty's Waist) in the bridge corridor in Shanghai's Yuyuan. It is the prelude to the greater rockery.

网师园冷泉亭内特置英石峰

The Yingshi Feng was specially placed inside Lengquan Ting (Note: Cold Spring Pavilion) in Wangshi Yuan.

扬州瘦西湖望春楼内室山采用"室内天井"的格局，楼上下可共享假山景观，室内产生山麓的野趣，满楼烟云，是江南园林中的孤例，可谓匠心运

In Yangzhou's Shouxi Lake, an open patio form is adopted in Wangchun Lou, where both the lower and upper story can enjoy the rockery. Wild nature interest is introduced indoors and with the foggy and misty atmosphere, it is the only such unique example in southern Chinese gardens.

（八）峭壁山

峭壁山者，靠壁理也。借以粉壁为纸，以石为绘也。理者相石皴纹，仿古人笔意，植黄山松柏、古梅、美竹，收之圆窗，宛然镜游也。

(8) Cliff Rockery

A cliff rockery is made by creating a rockery against a wall. This is like using the white paint wall as paper and rocks as the contents. During construction, one must pay attention to the textures of rocks and use ancient landscape painting theory to guide the work. Moreover, on the cliff, some Mount Huang pines, old plums, and pretty bamboos should be planted. Seeing these scenes from a round window is just like having a virtual tour in a mirror.

江南园林中以粉墙为纸，树石为绘的峭壁山精品颇多，主要是造园者多以古人条屏、扇面绘画作品为蓝本来完成营造的，加入门洞、方窗为框，所以十分入画。

There are many fine works in Southern Chinese gardens using trees and rocks as contents and white paint wall as paper. The garden designers mainly used ancient screen or fan-shape paintings as blueprints and added door and window frames. As a result, they became very picturesque.

拙政园海棠春坞庭院石景
Rockery scene in the yard of Haitang Chunwu in Zhuozheng Yuan

无锡二泉池前明代太湖石峰"观音石"，以粉墙为纸，藤石为绘，显有写意画境
In front of Wuxi's Erquan Chi (Note: Second Spring Pond), a Ming dynasty Tai Hu rockery "Guanyin Shi" uses white painted wall as the background and vine-climbed rockery as the objects to have mimicked an impressionist freehand brushwork in traditional Chinese painting.

狮子林海棠形洞门框景九狮峰
Crabapple flower shaped gate framing Jiushi Feng (Note: Nine Lions Peak) in Shizilin

怡园屏风三叠
The three fold rockery screen in Yiyuan

苏州怡园入口石景
Rockery scene in the entrance of Yiyuan

（九）山石池

山石理池，予始创者。选版薄山石理之。少得窍不能盛水，须知"等分平衡法"可矣。凡理块石，俱将四边或三边压报，若压两边，恐石平中有损。如压一边，即缚稍有丝逢，水不能注。虽做灰坚固，亦不能止，理当斟酌。

(9) Rockery Pond

To use rocks and stones to build a pond, I was the first. Using a thin rock slate for construction, if there is only a small hole or crack, we cannot guarantee the water would hold. One must understand "equal share equilibrium method" to build such a pond. If you want to use rocks to build a pond, you must hold down with rocks on the bottom slate at four or three sides. If you only stack at two ending sides, the flat slate at the bottom may break. If you only hold down at one side, then if only there is a slight crack at the seam, water cannot be filled. Even if you use lime mixed with China wood oil to seal the crack, it still could not prevent leaking. This is something that we should carefully think about when we use rocks to build a pond.

（十）金鱼缸

如理山石池法，用糙缸一只，或两只，并排作底。或埋、半埋，将山石周围理其上，仍以油灰抵固缸口。如法养鱼，胜缸中小山。

(10) Gold Fish Pond

To build a gold fish pond, one could use similar method as that for a rock pond. Use one or two large crocks and place them together at the bottom. Bury them completely or half way into the ground and then lay rocks around them. Use limes mixed with China wood oil to seal the rim. To raise fish, this is a better way than building a rockery from within a crock.

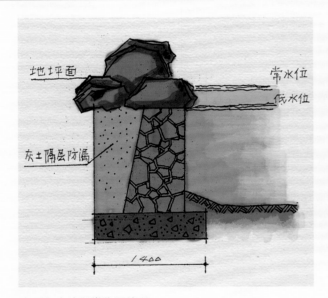

大面积水池驳岸防漏处理
Revetment leak prevention measures for large pond

水深不稳定状况山石驳岸的处理
Revetment treatment with unstable water levels

曲折有致，错落得体
Natural and irregular bank form appears appropriate

山石理池首先要解决水的渗、漏。另外要注意以下几点：

1. 驳岸要自然嶙峋，山石依皱拼结。
2. 水位不稳定时的综合效果。
3. 山石水池与建筑、道路、植物的艺术结合。主人亲水、近水的设施。
4. 水源的来无踪去无影的艺术处理。
5. 桥、岛、堤、岸、矶、岫、台等的综合艺术创作。

The first thing to do for building a pond with rocks is to address the leaking issue. In addition, one needs to pay attention to the following:

1) The bank should look rugged naturally and rocks should be connected according to their natural textures.
2) The complex effect when the water level varies.
3) The artistic integration of rock pond with buildings, paths, and plants. Outdoor furniture and facilities for the owner when he approaches water.
4) The artistic handling to conceal the origin of the water.
5) The synthetic artistic effect creation combining bridges, islets, dykes, banks, piers, caves, and terraces.

山石驳岸与道路、植物、水体组合野趣盎然（一）
Wild interest arises when rockery revetment, plants, and water body integrate well 1

咫尺山林，多方景胜
Miniature landscape with various scenes

山石驳岸与道路、植物、水体组合野趣盎然（二）
Wild interest arises when rockery revetment, plants, and water body integrate well 2

高低观之多致
A well-mixed rockeries with various elevations

做假成真的矶岸
A man-made but naturally looking rocky shore

似岛似矶，尤特致意
An islet or a pier? A unique treatment

瞻园石矶，以假乱真
The rock pier in Zhan Yuan was made almost like real

水边点石亦可供人小憩
Spotted rocks by water serve as great resting place for people

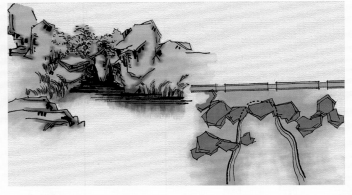

墙前水岫，深奥莫测
The water cave in front of a wall appears unfathomable

上海豫园崖矶与水门结合，别具风韵
In Shanghai's Yuyuan, combine rockery work with a water gate is uniquely charming

嘉兴南湖鱼乐园，其碑刻系明朝万历三十年董其昌手书，鱼池湖石嶙峋，游鱼嬉戏，此乃典型金鱼缸的章法
Yule Yuan (Note: Fish Pleasure Garden) in South Lake, Jiaxing. The inscriptions were from Dong Qichang in Wanli 30th year in the Ming dynasty. There are rugged lake rocks in the fish pond where fish frolic. This is a typical fish pond handling.

（十一）峰

峰石一块者，相形何状，选合峰纹石，令匠凿笋眼为座，理宜上大下小，立之可观。或峰石两块三块拼掇，亦宜上大下小，似有飞舞势。或数块掇成，亦如前式，须得两三大石封顶。须知平衡法，理之无失。稍有欹侧，久则逾欹，其峰必颓，理当慎之。

峰石的鉴赏标准自古至今均为：漏、透、瘦、皱、丑五个方面造型，具体如下图表示。用岩石多块拼掇的峰石首先是考虑造型艺术。由于多姿动感的姿态，所以拼接的牢固不容忽视。

(11) Peaks

If it is a single piece rock, one should observe its shape and select compatible rock to let craftsmen make a base rock with a mortise on top. Then place the main rock, with its larger part on top and smaller part at bottom, onto the base rock, in the preferred upright position. Or assemble two or three pieces, still larger upper and smaller lower, and place them on the base, to form a flying gesture. Or even use more pieces doing the same and use two or three larger rocks to end on top. One must know about the balancing art to prevent failure. If it is tilted, even if slightly, over time it could get more tilted and eventually it would fall. So we should practice caution during construction.

The appreciation standard of rocks since ancient times has been Lou (Note: vertically through), Tou (Note: horizontally through), Shou (Note: lean), Zhou (Note: textured and grained), and Chou (Note: queer) five aspects, which are illustrated below. For rockeries using multiple pieces, the first factor to consider is their formative art values. Then, because of their dynamic gestures, the security of their connection cannot be ignored.

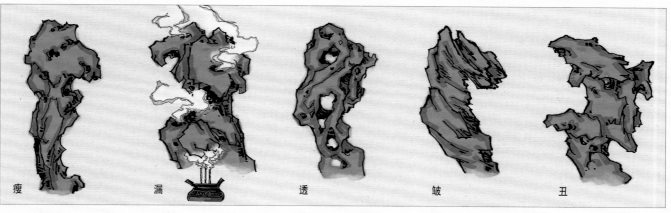

| 瘦 | 漏 | 透 | 皱 | 丑 |

峰石下小上大
Shou, the rock is smaller at its lower part and bigger upper

峰石上下有洞
Lou, the rock has holes running through vertically

峰石水平向有洞
Tou, the rock has holes running through horizontally

皱折重叠
Zhou, the rock has rich and repeated textures and grains

似有飞舞势
Chou, the rock seems to pose a dancing gesture

五石拼掇组合图　五石拼掇实景（苏州怡园）
Five rock combination illustration　Five rock combination actual example (Yiyuan in Suzhou)

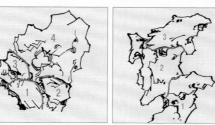

四石拼掇　三石拼掇　三石拼掇
Four rock combination　Three rock combination　Three rock combination

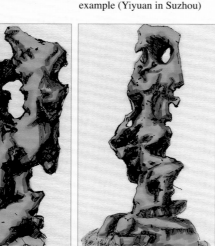

苏州瑞云峰　苏州冠云峰
Ruiyun Feng (Note: Propitious Cloud Peak) in Suzhou　Guanyun Feng (Note: Cloud Tower Peak) in Suzhou

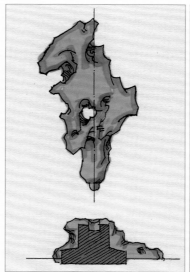 

峰石的安装　三石拼掇成石峰一块
Peak rock installation　Three pieces of rocks combine to form one peak rockery

江南园林中名石集锦

中国园林特有的"石文化",不仅是欣赏奇峰怪石,其隽永的意蕴传承至今,耐人寻味。

Well-known rockery collection in southern Chinese gardens

The unique "rock culture" in Chinese garden art lies not only in the individual appreciation of rare rocks but also in relishing its deep meaning collectively to future generations.

豫园玉玲珑,为江南三大名石之一,豫园镇园之宝,相传是宋徽宗"花石纲"遗物。此峰堪称石中上品,姿态婀娜,极具漏、透、皱、瘦之美

Yuyuan's Yu Linglong (Note: Exquisite Jade) is one of the three celebrity rocks in south China. It is said to be a relic of the Song dynasty's emperor Huizong's "Huashi Gang" and as a result, the treasure of the garden. It is considered as a top notch rockery with a gracefully charming figure excelling in all four categories of Lou, Tou, Zhou and Shou.

留园岫云峰,为江南名石之一

Liu Yuan's Xiuyun Feng is another well-known rockery piece in southern China.

留园瑞云峰,为江南名石之一

Liu Yuan's Guanyun Feng is another famous one.

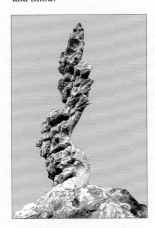

留园皱云峰,为江南三大名石之一

Liu Yuan's Zhouyun Feng is one of the three celebrity rockeries in southern China.

扬州船形钟乳石云盆,北宋"花石纲"遗物。此为大自然经过漫长时间,天然造就的玲珑巧石,看去如天然山水盆景,极富自然情趣,十分珍贵。

The stalactite evolved and boat shaped Yunpen (Note: Cloud Basin) in Yangzhou is a relic of the North Song dynasty's "Huashi Gang". It is a naturally formed exquisite rock after a long period of time in the wild. It appears like a real bonsai and is full of natural interests. How precious!

扬州清代九峰园遗石

A relic from Jiufeng Yuan (Note: Nine Peak Garden) from the Qing dynasty in Yangzhou

扬州九峰园峰石流泉

Yangzhou Jiufeng Yuan's Fengshi Liuquan (Note: Peak Rock Flowing Spring)

湖州莲花庄内的莲花峰,含苞欲放,雄奇俊秀

Lianhua Feng (Note: Lotus Peak) from Lianhua Zhuang (Note: Lotus Village) in Huzhou. Its budding gesture shows its magnificent beauty.

西湖孤山美人峰,亭亭玉立

Meiren Feng (Note: Beauty Peak) from Gushan (Note: Lone Mountain) in West Lake stands gracefully like a fair lady.

梅园"米襄阳拜石",凝结着中国园林特有的"石文化",其隽永的意蕴,耐人寻味

"Mi Xiangyang Bai Shi" of Mei Yuan is packed with Chinese unique "rock culture". It's intriguingly relishing.

绍兴沈园诗境石,沈园的诗境园中矗立着一块高达四米的太湖石,上刻陆游手迹"诗境"

Shijing Shi (Note: Poetic Conception Rock) of Shen Yuan in Shaoxing. It is a Tai Hu rockery with a towering four meter height, on which there is a calligraphic writing "Shi Jing" (Note: Poetic Conception) by Lu You.

胡雪岩故居，天井中的景石，玲珑剔透
At Hu Xueyan's former residence, the scenic rockery in the patio is dainty and exquisite.

浙江省海盐县绮园美人照镜石峰，石颈项处有一圆孔，朝阳从孔中穿过，光耀四射，流霞如镜
Meiren Zhaojing (Note: Beauty Mirror) peak is in Qiyuan in Haiyan County of Zhejiang Province. At its neck area there is a round hole, through which morning sun shines with bright light and forms rosy clouds of dawn.

南京煦园依云峰，传为北宋时期花石纲遗物。独峰突兀，挺拔壮观
Nanjing Xuyuan's Yiyun Feng (Note: Cloud Leaning Peak) is said to be a relic of Huashi Gang from the North Song dynasty. It's grand, precipitous, tall and straight.

狮子林名峰之一
A well-known peak in Shizilin, 1

狮子林名峰之二
A well-known peak in Shizilin, 2

狮子林名峰之三
A well-known peak in Shizilin, 3

狮子林名峰之四
A well-known peak in Shizilin, 4

狮子林名峰之五
A well-known peak in Shizilin, 5

狮子林名峰之六
A well-known peak in Shizilin, 6

太仓南园簪玉峰，峰顶翘出状如发簪的一支小峰，故名
Zanyu Feng (Note: Hairpin Peak) of Nan Yuan in Taicang. A branch sticking out on top resembling a hairpin, hence the name.

"鼋渚春涛"，言简意赅，发人遐想
"Yuanzhu Chuntao" (Note: Turtle Inslet Spring Billow), a laconic name, invokes imaginations.

绍兴沈园断云石，位于沈园大门西首，原本是一拙朴的卵形石块，后被分成两半，上刻陆游手迹"断云"二字。此石匠心独运，寓意深刻，蕴涵着陆游与唐琬的爱情悲剧
Shaoxing Shen Yuan's "Duanyun Shi" (Note: Cloud Breaking Rock) is located at the west side of the garden's entrance. It used to be a dull and simple egg-shaped rock but was split into two halves with a Lu You's hand written callifgraphy "Duan Yun" two characters. It shows its ingenuity and carries a deep meaning alluded to Lu You and Tang Wan's tragedy in their romantic lives.

（十二）峦

峦，山头高峻也，不可齐，亦不可笔架式，或高或低，随致乱摆，不排比为妙。

(12) Luan

Luan means high and steep peaks. It is a no-no to have equal heights, nor can it be like a writing brush holder. There must be variations in heights and the formation should be determined by the rocks' natural looks. Asymmetric arrangement is preferred.

峦是比主峰稍小与矮的山头。由于园林中假山规模有限，难觅实例图释，故在古代画中选取佳品。

Luan is only secondary to a main peak in its size and elevation. Because there are few examples in real gardens so it is difficult to show them. Here we use some ancient paintings for illustration purposes.

北宋－郭熙－山水图
Shanshui Tu (Note: Mountain and Water Drawing), by Guo Xi, the North Song dynasty

清－渐江－山水图
Shanshui Tu (Note: Mountain and Water Drawing), by Jian Jiang, the Qing dynasty

明－王绂－潇湘秋意图
Xiaoxiang Qiuyi Tu (Note: Autumn Scene in Hunan), by Wang Fu, the Ming dynasty

清－渐江－林泉图
Linquan Tu (Note: Forest and Spring Drawing), by Jian Jiang, the Qing dynasty

南宋－佚名－松风高卧图
Songfeng Gaowo Tu (Note: Pine Breeze Rest High Drawing), anonymous, the South Song dynasty

北宋－佚名－山水图
Shanshui Tu (Note: Mountain and Water Drawing), anonymous, the North Song dynasty

（十三）岩

如理悬岩，起脚宜小，渐理渐大，及高，使其后坚能悬。斯理法古来罕有，如悬一石，又悬一石，再之不能也。予以平衡法，将前悬分散后坚，仍以长条堑里石压之，能悬数尺，其状可骇，万无一失。

(13) Yan

A cliff rockery construction should start with a small base and grow larger as it stacks up. When reaching certain height, it becomes necessary to enforce its backend so that it could balance an overhang at the front. However, it has rarely been seen since the ancient times. Normally, only one rock sticks out, or at most another one can be added. If one wants to add the third one, it becomes almost impossible. I adopt a technique for balancing to distribute the weight from the front to the back. Using a long chunk of stone as a counter weight, the overhang can stick out several feet. Although this looks a bit scary, it is actually very safe.

理悬岩已在掇山篇图释。

崖的图释在实例中难以表达清楚，故选择古画中的作品，画家亦是时代相近且为计成赏识的画家，所选作品亦可作为今后理崖实践的蓝本。

The handling of cliff is already illustrated and explained in the rockery section.

The illustrations about Ya is hard to express with real examples, so here I selected some ancient paintings to help. The painters lived in close time frame as that of Ji Cheng, who also appreciated their works. The selected works can also serve as blueprints for future cliff rockery constructions.

《辞源》

岩：岩石的简称。

崖：高峻的山边或岸边。

故"岩"应为"崖"，可能因为《园冶》一书在国内外流传时有印刷错误所致。

From *"Etymology"*:

Yan: An abbreviation for rocks.

Ya: High and steep mountain side or water bank.

Therefore, "Yan" should actually be "Ya". This is probably a typo from *Yuan Ye* when it was circulated.

元－黄公望－快雪时晴图
Kuaixue Shiqing Tu (Note: Fast Snow Occasion Clear Drawing), by Huang Gongwang, the Yuan dynasty

元－马琬－乔岫幽居图
Qiaoxiu Youju Tu (Note: Tall Peak Seclude Living Drawing), by Ma Wan, the Yuan dynasty

现代－溥儒－白云青嶂图
Baiyun Qingzhang Tu (Note: White Cloud Green Mountain Drawing), by Pu Ru, a modern contemporary

南宋－夏圭－清远图
Qingyuan Tu (Note: Clear Far Drawing), by Xia Gui, the South Song dynasty

清－王鉴－仿古山水图
Fanggu Shanshui Tu (Note: Antique Model Mountain and Water Drawing), by Wang Jian, the Qing dynasty

明－王锷－雪岭风高图
Xueling Fenggao Tu (Note: Snow Ridge High Wind Drawing), by Wang E, the Ming dynasty

（十四）洞

理洞法，起脚如造屋，立几柱著实，掇玲珑如窗门透亮，及理上，见前理岩法。合凑收顶，加条石替之，斯千古不朽也。洞宽丈余，可设集者，自古鲜矣！上或堆土植树，或作台，或置亭屋，合宜可也。

(14) Caves

The construction method of caves is similar to that of a building. First a few columns need to be erected and stabilized. Select some exquisite rocks and stones with holes so that they can be used as windows for lighting. The construction method on top is similar to that for Yan. When it is ready to enclose on top, use strips for stability. In this way it can last forever. The cave can be as wide as one Zhang (Note: about 10 feet) and can be used to hold parties, which has been rare since the ancient time! On top of the cave, trees can be planted, or a terrace or pavilion can be built, too, as long as it is appropriate.

个园夏山洞中石柱与造景相结合
In Ge Yuan, the rock column in the Summer Mound's cave blends well with the scene

条石压石收定法
Slate weight securing the top of a rockery

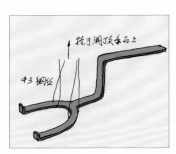

熟铁吊架，固定条石下挂山石
Wrought iron connector for securing the hanging rock of a beam

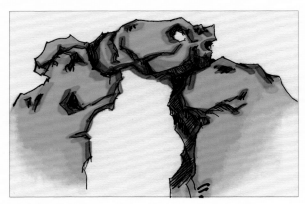

岩石合凑收定法
How rocks meet and lock on top

寄啸山庄　下洞上台山石景
A case below and terrace on top rockery treatment in Jixiao Shanzhuang

环秀山庄水岫一
Water rockery 1 in Huanxiu Shanzhuang

环秀山庄水岫二
Water rockery 2 in Huanxiu Shanzhuang

个园夏山下洞上亭台
Ge Yuan's Summer Mound has its cave at its lower part and a terrace and pavilion are on top

（十五）涧

假山依水为妙，倘高阜处不能注水，理涧壑无水，似少深意。

(15) Jian

A rockery by water can create wonderful scenes. However if it is hard to get water because of higher elevation it would have a dry gully, resulting a rather deficient scene.

涧，两山间的流水或离山之水。在园林中是指高耸驳岸夹峙的带状水体，往往寓意深刻，深奥莫测。

Jian, or gully, refers to a water stream between two mountains or water flowing away from a mountain. It is a linear water body sandwiched between two tall banks, often pregnant with meanings and seems unfathomable.

苏州耦园引城河水入园，并利用高差堆叠黄石山成涧谷造景。架折桥串联两壁，远处听橹楼屹立城河边借景，是巧妙将园融入环境的佳作，并是江南园林最杰出涧的实例。

Suzhou's Ou Yuan introduced city moat water into the garden. It uses yellow stones and leverages various elevations to form a gully scene. A zigzagged bridge connects the two sides of the gully. Tinglu Lou in the far side stands by the city moat, leveraging the outside scenes and scull sound. It is an excellent example of scene leveraging and is a great work of Jian scene in southern Chinese gardens.

八音涧平面图
Plan of Bayin Jian

八音涧是利用流经墙外的"天下第二泉"伏流，落石盂入潭，化为清溪曲涧，蛇行斗折于崖根石隙，曲曲而下，呈现了忽断忽续，忽隐忽现，忽急忽缓，忽聚忽散的不同水景，因而产生音响，水音与岩壑共鸣，不啻八音齐奏，故名"八音涧"

Bayin Jian (Note: Octatonic Gully) takes advantage of outside undercurrent "Tianxia Dierquan" (Note: Second Spring on Earth) and channels it into the garden stream. The water flows in a meandering way through rock base and cracks, sometime visible and others not, sometimes continuous and others disconnected, sometimes rapid and others slow, sometimes converging and others diverging, forming various water scenes and sound. It resonates with the rock gully, making a myriad of different sounds, hence the name.

谐趣园"玉琴峡"系人工琢石成涧溪，引后湖水形成多层跌落

In Xiequ Yuan's "Yuqin Xia" (Note: Jade Musical Instrument Gorge) is a man-made rock gully stream. It channels the water from the back lake to form cascading water falls.

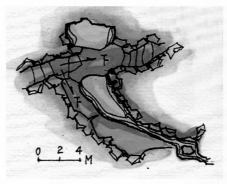

源泉平面图
Plan of the spring source

引"天下第二泉"从石隙流出，水经石盂跌落至小潭，化为清溪曲涧

The water comes from "Tianxia Dierquan" as it flows from rock cracks and falls into a broad-mouthed stone jar before falling further into the small pond, which becomes Qingxi Qujian (Note: Clear Stream Curved Gully).

常州兴福寺涌泉"空心潭"
"Kongxin Tan" (Note: Hollow Spring) is a gushing spring of Xingfu Temple in Changzhou.

羡园云墙之下，水门贯通二园内外盘曲的涧水，故题"槃涧"

Under the Yun Qiang (Note: cloud wall) of Xian Yuan, a water gate connects the meandering water outside and inside of the two gardens; therefore its name is entitled as "Pan Jian" (Note: Meandering Gully).

(十六) 曲水

曲水，古皆凿石槽，上置石龙头喷水者，斯费工类俗，何不以理涧法，上理石泉，口如瀑布，亦可流觞，似得天然之趣。

(16) Meandering Streams

To build a meandering stream, the ancients liked to build a stone cistern, on which water from a stone-carved faucet flowed out. This is labor intensive and looked tacky. Why not learn from the system on how streams work, build a spring from the rocks and let water overflow to form a waterfall? We can also let wine cups float on water for fun and enjoy the elegant taste with nature.

曲水，多指带状弯曲的溪水。自古文人就有绍兴兰亭曲水流觞的故事，园林中盘曲溪流是充满着诗情画意的。

Meandering streams refer to linear and winding streams. Since ancient times, there had stories of literati gathering at Orchid Pavilion in Shaoxing for refined pleasure of floating wine cups in the stream. So having meandering streams in a garden helps create a poetic and enjoyable scene.

绍兴兰亭曲水流觞景点，一弯流水蜿蜒、九曲回环，为当年兰亭修禊觞咏之地
Qushui LiuShang (Note: Meandering Water Floating Wine Cup) of the Orchid Pavilion in Shaoxing. A meandering water flows through many turns and it is where literati met to pray and played poem games.

盘曲溪流，野趣横生
Tortuous stream full of wild nature interest

山石驳岸与道路、植物、水体组合野趣盎然
Rocky revetment integrated with paths, plants, and water to form a scene with rich wild interest.

山石使带状水面产生深远的艺术效果
The rocks flanking linear water body to create a deep and far artistic effect.

曲水岸石悬挑，产生深奥莫测效果
Tortuous water bank with overhung embankment, creating a deep and unfathomable feeling.

留园西部曲流潺流，有桃花源意境，水尽头为活泼地水阁
A meandering stream flows in the west of Liu Yuan, alluding a Taohua Yuan (Note: an ideal and pastoral utopia) and ending with Huopodi water hall.

（十七）瀑布

瀑布如峭壁山理也。先观有高楼檐水，可洞至墙顶作天沟，行壁山顶，留小坑，突出石口，泛漫而下，才如瀑布。不然，随流散漫不成，斯谓"坐雨观泉"之意。

夫理假山，必欲求好，要人说好，片山块石，似有野致。苏州虎丘山，南京凤台门，贩花扎架，处处皆然。

(17) Waterfalls

The way to construct a waterfall is the same as that for a cliff wall. First one needs to observe whether there is a tall building eave that can hold rain water source. Water can also be channeled from a stream along a gutter on top of a wall to a wall rockery into a small reservoir on its top. Then the water would flow down through a small stone opening to simulate a waterfall from a cliff. Otherwise, water would drip down from everywhere failing to form a waterfall. This is the so-called artistic concept of "enjoying a waterfall when it rains".

To build a garden rockery, one must pursue a certain artistic state and create an amazing and admirable work. Even with one piece of rock, a wild taste of nature should be sought after and obtained. Nowadays at Mount Huqiu in Suzhou or Fengtai Men in Nanjing, flower peddlers make bonsais that are so artificially twisted into various forms. This handling of plants is against nature and should not be modeled after.

古代瀑布水源仅靠天然雨水，故奇思妙想汇集屋面雨水至天沟，水口落到假山顶然后泛漫而下形成瀑布。现今水源充足，就可塑造出：线瀑、布瀑、悬瀑、分段瀑、分层瀑等众多形式。江南园林中以泉造景更为普遍，故在此补充泉景的实例。

在园林里掇山叠石的关键是片山块石，要有野趣，即"以真为假，做假成真"，也就是古代画家强调的"造化为师""搜尽奇峰打草稿"，这样就能把人工的假山达到宛若真山的艺术效果。

In ancient times, a waterfall depended on natural rain water as its source, so people thought of all kinds of methods to get water from roof and to collect it through gutter and direct to the top of a rockery before it fell down as a waterfall. Nowadays we have plenty of water sources and so we could create various forms of waterfalls, such as streamed, screened, segmented, layered etc. Using waterfalls to create scenes is even more popular; therefore more spring scene examples are shown here.

The key to build a rockery in a garden is on the rocks. It must carry a wild interest, that is, "to use real as fake and make fake as real". This is also what the ancient landscape painters said "treat nature as our mentor", and "research exhaustively on all the exotic peaks and rocks". Only through these efforts can we achieve the artistic state of making artificial rockeries appear as real ones.

南京瞻园假山洞壑瀑布
Cave and gully waterfall in Zhan Yuan, Nanjing

扬州寄啸山庄高楼檐水形成的瀑布
A waterfall formed by falling water from the eaves of a tall building, Jixiao Shanzhuang, Yangzhou

狮子林问梅阁北，瀑布自山巅飞泻而下，直注涧底，潺然有声，有做假成真效果
North of Wenmei Ge (Note: Ask Plum Building) of Shizilin, the waterfall runs all the way down from the rockery top, into the pond, making splashing sound, like a real one.

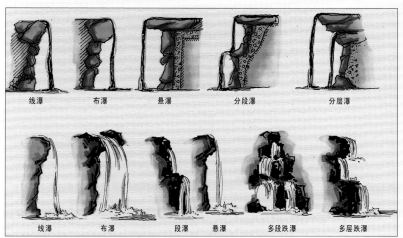

汇集归纳众多的瀑布形式，可供造园时借鉴
A collection of different waterfalls, for garden design reference

北京香山饭店庭院的瀑布
A waterfall in a yard in Xiangshan Hotel, Beijing

水源以跌水与瀑布两种形式注入，可谓匠心独运
An interesting scene combining a cascade and waterfall

山水有清音
Natural music

苏州虎丘天平山"白云泉"
Baiyun Quan (Note: White Cloud Spring) of Mount Tianping in Huqiu, Suzhou

芜湖翠明园庭院瀑布
A yard waterfall in Cuiming Yuan, Wuhu

依皴合掇的水口，真假难分
A water source following the nature's rule making it hard to tell a fake from a real one

苏州虎丘下"天下第三泉"
"Tianxia Disan Quan" (Note: The 3rd Spring on Earth) in Huqiu, Suzhou

庭院中的盂泉，用毛竹产生二级跌落
Yu Quan (Note: Receptacle Spring) in the yard uses bamboo ducts for two level cascade

毛竹从石壁引泉入盂，并点缀石灯笼
A bamboo duct introduces water from a rock wall. A stone lantern is dotted as a decoration

庭园中的山石几案
A rock table in a garden yard

山石小品既遮掩阴井，又成建筑抱角
A rockery opusculum serves a dual purposes - a cover for hidden well and a building base ornament

苏州虎丘山"片山块石，似有野致"，是天然佳作，可以借鉴
At Mount Huqiu in Suzhou, "a few rocks create a wild nature feel". This nature oriented work can serve as a great example

园冶
图释

Yuan Ye Illustrated
——Classical Chinese Garden Explained

十 选石

10 Rock Selection

夫识石之来由，询山之远近。石无山价，费只人工，跋蹩搜巅，崎岖究路。便宜出水，虽遥千里何妨；日计在人，就近一肩可矣。取巧不但玲珑，只宜单点；求坚还从古拙，堪用层堆。须先选质无纹，俟后依皴合掇。多纹恐损，无窍当悬。古胜太湖，好事只知花石；时遵图画，匪人焉识黄山。小仿云林，大宗子久。块虽顽夯，峻更嶙峋，是石堪堆，便山可采。石非草木，采后复生，人重利名，近无图远。

Before excavating rocks, one must study its source, and find out the distance. When it is still in the mountain, a rock does not have any value; the cost is on labor. To find exotic rocks, one must endure the travail of climbing and wading in the wild. If water transport is feasible, then even if it is hundreds miles away it is still manageable. If it only takes a few days to move the rocks, then even if it is carried by human labors it is all right. To select a rock, one must not only pick those exotic ones for single rock construction, he also needs to find those queer but hard ones so that he could lay and stack them. One must select those rocks with excellent texture but no cracks, and then base on the ancient landscape paintings to guide the rock texture building. If there are too many cracks then one need to practice caution to prevent it from being broken. If a rock has no holes then it should be used on an overhung position. The ancient people argued that a lake rock was the best and people went after it simply after hearing about it. Not many people understand the wonders of rocks from Mount Huang. For smaller rockery, one can model after Ni Yunlin's simple yet meaningful style, and for larger rockery, Huang Zijiu's majestic strokes can be applied. Although rocks are dull and insensate, when it is made high and steep it could look really like rugged rocks with wild nature feel. This kind of rocks can all be used for rockery construction and they are available in nearby mountains. Unlike plants, rocks cannot regenerate. People want rare and exotic rocks with convenience, so if there is none available nearby they would travel far to get it.

斧劈皴　Axe cut texture

马牙皴　Horse teeth texture

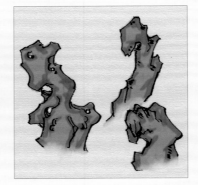

湖石类形态　Lake rock appearance

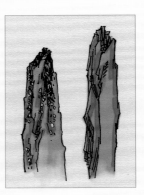

石笋类　Rock shoot category

云头皴　Cloud texture

解索皴　Untie texture

黄石类形态　Yellow stone appearance

披麻皴
依皴合掇：掇山要按山石纹理堆叠　Hemp texture

折带皴　Turning belt texture

园林中掇山叠石的山石基本上是湖石、黄石、石笋三种类型
There are basically three types of rocks in gardens – lake rock, yellow rock, and rock shoot.

黄子久山水画
Landscape Paintings of Huang Zijiu

富春山居图（局部）1347-1350 年
Fuchun Mountain Dwelling Drawing (a portion) 1347-1350

黄公望（1269—1354 年）

元代画家，字子久，号大痴、大痴道人，又号一峰道人、净竖，晚号井西道人，平江常熟（今江苏常熟）人。从事绘画，以山水为主，经常优游于名山胜水之间，促进了他写实山水画风的形成，与倪云林、吴镇、王蒙合称"元代四大家"，其诗书画三绝，为推元四家之首。

Huang Gongwang (1269 – 1354)

Huang Gongwang is an art painter of the Yuan dynasty. His Zi (Note: a second name) is Zijiu and his Hao (Note: a third name) Dachi Daoren. He was also referred to as Yifeng Daoren, or Jingshu. In his late years, he got another Hao as Jingxi Daoren, and he is from Changshu (nowadays Changshu in Jiangsu). His paintings mostly use mountain and water as the main theme. He often toured and traveled to famous mountains and rivers and that helped him in his realistic landscape painting style. He and Ni Yunlin, Wu Zhen and Wang Meng together are referred to as the "four great masters of the Yuan dynasty", and his accomplishments in poetry, calligraphy, and painting top the group.

黄石的石皴与山皴一致
Yellow stone's rock texture is compatible with mountain's

水阁清幽图（1349 年）
Shui Ge Qing You Drawing (1349)

剡溪访戴图（1349 年）
Shanxi Fangdai Drawing (1349)

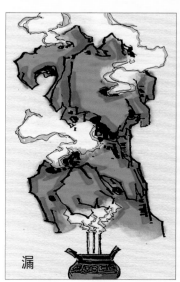
漏

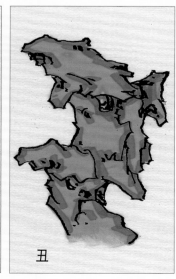
丑

瘦

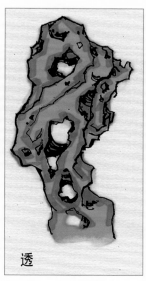
透

皱

适合单点的玲珑石形
Exquisite Rocks Suitable for Individual Display

倪云林山水画
Landscape Painting of Ni Yunlin

幽润寒松图（1372年）Youjian Hansong Drawing (1372)

倪瓒（1301—1374年）

元代画家，初名珽，字元镇，号云林、云林子、荆蛮民等，自称倪迂、懒瓒等，江苏无锡人。家豪富，后卖田散财，浪迹太湖、泖湖一带。倪瓒诗书画三绝，绘画开创了水墨山水的一代画风，与黄公望、吴镇、王蒙合称"元代四大家"。画法疏简，格调天真幽淡，以淡泊取胜。作品多画太湖一带山水，构图多取平远之景，善画枯木、竹石茅舍。其画多以干笔皴擦，笔墨极简，所谓"有意无意，若淡若疏"，形成荒疏萧条一派。

Ni Zan (1301–1374)

Ni Zan is an art painter of the Yuan dynasty. His initial name is Ting, his Zi is Yuanzhen, and his Hao is Yunlin, Yunlinzi, Jingmanmin etc. He called himself Ni Yu, Lan Zan and so on, and he is from Wuxi. He was born in a wealthy family, but later his land was sold and his assets were distributed or given away before he started traveling around Tai Hu and Mao Hu areas. Ni Zan's poetry, calligraphy, and paintings are marvelous, and he created the Chinese Ink Painting School. He and Huang Gongwang, Wu Zhen, and Wang Meng together are referred to as the "four great masters of the Yuan dynasty". His painting strokes are simple and his style is innocent and light. He excels in indifference. He mostly drew and painted the mountains and waters in Tai Hu area. The compositions of his paintings are mostly flat and far type and he is really good at drawing dead wood, bamboos, rocks, and cottages. He often used dry brushes to draw textures and his drawings are extremely simple and minimalized. This is so-called "intentional or not, light and sparse" genre and as such he had formed a school of the rusty and desolate.

江亭山色图（1372年）
Jiangting Shanse Drawing (1372)

梧竹秀石图（1337年）
Wuzhu Xiushi Drawing (1337)

容膝斋图（1372年）
Rongxizhai Drawing (1372)

竹石乔柯图（1357年）
Zhushi Qiaoke Drawing (1357)

江亭山色（局部）
Jiangting Shanse (a portion)

（一）太湖石

苏州府所属洞庭山，石产水涯，惟消夏湾者为最。性坚而润，有嵌空、穿眼、宛转、险怪势。一种色白，一种色青而黑，一种微黑青。其质文理纵横，笼络起隐，于石面遍多坳坎，盖因风浪中冲激而成，谓之"弹子窝"，扣之微有声。采人携锤錾入深水中，度奇巧取凿，贯以巨索，浮大舟，架而出之。此石以高大为贵，惟宜植立轩堂前，或点乔松奇卉下，装治假山，罗列园林广榭中，颇多伟观也。自古至今，采之已久，今尚鲜矣。

（二）昆山石

昆山县马鞍山，石产土中，为赤土积渍。既出土，倍费挑别洗涤。其质磊块，嶙岩透空。无耸拔峰峦势，扣之无声。其色洁白，或植小木，或种溪荪于奇巧处，或置器中，宜点盆景，不成大用也。

（三）宜兴石

宜兴县张公洞、善卷寺一带山产石，便于竹林（祝陵）出水，有性坚、穿眼，险怪如太湖者。有一种色黑质粗而黄者，有色白而质嫩者，掇山不可悬，恐不坚也。

（四）龙潭石

龙潭金陵下七十余里，沿大江，地名七星观，至山口、仓头一带，皆产石数种，有露土者，有半埋者。一种色青，质坚，透漏文理如太湖者；一种色微青，性坚，稍觉硬夯，可用起脚压泛；一种色纹古拙，无漏，宜单点；一种色青如核桃纹多皴法者，掇能合皴如画为妙。

（五）青龙山石

金陵青龙山石，大圈大孔者，全用匠作凿取，做成峰石，只一面势者。自来俗人以此为太湖主峰，凡花石反呼为"脚石"。掇如炉瓶式，更加以劈峰，俨如刀山剑树者，斯也。或点竹树下，不可高掇。

（六）灵璧石

宿州灵璧县地名"磬山"，石产土中，岁久，穴深数丈。其质为赤泥渍满。土人多以铁刃遍刮，凡三次，既露石色，即以铁丝帚或竹帚兼磁末刷治清润，扣之铿然有声，石底多有渍土不能尽者。石在土中，随其大小具体而生，或成物状，或成峰峦。嶙岩透空，其眼少有宛转之势，须借斧凿，修治磨砻，以全其美。或一两面，或三面，若四面全者，即是

(1) Tai Hu Rock

The lake by Mount Dongting, which belongs to Suzhou city, is rich in Taihu Rocks. The best ones are located in Xiaoxia Bay area. Taihu rocks are hard, moist, and shiny, and have various looks such as hollow, twisted, and queer etc. One type of Taihu rock is white, and another is dark green or light dark green. It is rich in inter-twining textures, forming protruding and denting surfaces, which is referred to as "slingshot dents", caused by wind and water currents. When it is knocked on, one can hear a faint sound. Rock excavators carried hammers and chisels and dove into the water before they selected exquisite rocks and separated them by chisels. Thick ropes were then used to tie them up before they were hoisted out of water by a wooden structure on large boats. The taller and larger a Taihu rock is, the more valuable it is. It is particularly appropriate to have one erected in front of a Xuan or Tang, or stand by a fine pine and exotic flowers, or made into a rockery. If it is placed in a broad Xie it would feel much grander. Since Taihu rocks have been excavated since the ancient times, there are few left.

(2) Kunshan Rock

At Mount Ma'an in Kunshan County, there is a rich deposit of Kunshan rocks buried in the red dirt in the mountains. Because they are tainted, after they are excavated, much effort is needed for selection and cleaning. The texture is rough and its form is protruding and hollowing. There is no distinguished resemblance of a mountain peaks or ranges. When it is knocked on there is no sound. A Kunshan rock is pure white and is best to be used along with a small tree or with some shallow water plants, or in a container as bonsais. It is not suitable for a larger scale display.

(3) Yixing Rock

In Yixing County, there is much supply of Yixing rocks in Zhanggong Cavern and Shanjuan Temple areas. They are particularly rich in areas where water flows out of a bamboo grove. One kind of Yixing rock is hard with a queer form of hollowness, similar to Taihu rocks. One type is dark colored with some yellow tint on rough grains. Another type has fine textures and appears white. Yixing rocks cannot be used as a protruding hangover in rockeries since it is not sturdy enough.

(4) Longtan Rock

Longtan (Note: Dragon Pool literally) is located seventy kilometers east of Nanjing. Along the Yangtze River, starting from Qixing Guan (Note: Seven Star Taoist Temple) to Shankou, Cangtou areas, there are several kinds of Longtan rocks that can be found. Some are above ground while others are buried half-way. One kind is bluish green colored, hard, hollow, with a texture resembling that of Taihu rocks. Another kind is light bluish green, hard and insensate. It can be used as a foundation of a rockery or a weight pressed on top of a piling. Some Longtan rocks have unadorned textures but no holes and they are appropriate for individual embellishments. There is another blue-green colored Longtan rock with walnut texture, which is very close to what some ancient landscape paintings show. So when building a rockery, if one can incorporate this kind of rocks well, the effect would be much like that of an old wild landscape painting.

(5) Mount Qinglong Rock

The Mount Qinglong rocks from southeast of Nanjing with large rings and holes are all chiseled out by masons. If they are used to make a rockery, then only one side has a peak appearance. In the past some used it as the main peak and used Taihu rocks as "foot rocks", or subordinate ones. The rockeries they made looked like incense burner or vase on a worship table, and when two side sharp rocks were added, they really resembled "mountain of knives and grove of swords", a very low taste. Mount Qinglong rocks can be used as decorations under bamboo trees and should not be built too high.

(6) Lingbi Rock

In Lingbi County in Suzhou, there is a place called "Pan Shan" (Note: Huge Rock Mountain), where there is a rich deposit of Lingbi rocks. However, due to years of excavations, some pits are already several Zhangs deep. Because Lingbi rocks have long been buried in red clay, they are full of silt stains. The locals used iron tools to scrape them twice or three times in order to reveal the original rock color. Afterwards, iron or bamboo brooms were used to sweep them clean with porcelain powder to make it smooth and sleek. When it is knocked on it makes some clanging sound. At the bottom of a rock, there would always be deposit or stains that cannot be cleaned up completely. Lingbi rocks stay in the ground and thus

从土中生起，凡数百之中无一二。有得四面者，择其奇巧处镌治，取其底平，可以顿置几案，亦可以掇小景。有一种扁朴或成云气者，悬之室中为磬，《书》所谓"泗滨浮磬"是也。

（七）岘山石

镇江府城南大岘山一带，皆产石。小者全质，大者镌取相连处，奇怪万状。色黄，清润而坚，扣之有声。有色灰青者，石多穿眼相通，可掇假山。

（八）宣石

宣石产于宁国县所属，其色洁白，多于赤土积渍，须用刷洗，才见其质。或梅雨天瓦沟下水，冲尽土色。惟斯石应旧，愈旧愈白，俨如雪山也。一种名"马牙宣"，可置几案。

（九）湖口石

江州湖口，石有数种，或产水中，或产水际。一种色青，浑然成峰、峦、岩、窦，或类诸物。一种扁薄嵌空，穿眼通透，几若木版以利刃剜刻之状，石理如刷丝，色亦微润，扣之有声。东坡称赏，目之为"壶中九华"，有"百金归贵小玲珑"之语。

（十）英石

英州含光、真阳县之间，石产溪水中，有数种：一微青色，间有通白脉笼络；一微灰黑；一浅绿。各有峰、峦、嵌空穿眼，宛转相通。其质稍润，扣之微有声。可置几案，亦可点盆，亦可掇小景。有一种色白，四面峰峦耸拔，多棱角，稍莹彻，而面面有光，可鉴物，扣之无声。采人就水中度奇巧处凿取，只可置几案。

（十一）散兵石

"散兵"者，汉张子房楚歌散兵处也，故名。其地在巢湖之南，其石若大若小，形状百类，浮露于山。其质坚，其色青黑，有如太湖者，有古拙皴纹者，土人采而装出贩卖，维扬好事，专买此石。有最大巧妙透漏如太湖峰，更佳者，未尝采也。

（十二）黄石

黄石是处皆产，其质坚，不入斧凿，其文古拙。如常州黄山，苏州尧峰山，镇江圌山，沿大江直至采石之上皆产。俗人只知顽夯，而不知奇妙也。

have various forms and appearances, some resembling mountain peaks and ranges. Shrillest and perforated with small holes, it is mild and tortuous. Some axing, chiseling, and polishing works are needed to make it perfect. It is rare to find a rock from the ground that has a perfect form in multiple or even one face. If a Lingbi rock looks great on all of its four sides, then we could focus on its unique features and perform some refinements. We would polish its bottom to make it flat so that it could stand steadily on a table forming a miniature scene on its own. There is an ancient, flat or cloud shaped Lingbi rock that can be hung indoors as a Pan, a large rock decoration, and this is what is said in "Shang Shu" as "floating rock in Si River".

(7) Xianshan Rock

Xianshan rocks are produced in Mount Daxian south of Zhenjiang. Smaller ones can be excavated in whole but larger ones can only be chiseled out at the connection to the main mountain body. Xianshan rocks are shown in different weird forms and they are generally yellow colored. They are moist and hard. When being knocked on, they make a sound. Another kind of rock is beige or taupe colored and it has many holes that are inter-connected. It can be used to make a rockery.

(8) Xuan Rock

Xuan rocks are produced in Ningguo County. The rock is white but mostly have residual taint from red dirt. It must be cleaned and brushed before its real white color is shown. Or one may place it under house eave to let the dripping rain water to wash out the deposit stains. The older rocks are preferred. The older the rock is, the whiter it is. It resembles a snow mountain. One of the Xuan rock is called "Maya Xuan" (Note: Horse Tooth Xuan), and it can be placed on tables for visual pleasure.

(9) Hukou Rock

In Hukou area in Jiangzhou, there are many kinds of rocks. Some are from the lake and others from the bank. There is one bluish green colored Hukou rock that resembles naturally a peak, crest, crag, gully or other forms. There is another kind that is flat and thin with perforations running through. It can be as thin as a plank of wood, as if it is cut by a sharp knife or sculptured by a chisel. Its grains and textures are as if it has been just brushed. It looks a bit moist and smooth and it makes a sound when knocked on. It was highly praised by Su Dongpo as he viewed it as "Mount Jiuhua in a kettle". He also wrote a poem that goes, "I'd rather spend much money for a small and exquisite rock."

(10) Ying Rock

Between Hanguang county and Zhenyang county in Yingzhou, there are Ying rocks in its streams. There are different kinds. One is slightly bluish green, with vein-like white grain. Another one is dark gray and the third one is light green. Ying rocks can have different forms of peaks and ranges and have mild and indirect holes that connect to each other. Its textures are smooth, moist and shiny and when it is knocked on, it emits faint sound. A Ying rock can be placed on a table, or used in a bonsai for decorations, or can be made into a small rockery in a garden. It can have multiple edges and corners and can be crystal clear. There could show a sheen from inside and reflect images on its body. There is little sound when it is knocked on. Exquisite pieces are excavated in water and they could be placed on tables for enjoyment.

(11) Stragglers Rock

The reason it is called "stragglers" is that the rocks are from the place where Xiang Yu was surrounded and Zhang Liang, who offered an advice to let soldiers sing Chu songs to disintegrate Xiang Yu's troops. The location is south of Chao Lake. The stragglers rocks have various sizes and forms and they are exposed on the mountain ground. Stragglers rocks are hard and their color is dark bluish green. It looks a little like Taihu rocks and has unadorned grain resembling that in a landscape painting. The locals excavated them for sale. Enthusiasts from Yang zhou loved buying only this kind of rocks. The current largest stragglers rock, much like a Taihu peak rock, is exquisite and with twisted perforations. The ones better than this one have not been excavated yet.

(12) Yellow Rock

Yellow rock is produced in many places. It is hard and difficult to deal with by tools such as axes and chisels. Its grain looks old and unadorned. It can be obtained along the Yangtze River all the way up to Caishi Ji (Note: Quarry Rock), such as Mount Huang in Changzhou, Mount Raofeng in Suzhou, and Mount Chui

(十三) 旧石

世之好事，慕闻虚名，钻求旧石。某名园某峰石，某名人题咏，某代传至于今，斯真太湖石也，今废，欲待价而沽，不惜多金，售为古玩还可。又有惟闻旧石，重价买者。夫太湖石者，自古至今，好事采多，似鲜矣。如别山有未开取者，择其透漏、青骨、坚质采之，未尝亚太湖也。斯岂古露风，何为新耶？何为旧耶？凡采石惟盘驳，人工装载之费，到园殊费几何？予闻一石名"百米峰"，询之费百米所得，故名。今欲易百米，再盘百米，复名"二百米峰"也。凡石露风则旧，搜土则新，虽有土色，未几雨露，亦成旧矣。

(十四) 锦川石

斯石宜旧。有五色者，有纯绿者，纹如画松皮，高丈余，阔盈尺者贵，丈内者多。近宜兴有石如锦川，其纹眼嵌石子，色亦不佳。旧者纹眼嵌空，色质清润，可以花间树下，插立可观。如理假山，犹类劈峰。

(十五) 花石纲

宋"花石纲"，河南所属，边近山东，随处便有，是运之所遗者。其石巧妙者多，缘陆路颇艰，有好事者，少取块石置园中，生色多矣。

(十六) 六合石子

六合县灵居岩，沙土中及水际，产玛瑙石子，颇细碎。有大如拳、纯白、五色者，有纯五色者；其温润莹彻，择纹彩斑斓取之，铺地如锦。或置涧壑及流水处，自然清目。

夫葺园圃假山，处处有好事，处处有石块，但不得其人。欲询出石之所，到地有山，似当有石，虽不得巧妙者，随其顽夯，但有文理可也。曾见宋杜绾《石谱》，何处无石？予少用过石处，聊记于右，余未见者不录。

in Zhenjiang. People only learn about its unruliness but rarely know it can be made into a wonderful rockery.

(13) Old Rock

Some arty people seeking undeserved reputations single-mindedly looked for old rocks. There are some peak rocks in a certain famous garden with an inscription by some famous person and those rocks are handed over since certain dynasty and they are real Taihu rocks and so on. Now the garden is in neglect and its rocks are up for sale. Because of this, the arty guy wants to spend much to buy the rocks, which is okay to collect as antiques. There are some people who would buy them as long as they hear that they are old rocks. However real Taihu rocks have been sought and excavated since ancient times by enthusiasts and as a result, there isn't much left. If there are other unexcavated areas with rocks that are hard, dark, and perforated, they would not be necessarily inferior to Taihu rocks. These rocks are exposed and weathered over thousands of years. How can you tell the difference between the old and the new? An excavation work involves land and water transportation and manual installation costs and the total cost could hardly accurately measured. I heard that there is a rare rock called "Baimi Feng" (Note: Hundred Dan Peak). After a careful inquiry I got to know that the rock bought after spending money is worth a hundred Dans (Note: An old Chinese measure, which is close to sixty kilograms) of rice. So if we bought a rock at this price and would spend the same amount transporting it, then it could be referred to as "Two Hundred Dan Peak", right? No matter what kind of rocks, as long as it is exposed and weathered, it would turn old. If a rock is just excavated from the ground, it may look new, although with the color from the dirt; however as long as it's exposed for long, it would turn old for sure.

(14) Jinchuan Rock

For Jin chuan rocks, the older it is, the higher the grade is. Some Jinchuan rocks have interwoven multi-colors while others just pure green. Its grain looks like pine barks painting. The most precious ones are those that are one Zhang (Note: about ten feet long) tall and one Chi (Note: about a foot long) wide; however most rocks are shorter than one Zhang. Some recently produced rocks from Yixing look like Jinchuan rocks but there could be pebbles embedded in their grainy holes and their colors are not that good looking either. Old Jinchuan rocks have hollow grainy eyes with clear, fresh colors and smooth and moist textures. They could form beautiful scenes among flowers or under a tree. If they are used in a rockery, then they are good for steep cliffs.

(15) Huashigang

"Huashigang" in the Song dynasty is located in Hanan where it borders Shandong. They can be found anywhere because they were abandoned during transportation in those years. A lot of Huashigang rocks have very exquisite forms. Because of difficulties in ground transportation, some enthusiasts just picked some smaller amount to decorate their gardens, making the garden scene more lively.

(16) Liuhe Cobblestone

At Lingju Yan in Liuhe County, in the sand or by a river bank, there are a lot of small agate pebbles. Some can be as big as a fist. They show either white interwoven with five colors or five pure colors. They are moist, shiny, and crystal clear. Normally people select the most colorful and pattern rich ones to use, and if it's used as a paver, then the path would feel and look like a brocade. They could aslo be placed by a stream or flowing water, highlighting its natural and crystal clear beauty.

For garden rockery building, there are enthusiasts everywhere, and rocks can be found everywhere; however, there are very few people who truly understand the art of rockery. Talking about the source of rocks, we know there are mountains everywhere and there should be rocks in any mountain. If exquisite rocks cannot be found, even if it's an unadorned one, as long as it has some grains or textures, they can be used for a rockery. I once read *Yunlin Rock Guidebook*, written by Du Wan of the South Song dynasty. Apparently we could find rocks almost anywhere. The rocks that I used originated from a few places and I have listed them as aforementioned. I did not list the rocks that I did not see.

为确保识石准确，特请赏石大师《中国赏石文化发展史》巨著主编，学兄贾祥云教授提供各类石种如下：

To ensure the accuracy of rock recognition, I specially invited Professor Jia Xiangyun, a great rock connoisseur and author of *An Evolutionary History of Rock Appreciation in China*, to list various types of rocks below:

蒲松龄 海岳石－灵璧石
Haiyue Rock – Lingbi Rock, by Pu Songling

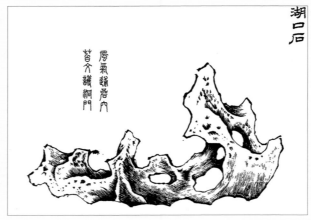
素园石谱－湖口石
Hukou Rock, from "Suyuan Shipu"

素园石谱－壶中九华
Huzhong Jiuhua (Note: Mountains in a Kettle), from "Suyuan Shipu"

苏州网师园鹰石－英石
Ying Rock, in Wangshi Yuan, Suzhou

壶中九华石
Huzhong Jiuhua Rock

一代名石－皱云峰－英石
A celebrity rock – Zhouyun Feng – Ying Rock

扬州个园四季假山（秋）－黄石
Yangzhou Geyuan's Four Season Rockery (Autumn) – Yellow Rock

趵突泉待月峰－太湖石1－旧石
Baotu Spring's Daiyue Feng (Note: Moon Waiting Peak) – Taihu Rock 1 – Old Rock

万竹园白云峰－太湖石2－旧石
Wanzhu Yuan's Baiyun Feng (Note: White Cloud Peak) – Taihu Rock2 – Old Rock

江苏昆山石－雪山悠悠
Xueshan Youyou (Note: Snow Mountain) – Jiangsu Kunshan Rock

锁云－灵璧石
Suoyun (Note: Locking Cloud) – Lingbi Rock

墨玉通灵－灵璧石
Moyu Tongling – Lingbi Rock

江苏昆山石
Jiangsu Kunshan Rock

江苏昆山石－仙骨寒香
Xiangu Hanxiang – Jiangsu Kunshan Rock

上海玉玲珑－太湖石
Shanghai Yulinglong – Taihu Rock

苏州瑞云峰－太湖石
Suzhou Ruiyun Feng – Taihu Rock

扬州个园四季假山（冬）－宣石
Yangzhou Geyuan Four Season Rockery (Winter) – Xuan Rock

雨花石－庐山夕照
Lushan Xizhao (Note: Mount Lu Sunset) – Yuhua Rock

雨花石－九寨风情－六和石子
Jiuzhai Fengqing – Yuhua Rock – Liuhe Pebble

安徽无为县－石丈－巢湖石
Anhui Wuwei County – Shi Zhang – Chaohu Rock

醉道士石－太湖石－旧石
Zuidao Shishi – Taihu Rock – Old Rock

石器时代－雨花石－六和石子
Stone Age – Yuhua Rock – Liuhe Pebble

宋代花石纲遗物－钟乳石云盆－存于扬州
Relics of Huashigang of the Song dynasty – Stalactite Yunpen (Note: Cloud Basin), kept in Yangzhou

趵突泉龟石－宋代花石纲遗物
Relics of Huashigang of the Song dynasty – Baotu Spring's Gui (Note: Turtle) Rock

园冶图释

Yuan Ye Illustrated
——Classical Chinese Garden Explained

十一 借景

11 Scene Leveraging

构园无格，借景有因。切要四时，何关八宅。林皋延竚，相缘竹树萧森；城市喧卑，必择居邻闲逸。高原极望，远岫环屏。堂开淑气侵人，门引春流到泽。嫣红艳紫，欣逢花里神仙；乐圣称贤，足并山中宰相。《闲居》曾赋，"芳草"应怜；扫径护兰芽，分香幽室；卷帘邀燕子，闲剪轻风。片片飞花，丝丝眠柳。寒生料峭，高架秋千。兴适清偏，怡情丘壑。顿开尘外想，拟入画中行。林阴初出莺歌，山曲忽闻樵唱，风生林樾，境入羲皇。幽人即韵于松寮，逸士弹琴于篁里。红衣新浴，碧玉轻敲。看竹溪湾，观鱼濠上。山容蔼蔼，行云故落凭栏；水面鳞鳞，爽气觉来欹枕。南轩寄傲，北牖虚阴。半窗碧隐蕉桐，环堵翠延萝薜。俯流玩月，坐石品泉。学衣不耐凉新，池荷香绾；梧叶忽惊秋落，虫草鸣幽。湖平无际之浮光，山媚可餐之秀色。寓目一行白鹭，醉颜几阵丹枫。眺远高台，搔首青天那可问；凭虚敞阁，举杯明月自相邀。冉冉天香，悠悠桂子。但觉篱残菊晚，应探岭暖梅先。少系杖头，招携邻曲。恍来林月美人，却卧雪庐高士。云幂黯黯，木叶萧萧。风鸦几树夕阳，寒雁数声残月。书窗梦醒，孤影遥吟；锦幛偎红，六花呈瑞。樽兴若过刻曲，扫烹果胜党家。冷韵堪赓，清名可并；花殊不谢，景摘偏新。因借无由，触情俱是。

Although there is no fixed rules in garden design, scene leveraging must have some grounds, which are mostly based on seasonal changes and have nothing to do with the eight residential orientations, a concept in Fengshui. In a forest or a place close to water, it is appropriate to pause for a leisurely recreation. This is because the trees help create a deep and quiet state of mind. One should avoid the hustle and bustle in an urban environment and select a quiet and leisurely place to live. Climbing up high one can look into the distance and the faraway mountain peaks and ranges are just like surrounding verdant screens. Mild breeze greets you when you are in an open hall and spring streams flow at the gate. Fairies seem to live in colorful flowers and living in seclusion, one could feel as the Prime Minister Tao Hongjing hiding in the mountain. Resigning from an official post one can follow Pan Yue in composing and intoning "Xian Ju" (Note: Leisure Living), or performing self-cultivation like Qu Yuan in taking pity on "fragrant grass". Cleaning flower paths and caring for orchid buds to enjoy the fragrance in a quiet room. Rolling up a window curtain to invite in spring swallows and their flying movements in the air are just like scissors cutting the breeze. Falling flower petals flying in the air; low-hanging willow branches waving in the breeze. It is still a bit chilly in spring while in autumn a high flying swing is already in full swing. In leisure time, one feels in an aesthetics and elegance seeking mood while reposing personal feelings in natural mountains. It is as if a land of idyllic beauty and a wonderland like that in a beautiful landscape picture. Just enjoy birds happily chirping in the forest and here comes the singing of a woodsman from a hill bend. With cool mountain breeze blowing into my face, it seems that I entered the immemorial pristine time. Humming poem verses, one lives in seclusion in a cottage in a pine forest; playing musical instruments, a leisurely minded folk enjoys the serenity of a deep bamboo forest. A lotus flower sticking out of water is like a beauty just out of bath; rain drops falling on a Chinese parasol tree is just like jade beads hitting each other. What a pleasure it is to enjoy the beauty of bamboos at a creek bend and watching swimming fish in a pond! Looking into the distance by a railing seeing misty mountains, I feel that the clouds just fall in front of me; hearing rippling waters in the pond while I fall asleep brings a cool breeze to my pillow. I repose my pride and haughty spirit in the south hall while enjoying the afternoon shade behind the north window. Shady trees loom outside my window; verdant vines spread over the fence wall. I enjoy the moon's reflection by the water and savor the spring's gurgling sound when sitting still on a rock. In autumn, a thin dress would not protect you from the chilly weather; however, the fragrance of lotus flower in the pond still lingers. The falling leaves of parasol trees show the arrival of the fall season and the insects seem to lament the passing of summer. The calm and undisturbed water enhances the tranquility of the lake and the wildly beautiful landscape adds to the charm of the mountain. Here is a picture: a line of white egrets flying in the bright and blue sky; a few red maple trees stand by their reflections in clear and green water. Climbing onto a terrace and looking afar, I could talk to the blue sky; standing by a railing and looking up in the starry vault of heaven, I would toast to the bright moon. Intoxicating osmanthus flower fragrance permeates in the air in August. In winter, the chrysanthemum flowers by the fence appear withered and dying; we should go and find out whether the plum flowers on the mountain ridge is blooming towards the sun? Bring some cash, get some wine, and invite our good neighbors for a drink. Seeing plum flower bloom under the trees and bright moonlight, I wonder if a beauty would appear. In a snow caped cottage in the mountain, there might be extraordinary people resting. Dark clouds make it look like dusk and howling wind adds to the gloomy atmosphere. There are a few old crows wobbling in evening wind and some wild geese crying under the waning moon in the cold. Waking up in my studio, I chant poems lonely; with brocade hanging scrolls around a stove, I savor the snow in winter. Rowing a boat is just for the fun; using snow water to make some tea, the taste may be better than that in a tea ceremony. We could keep doing our literary and artistic pursuits in the cold winter, as we could rival those aloof and lofty people. Famous flowers are always available in all seasons but the best scenes seem always depend on the new and the fresh. One does not have to find a reason to leverage scenes on context but keep in mind that a scene and an emotion evoke each other.

"借景"一章是描写意境创作上的借景,并非以空间划分基础的"园区",园区是规划设计的单元,而意境则是设计构思艺术处理的范畴。前者是通过"体、宜、因、借"以处理园林的组成结构,后者则亦通过"体、宜、因、借"而赋予艺术处理的灵魂,这是它们相同与相异之处。而且景区划分,也要求有一个全区性的总意境,只是二者的重点范围确实互不相同。

兴造论中规划设计中的借景,是指景区构图上的借景,此章的借景主要是意境上的,比起"俗则屏之,嘉则收之",构图上的借景要灵活得多、变化得多、丰富得多。下面就用画幅的形式来图释意境创作上的借景。

The "Scene Leveraging" chapter is about that in artistic concept creation, not about "garden zone" idea used in spatial zoning. A garden zone is a unit for planning and design but an artistic concept belongs to the realm of design idea and artistic handling. The former is through "Ti Yi Yin Jie", (Note: fit, proper, extensible, and leveraging) to handle a garden's components and structure and the latter is still through the same four criteria to endow a soul and spirit of artistic handling. These are what they share in common and what they differ. Moreover, even for scene zoning, there should also be a general artistic concept for the whole garden; it is just that the focuses and extents of the two are different.

In the On Design and Construction opening section, the scene leveraging means the one for scene zoning composition; however, in this chapter, it mainly means the one on artistic concept, which is much more flexible, various, and richer than what is stated in "Nice views should be included; un-tasteful ones should be excluded." The following illustrations are aimed to help explain what it means by scene leveraging at the artistic concept creation level.

林皋延伫,相缘竹树萧森
(录自《邱思泽画集》"浓竹浓荫夏日长"图,略有剪裁)
In a forest or a place close to water, it is appropriate to pause for a leisurely recreation. (from *Painting Collection of Qiu Size*, with some editing)

城市喧卑,必择居邻闲逸
(录自《邱思泽画集》"春水"图)
One should avoid the hustle and bustle in an urban environment and select a quiet and leisurely place to live. (from "Spring Water" in *Painting Collection of Qiu Size*)

嫣红艳紫,欣逢花里神仙(录自《邱思泽画集》"秋韵"图)
Fairies seem to live in colorful flowers and living in seclusion. (from "Qiuyun Tu" in *Painting Collection of Qiu Size*)

高原极望,远岫环屏(录自《曹仁容画集》)
Climbing up high one can look into the distance and the faraway mountain peaks and ranges are just like surrounding verdant screens. (from *Painting Collection of Cao Renrong*)

堂开淑气侵人,门引春流到泽(录自《李树人画集》"深山之恋"图)
Mild breeze greets you when you are in an open hall and spring streams flow at the gate. (from "Mountain Love" in *Painting Collection of Li Shuren*)

借景 SCENE LEVERAGING 315

《闲居》曾赋,"芳草"应怜(录自《杨明义画扇》)
Resigning from an official post one can follow Pan Yue in composing and intoning *Xian Ju* (Note: Leisure Living), or performing self-cultivation like Qu Yuan in taking pity on "fragrant grass". (from *Yang Mingyi Fan Paintings*)

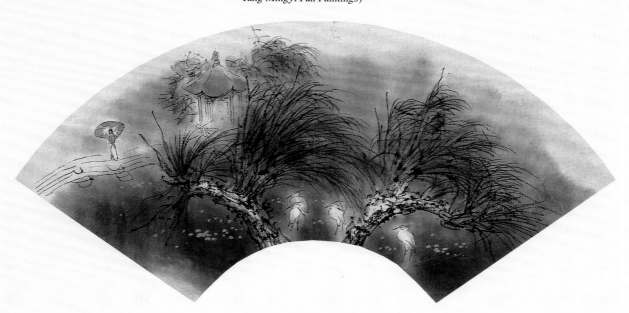

片片飞花,丝丝眠柳(录自《孙君良扇页小品画集》"春雨潇潇"图)
Falling flower petals flying in the air; Low-hanging willow branches waving in the breeze. (from "Spring Drizzle Painting" in *Sun Junliang Fan Painting Collection*)

乐圣称贤,足并山中宰相(录自《孙君良扇页小品画集》"春韵"图)
One could feel as the Prime Minister Tao Hongjing hiding in the mountain. (from "Chunyun Tu" in *Sun Junliang Fan Painting Collection*)

扫径护兰芽，分香幽室。（录自《李树人画集》"山花"图）
Cleaning flower paths and caring for orchid buds to enjoy the fragrance in a quiet room. (from "Mountain Flower Painting" in *Painting Collection of Li Shuren*)

怡情丘壑，拟入画中行。（录自《曹仁容画集》"上方雨意"图）
It is as if a land of idyllic beauty and a wonderland like that in a beautiful landscape picture. (from "Shangfang Yuyi Tu" in *Painting Collection of Cao Renrong*)

卷帘邀燕子，闲剪轻风。（录自《杨明义画扇》"雨后"图）
Rolling up a window curtain to invite in spring swallows and their flying movements in the air are just like scissors cutting the breeze. Falling flower petals flying in the air; low-hanging willow branches waving in the breeze. (from "After Rain Painting" in *Yang Mingyi Fan Paintings*)

观鱼濠上（录自《杨明义画扇》"秋池"图）
The pleasure to watch swimming fish in a pond (from "Autumn Pond" in *Yang Mingyi Fan Paintings*)

借景 SCENE LEVERAGING 317

兴适清偏，顿开尘外想（录自《黄澄钦画文集》"西湖难忘缘绮琴"图）
In leisure time, one feels in an aesthetics and elegance seeking mood while reposing personal feelings in natural mountains. (from "Unforgettable Music Plan on West Lake" in *Huang Chengqin Painting and Prose Collection*)

山曲忽闻樵唱（录自《戴敦邦新绘和谐自然集》）
Here comes the singing of a woodsman from a hill bend. (from *New Painting Collection on Harmonious Nature of Dai Dunbang*)

风生林樾，境入羲皇（录自《曹仁容画集》"汉柏四绝"图）
With cool mountain breeze blowing into my face, it seems that I entered the immemorial pristine time. (from "Four Han Cypress Wonders" in *Painting Collection of Cao Renrong*)

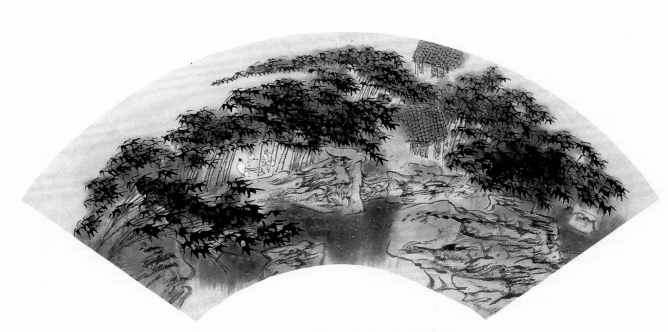

寒生料峭（录自《孙君良扇页小品画集》"竹深留客处"图）
Winter scene (from "Guest Reception Place in Deep Bamboo Grove" in *Sun Junliang Fan Painting Collection*)

林阴初出莺歌（录自《杨明义画扇》"秋荷"图）
Enjoying birds happily chirping in the forest (from "Autumn Lotus" in *Yang Mingyi Fan Paintings*)

红衣新浴，碧玉轻敲（录自《杨明义画扇》"荷塘清趣"图）
A lotus flower sticking out of water is like a beauty just out of bath; rain drops falling on a Chinese parasol tree is just like jade beads hitting each other. (from "Simple Pleasure in Lotus Pond" in *Yang Mingyi Fan Painings*)

借景
SCENE LEVERAGING 319

幽人即韵于松寮（录自《孙君良扇页小品画集》"只听松声好"图）
Humming poem verses, one lives in seclusion in a cottage in a pine forest. (from "Enjoy Wind Through Pines Painting" in *Sun Junliang Fan Painting Collection*")

逸士弹琴于篁里（录自《孙君良扇页小品画集》"晴光穿竹翠玲珑"图）
Playing musical instruments, a leisurely minded folk enjoys the serenity of a deep bamboo forest. (from "Light Through Bamboo Grove" in *Sun Junliang Fan Painting Collection*)

南轩寄傲，北牖虚阴，半窗碧隐蕉桐，环堵翠延萝薛（录自《孙君良扇页小品画集》"春雨天边"图）
I repose my pride and haughty spirit in the south hall while enjoy the afternoon shade behind the north window. Shady trees loom outside my window; verdant vines spread over the fence wall. (from "Spring Rain" in *Sun Junliang Fan Painting Collection*)

看竹溪湾（录自《黄澄钦画文集》"逍遥话逍遥"图）
The pleasure to enjoy the beauty of bamboos at a creek bend (from "Leisurely Talking about Leisure" in *Huang Chengqin Painting and Prose Collection*)

水面鳞鳞，爽气觉来欹枕（录自《曹仁容画集》"夕阳水榭"图）
Hearing rippling waters in the pond while I fall asleep brings a cool breeze to my pillow. (from "Shuixie at Sunset" in *Painting Collection of Cao Renrong*)

梧叶忽惊秋落,虫草鸣幽(录自《杨明义画扇》"晨"图)
The falling leaves of parasol trees show the arrival of the fall season and the insects seem to lament the passing of summer. (from "Morning" in *Yang Mingyi Fan Painings*)

苧衣不耐凉新,池荷香绽。
(录自《杨明义画扇》"湖塘清趣之三"图)
In autumn, a thin dress would not protect you from the chilly weather; however, the fragrance of lotus flower in the pond still lingers. (from "Lake Pond Simple Pleasure, Part III" in *Yang Mingyi Fan Paintings*)

俯流玩月,坐石品泉(录自《黄澄钦画文集》"水帘飞瀑"图)
I enjoy the moon's reflection by the water and savor the spring's gurgling sound when sitting still on a rock. (from "Cascading Waterfall" painting in *Huang Chengqin Painting and Prose Collection*)

湖平无际之浮光，寓目一行白鹭
（录自《曹仁容画集》"太湖晨曦"图）
A line of white egrets fly in the bright and blue sky over tranquil lake. (from "Morning in Taihu" from *Painting Collection of Cao Renrong*)

醉颜几阵丹枫
（录自《曹仁容画集》）
A few red maple trees stand by their reflections in clear and green water. (from *Painting Collection of Cao Renrong*)

眺远高台，搔首青天那可问
（录自《曹仁容画集》"灵岩冬翠"图，略有删减）
Climbing onto a terrace are looking afar, I could talk to the blue sky. (from adapted "Lingyan Dongcui" in *Painting Collection of Cao Renrong*)

冉冉天香，悠悠桂子
（录自《曹仁容画集》。"圣恩古刹"图）
Intoxicating osmanthus flower fragrance permeates in the air in August. (from "The Ancient Temple" from *Painting Collection of Cao Renrong*)

借景
SCENE LEVERAGING 323

但觉篱残菊晚，应探岭暖梅先
（录自《曹仁容画集》"邓尉香雪"图，略有删减）
In winter, the Chrysanthemum flowers by the fence appear withered ang dying,we should go and find out whether the plum flowers on the mountain ridge is blooming towards the sun. (from adapted "Dengwei Xiangxue" painting in *Painting Collection of CaoRenrong*)

恍来林月美人
（录自《黄澄钦画文集》"东坡到处有西湖"图，美人系录《华三川绘新百美图》）
I wonder if a beauty would appear. (The scene is from "There's a West Lake where Dongpo's been" in *Huang Chengqin Painting and Prose Collection*; The beauty is from *Hua Sanchuan Draws New Hundred Beauties*)

风鸦几树夕阳
（录自《黄澄钦画文集》"雁塔斜晖"图，略有删减）
Dark clouds make it look like dusk and howling wind adds the gloomy atmosphere. (from adapted "Yanta Xiehui" in *Huang Chengqin Painting and Prose Collection*)

寒雁数声残月
（录自《曹仁容画集》"石湖串月"图，将圆月改为月牙形）
There are a few old crows wobbling in evening wind and some wild geese crying under the waning moon in the cold. (from "Moon via Rock Lake", with the round moon changed into a crescent, in *Painting Collection of Cao Renrong*)

书窗梦醒，孤影遥吟（录自《戴敦邦新绘和谐自然集》）
Waking up in my studio, I chant poems lonely. (from *New Painting Collection on Harmonious Nature of Dai Dunbang*)

花殊不谢，景摘偏新（录自《李树人画集》"尘世之外"图）
Famous flowers are always available in all seasons but the best scenes seem always depend on the new and the fresh. (from "Beyond the Mortal World" from *Painting Collection of Li shuren*)

借景
SCENE LEVERAGING 325

棹兴若过剡曲（录自《李树人画集》"梦乡"图）
Rowing a boat is just for the fun. (from "Dreamland" in *Li Shuren Painting Collection of Li Shuren*)

锦幛偎红，六花（即雪花）呈瑞（录自《曹仁容画集》）
The brocade hangs scrolls around a stove, while I savor the snow in winter. (from *Painting Collection of Cao Renrong*)

因借无由，触情俱是
（录自《杨明义画扇》"春天来了"图）
One does not have to find a reason to leverage scenes on context but keep it in mind that a scene and an emotion evoke each other. (from "The Spring's Here" in *Yang Mingyi Fan Paintings*)

借景篇意境类画图摘录于以下画集，有部分画图由于文字内容需要进行了剪裁、删节、补充，特此说明。《戴敦邦新绘和谐自然集》、《邱思泽画集》、《孙君良扇页小品画集》、《杨明义画扇》、《李树人画集》、《吴中风光——曹仁容画集》《苏州园林名胜图——曹仁容》《黄澄钦画文集》。

The many paintings used in the Scene Leveraging Chapter are from the following sources, with some adaptations: New Paining Collection on Harmonious Nature of Dai Dunbang, Painting Collection of Qiu Size, Sun Junliang Fan Painting Collection, Yang Mingyi Fan Paintings, Painting Collection of Li Shuren, Painting Collection of Cao Renrong – Landscape Scenes in Wu Area, Suzhou Garden Attractions by Cao Renrong, Huang Chengqin Painting and Prose Collection.

夫借景，林园之最要者也。如远借，邻借，仰借，俯借，应时而借。然物情所逗，目寄心期，似意在笔先，庶几描写之尽哉。

Scene leveraging is the most important point to consider in garden design and construction. There are different ways of scene leveraging, such as leveraging from distant views, from adjacent views, from views up higher, views down below, and views from different times and seasons.

However because of attractions from various objects to our eyes and reactions in our minds and feelings, before any design, one must have a draft mental plan before a detailed plan and fully descriptive designs on paper.

此段所说的借景，先指规划设计空间划分范围内构图上的借景，遵循"俗则屏之，嘉则收之"的原则，要"巧于因借"。而"物情所逗"，"目寄心期"，"意在笔先"，则指意境上的借景。关于前者，江南古典园林中各类借景佳作均有实例，经选择图释如下。

The scene leveraging mentioned here means compositional leveraging in terms of spatial planning and design based on the principles of "including nice views and excluding un-tasteful ones" and "artful and clever extending and leveraging". However, "scene-emotion mutual evoking", "repose of the eyes and expectation of the heart", and "design idea before design paper" suggest scene leveraging at the artistic conception level. On the former connotation, there are examples of each type in classic southern Chinese gardens and some selected ones are shown below.

苏州拙政园远借北寺塔
Suzhou Zhuozheng Yuan leverages distant Beisi Ta (Note: North Temple Pagoda)

瘦西湖钓鱼台又名吹台，三面皆有圆形洞穴，西门洞借景五亭桥卧波，南圆门借景白塔耸立，十分巧妙
Shouxi Lake Fishing Terrace, with round doors at three sides. The west round door draws in the Five Pavilion Bridge while the south round door brings in the dominating white pagoda, a very clever handling.

瘦西湖熙春台邻借望春楼，远借五亭桥、白塔，视野甚广阔
Shouxi Lake's Xichun Tai borrows in nearby Wangchun Lou and faraway Five Pavilion Bridge and white pagoda, enjoying a broad view.

无锡寄畅园仰借锡山塔景
Wuxi's Jichang Yuan leverages the higher Mount Xi's pagoda and its reflection in the pond.

瘦西湖五亭桥上俯借凫庄景色，远借钓鱼台、小金山诸景
On top of Shouxi Lake's Five Pavilion Bridge, one can overlook nearby Fuzhuang and distant Fishing Terrace and Xiao Jinshan. (Note: Little Gold Mountain)

小李将军画本借隔湖熙春台景，收入扇面窗内，酷似一幅装裱的山水画
Xiaoli Jiangjun Huaben is drawn by cross lake Xichun Tai and contained in its fan-shaped window, just like a framed landscape painting.

《园冶》自识

崇祯甲戌岁，予年五十有三，历尽风尘，业游已倦，少有林下风趣，逃名丘壑中，久资林园，似与世故觉远，惟闻时事纷纷，隐心皆然．愧无买山力，甘为桃源溪口人也。自叹生人之时也，不遇时也。武侯三国之师，梁公女王之相。古之贤豪之时也，大不遇时也！何况草野疏遇，涉身丘壑，暇著斯"冶"，欲示二儿长生、长吉，但觅梨栗而已。故梓行，合为世便。

Postscript of *Yuan Ye*

In Chongzhen 7th year (1634), I was fifty three years old. After much hardships and uncertainties in an unstable society, I became tired of distant migration due to the pressure for survival. I nurtured a love for natural landscapes from my childhood, so I escaped to wild mountains and waters to avoid the fetters of fame and have been engaged in making gardens, feeling remote to mundane life. During chaotic and troubled times, people try to live in seclusion for safety reasons; however, I could not afford to buy a large property in the wild and had to depend on someone who could. I lament that I am not born at the right time with great opportunities. The ablest people in ancient times include Zhu Geliang, who only served as a senior consultant for the state of Shuhan, and Di Renjie, who only worked as the Prime Minister for Wu Zetian. They were both limited by their time and opportunity and could not find more favorable opportunities. Much less for me, a mountain living and leisure guy in wild country! I took time to write this book, with a purpose to guide my two sons Changsheng and Changji. However, they are still young and ignorant, so I publish this book to share with the world.

园冶图释

Yuan Ye Illustrated
— Classical Chinese Gardens Explained

海外交流

Overseas Exchange Activities

随着时代的发展，中西方文化亦相互渗透，风景园林行业也同样如此。如德国学者玛丽安娜早在20世纪90年代就著有《中国园林》一书（中文、德文两种版本）。众多的西方同行对中国造园充满诗情画意感兴趣，也开始效仿和借鉴，并咨询笔者学习的方法。笔者认为《园冶图释》就是较全面综合的教科书，在读懂该书的前提下，还需要多动手画。画图不仅仅是手头表现功夫，更重要的是在画的过程中能渐渐悟到诗情画意造园的真谛和具体做法，故选择几十年部分传统园林设计的手工图习作，可供参考借鉴。

为供观赏、参考与借鉴方便，按以下八类编排：

（1）山石小品类
（2）山水结合类
（3）古建筑类
（4）山、水、建筑、植物结合类
（5）鸟瞰图类
（6）仿古图类
（7）彩色综合类
（8）常用印章

As we progress, we see mutual infiltrations between the oriental and occidental culture. It also exists in the profession of landscape architecture, and one example is the book Chinese Gardens (in both Chinese and German editions), written by German scholar Marianna in the 1990s. Peers from the west, fascinated by poetic and picturesque aspects of Chinese landscape gardening, started learning and emulating. They consulted me about the methodologies on Chinese garden design. The authors believe that Yuan Ye Illustrated is a comprehensive textbook for learning. Besed on the premise that the book is well understood, one also needs to sketch, draw, and paint more. Sketching and drawing not only serve the purpose of enhancing the representation skills, more importantly, one could in the process gradually appreciate the gist and specific practices of the poetic and picturesque landscape gardens. Therefore, I selected some of our drawings and paintings during decades of traditional garden design work for your reference and enjoyment.

For your convenience, we have listed seven categories below:

(1) Rock Decorations/Opuscula
(2) Integration of Mountain and Water
(3) Ancient Architecture
(4) Integration of Mountain, Water, Building and Plant
(5) Perspective or Bird's-Eye Views
(6) Antique Modeling
(7) Comprehensive Color Representations
(8) Seals Frequently used

山石小品类　Rock Decoration/Opuscula

山水结合类　Integrtion of Mountain and Water

海外交流
OVERSEAS EXCHANGE ACTIVITIES 339

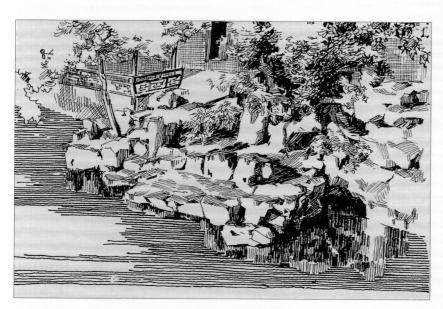

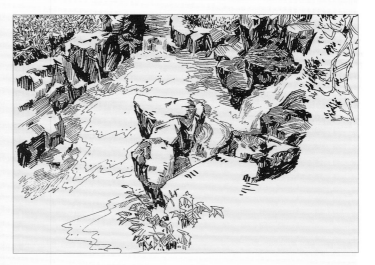

海外交流
OVERSEAS EXCHANGE ACTIVITIES

古建筑类　Ancient Architeture

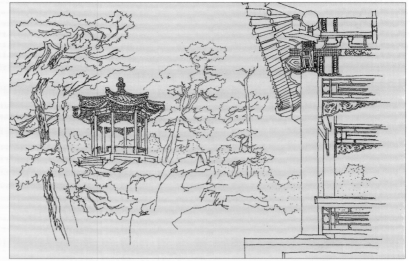

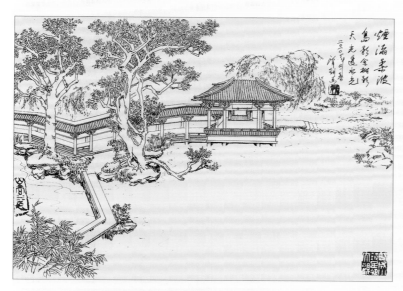
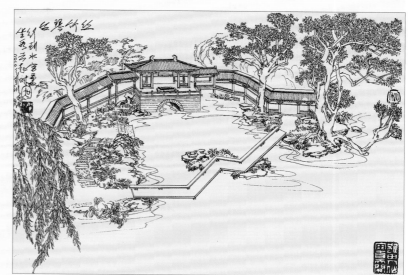

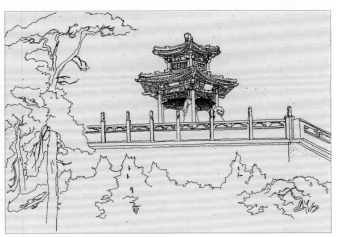

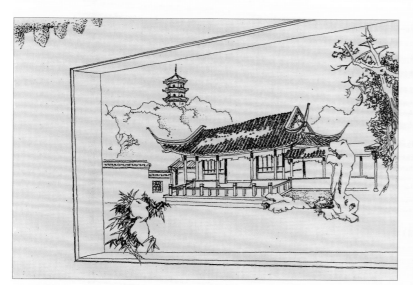
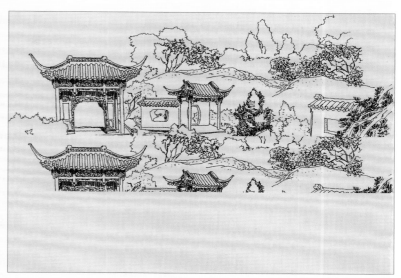

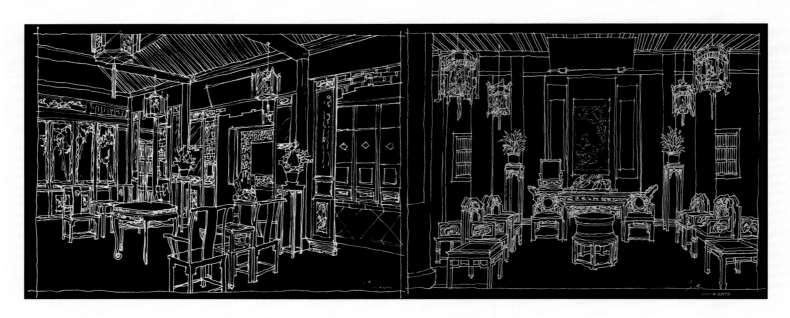

山、水、建筑、植物综合类 Integration of Mountain, Water, Building & Plant

海外交流
OVERSEAS EXCHANGE ACTIVITIES 349

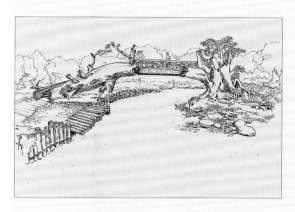

鸟瞰图类　Perspective or Bird's-Eye Views

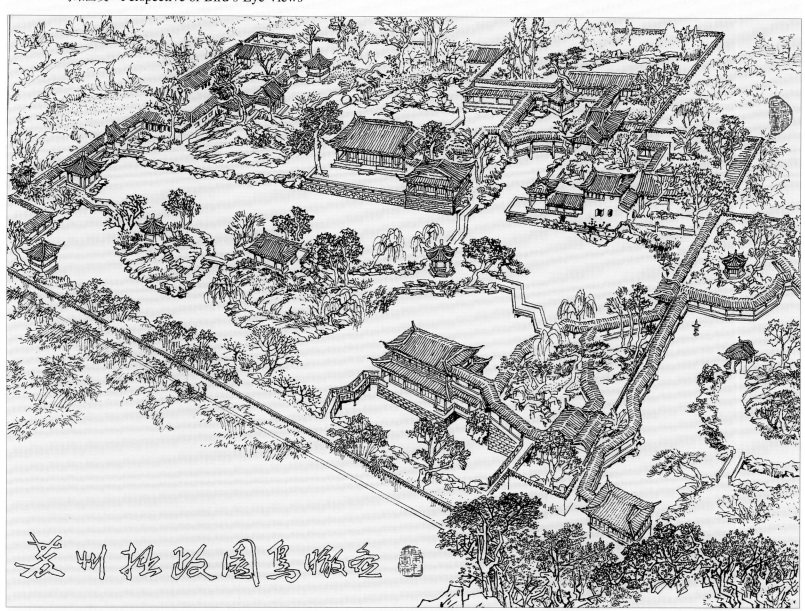

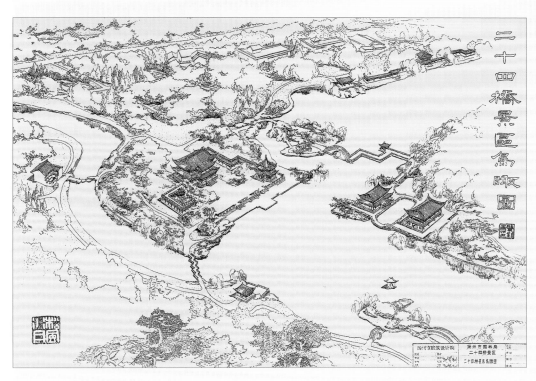

仿古类　Antique Modeling

彩色综合类　Comprehensive Color Representations

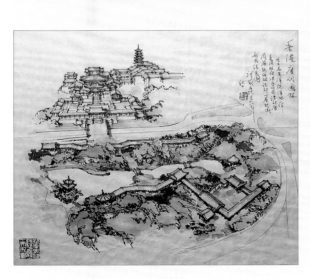

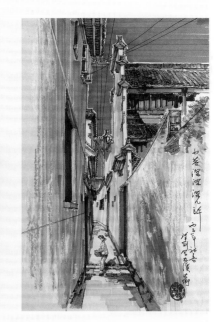

常用印章　Seals Frequently Used

 路漫漫其修远兮
吾将上下而求索

 造化为师

 有节无心

 石瘦梅清铁为骨

 虽由人作
宛自天开

 城市山林

 略成小筑
足征大观

 心淡若无

 借山造景

 穿池叠石写蓬壶

 游于艺

 听松看涛

 天坠地出

 借古开今

 古为今用

 先有古人勿忘我

 寄情

 寄情

 山高水长

 清风明月

 春

 夏

 秋

 冬

 风月无价

 山水有情

 怡情丘壑

 借古开今

 名园多依水

 得趣

 山高水长

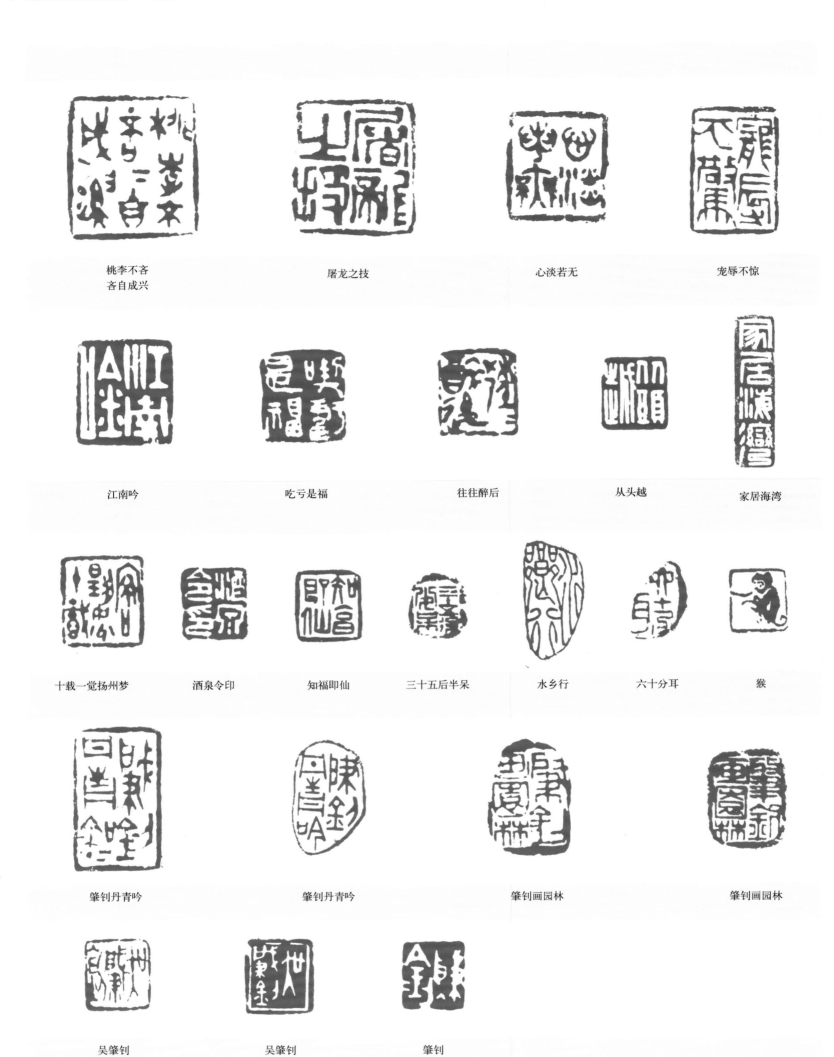

主要参考文献 / Major Reference

1. 陈植. 园冶注释. 北京：中国建筑工业出版社，1988.
2. 刘敦桢. 苏州古典园林. 北京：中国建筑工业出版社，1979.
3. 潘谷西. 江南理景艺术. 南京：东南大学出版社，2011.
4. 陈从周. 江南园林. 上海：上海科学技术出版社，1983.
5. 江苏省志·风景园林志. 南京：江苏古籍出版社，2000.
6. 江苏省基本建设委员会. 江苏园林名胜. 南京：江苏科学技术出版社，1982.
7. 罗哲文，陈从周. 苏州古典园林. 苏州：古吴轩出版社，1999.
8. 杨瑾，薛德震. 中国园林之旅. 石家庄：河北教育出版社，2006.
9. 孟兆祯. 孟兆祯文集. 天津：天津大学出版社，2011.
10. 杨鸿勋. 江南园林论. 上海：上海人民出版社，1994.
11. 彭一刚. 中国古典园林分析. 北京：中国建筑工业出版社，1986.
12. 瘦西湖. 瘦西湖风景区编印
13. 陈从周. 中国厅堂·江南篇. 上海：上海画报出版社，2003.
14. （明）文震亨. 长物志. 北京：中国建筑工业出版社，2010.
15. （明）李渔. 一家言. 北京：中国建筑工业出版社，2010.
16. 朱有玠. 岁月留痕——朱有玠文集. 北京：中国建筑工业出版社. 2010.
17. 董其昌书画集. 天津：天津人民美术出版社，1996.
18. 石涛书画全集. 天津：天津人民美术出版社，2010.
19. 戴敦邦. 戴敦邦新绘全本红楼梦. 上海：上海古籍出版社，2000.
20. 王绣. 山水. 郑州：河南美术出版社，2010.
21. 王绣. 青绿山水. 郑州：河南美术出版社，2010.
22. 李明. 山水画特写. 郑州：河南美术出版社，2010.
23. 杨明义. 杨明义画扇. 香港：心源美术出版社，1996
24. 孙君良. 孙君良扇面小品画集. 香港：心源美术出版社，1996.
25. 邱思泽. 邱思泽画集. 万宝阁艺术中心，1992.
26. 李树人. 李树人画集. 深圳：海天出版社，1993.
27. 曹仁容. 吴中风光——曹仁容画集. 苏州：古吴轩出版社，1999.
28. 曹仁容. 苏州园林名胜园. 苏州：古吴轩出版社，2010.
29. 黄澄钦. 黄澄钦画文集. 哈尔滨：哈尔滨出版社，2000.
30. 崔卫. 黄公望. 石家庄：河北教育出版社. 2006.
31. 盛东涛. 倪瓒. 石家庄：河北教育出版社. 2006.
32. 刘墨. 石涛. 石家庄：河北教育出版社. 2006.
33. 何延喆，刘家晶. 刘子久. 石家庄：河北教育出版社. 2008.
34. 马鸿增，马晓刚. 荆浩关仝. 石家庄：河北教育出版社. 2006.
35. 杨建峰. 文徵明. 南昌：江西美术出版社，2009.
36. 袁江，袁耀. 袁江袁耀画集. 天津：天津人民美术出版社，1996.
37. 戴敦邦. 和谐自然集. 上海：上海辞书出版社，2006.
38. 华三川. 华三川绘新百美图. 上海：上海古籍出版社，2003.

（注：屋宇、装折、门窗、墙垣篇墨线图多摘录于《苏州古典园林》、《江南理景艺术》、《扬州园林》）

跋一

历经艰辛成大业，风雨过后见彩虹。

吴肇钊先生大学毕业后一直在扬州园林部门工作。自调入中外园林建设有限公司任总工程师并深圳市中外园林建设有限公司任总经理工作二十多年来，他的作为和事业如日中天青云直上，足迹遍布国内，并到过世界诸多国家，而他心里时刻都忘不了是扬州这座名城为他提供锤炼、成长和施展才华的实践舞台，是扬州厚重的文化积淀给予他丰富的创作源泉和艺术的升华，是扬州众多的友人给了他真情的关心爱护与热忱相助！

吴肇钊先生第四部力作《园冶图释》即将付印前，他怀着对古城扬州以及对挚友深厚的情谊，刻意安排回到扬州，邀约昔日好友相聚于美丽的瘦西湖畔。在其亲自设计的《吟月茶楼》内置酒筵，大家席间畅叙情怀，频频举杯为新作《园冶图释》即将出版面世庆贺。肇钊先生谈吐潇洒，一再表露对扬州和对扬州友人的诚挚之情，乃新著得以问世的渊源。

《园冶图释》这部园林专著，是吴肇钊先生一生理论研究与实践的总结。肇钊先生的业绩是从扬州开始和发展的。扬州园林早在隋唐就盛名天下，有"园林多是宅，车马少于船"之誉。历经沧桑，扬州园林多遭破坏。自十年动乱后，扬州园林迎来发展的大好机遇，吴肇钊先生先后主持复建了白塔晴云、片石山房、卷石洞天、二十四桥、静香书屋等处景区景点，荣获过国际金奖等多项奖励。吴肇钊先生尊重历史，继承传统，面向未来，勇于创新，不但再现了扬州园林昔日的辉煌，又赋予它新的生命，在新著中，其不少篇章可为佐证，且已成为修复中国传统园林的经典之作。

新著《园冶图释》可谓是世界最古造园名著《园冶》的图画诠释，难度可想而知，但其得力于对扬州园林千年文化积淀的潜心研究，深深领悟到古典园林的精髓所在：注重诗情，讲究画意，更特别追求意境，这正是他的过人之处，在其专著中得到充分体现。如相地篇中，他选择同一地区不同地形、环境，因地制宜规划设计原则与手法的瘦西湖古园林群，他仿画的宫廷瘦西湖古图形象而又具体地图释园冶阐述的山林地、城市地、村庄地、郊野地、傍宅地、江湖地园林选址，不同的园基各异的创作手法，营造出独具特色的园林；然后又用江南现存园林实例进一步论证，可让读者清晰明了，方便理解与借鉴，足见其扎实的研究功底。又立基篇中，则借助参与园林界泰斗汪菊渊先生《中国古典园林史》的研究成果，根据史料记载复原计成在扬州营造实例"影园"，反过来用复原后的"影园"来验证其立基的理论，发现立基篇的论述均系其实践的总结与提升，诸如书中：动"江流天地外"之情，合"山色有无中"之句，均系实景的写照。难怪清华大学已故著名学者周维权先生评价肇钊先生的《计成与影园》论文是开实例论证理论的"先河"。"园者，画之见诸行事也"，"造园是将二维空间的画变成三维空间的实景"这是肇钊先生造园时严格遵循的原则。如在复建明代大画家石涛园林作品"片石山房"时，他首先将假山组成元素：峰、峦、洞、壑、瀑、潭、坡、矶、池、沼等均从石涛的山水画中找到画本，以此为依据画出设计图，其结果施工过程清理倒塌山石泥土时，原深潭、池岸、谷道位置与设计图不谋而合，使得复建工作十分成功，获国家、省级奖项。故陈从周老先生欣然书写碑记"近吴君肇钊……细心复笔，画本再全，功臣也。石涛有知，亦当含笑九泉。"陈老先生将活人写进碑记尚是唯一孤例。更令人赞赏的是新著中一幅幅精美的绘画典雅清秀，风格脱俗，耐人寻味。肇钊先生的绘画功底除得力于"文革"期间去中央美院进修油画外，他刚来扬州时，别人忙于"派性"斗争，他却专心于美术创作，他所绘制的众多巨幅毛主席油画像，与人合作完成的"收租院"泥雕群雕，令很多扬州人至今记忆犹新。记得他在德国国际园林博览会主持中国参展"清音园"建设期间，德方还专为他举办个人画展，有的画作售价达二万马克（相当人民币十二万元）；所设计的"清音园"亦荣获"大金奖"，德国政府还颁发荣誉勋章，为国争了光，我们为之欣喜万分。

吴肇钊先生在扬州工作期间，热忱待人，与单位领导、同事或下属建立了深厚的友谊，他夜以继日勤奋工作精神与注重办事效率，敏捷的才思与聪慧，为扬州园林作出了很大贡献，留下了众多佳作。他始终铭记着有关领导对他的信任，珍惜放手给他创作与实践的机会，不辜负同仁们对他的关心支持。随着时间的推移，友谊越发显得珍贵。在《园冶图释》出版之际，大家相聚同贺同庆，品尝杯中美酒，更品尝比美酒更浓更香的友谊，大家谨此热情恭祝肇钊先生今后事业取得更大成就，艺术青春永驻。

遵照吴肇钊先生意见，聚会友人一同签名留念：他们是原扬州市建委领导赵明先生、扬州市园林管理局原领导陈景贵先生、杨文祥先生，管委会规划局长周康女士、副局长韩涛先生，扬州市建筑设计研究院有限公司副院长季文彬先生，得意弟子扬州市天翼园林设计研究院院长王峰先生。

是为记。

赵明

2011 年 12 月 07 日

杨文祥　　陈景贵　　周康　　韩涛

李文彬　　王峰

Postscript 1

Mr. Wu Zhaozhao had been working in the organizations of landscape architecture in Yangzhou since his graduation from college. After that, he has been working as the chief engineer for Chinese and Foreign Landscape Architecture Construction Company and the director for Shenzhen branch for more than 20 years. During this period of time, he has achieved great success in his career and travelled a lot at home and abroad. Yangzhou as a historic city has always been on his mind, which has provided opportunities for his artistic practice, the rich culture of which has served as the source of his artistic creation, and where he has made a lot of friends who have consistently and enthusiastically helped him.

Before the publication of his works *Yuan Ye Illustrated*, he held a banquet at Yin Yue Teahouse, which was designed by himself, for the purpose of expressing his appreciation for the friends in Yangzhou. During the banquet, Mr. Wu expressed his gratitude for his friends and explained how his new book came into being.

His new book *Yuan Ye Illustrated*, which he has taken pains to compose, is an elaboration of *Yuan Ye*, the oldest book of landscape architecture. The book reflects his comprehensive research on the culture of Yangzhou gardens and his thorough understanding of the essence of classical gardens: poetic, picturesque, and imaginary. For example, in the chapter of site selection, he has explained in details how the gardens in Slender West Lake Park should be laid out and constructed in harmony with their surrounding environment. Then he has further elaborated on this point with existing gardens, which makes it easy for readers to follow and which displays his expertise in this field. In addition, in the chapter of foundation, with the help of research done by Wang Juyuan, author of *the History of Classical Gardens in China* and with the help of historical documents, he has successfully duplicated Ying garden built by Ji Cheng in Yangzhou. Ying garden duplicated by him has served best to prove the theory proposed by him, which is based on years of practice. Mr. Zhou Weiquan, the late scholar of Tsinghua University, remarked that his thesis *Mr. Ji Cheng and Ying Garden* was a pioneer of case study.'Gardens are a mirror to paintings'. 'Garden building is to transform the two dimensioned painting into the three dimensioned landscape'. The two are the principles strictly observed by Mr. Wu when he is building a garden. A case in point is the restoration of "Pian Shi Shan Fang", a garden in Yangzhou which was designed by Shi Tao, a famous painter of Qing Dynasty. He has carefully studied Shi Tao's paintings before drawing the blue print, which agrees perfectly with the original design made by Shi Tao himself. This restoration has earned a lot prizes for him. Mr. Chen Congzhou has unprecedented inscribed his name on the stone. A painting, which has made its first appearance in his book, is elegant, unworldly and subtle. His expertise in painting owns a lot to his study in Central Institute of Fine Arts. When in Yangzhou, he was deeply immersed in painting while others were involved in political activities. His portraits of Chairman Mao and his sculptures, which have been created in this period of time, are still cherished by a lot of people in Yangzhou. When he was presiding at the construction of Qing yin Garden of China for the International Exposition for Gardens held in Germany, an exhibition in his name was being given. Some of his paintings were sold at the price of 20,000 marks. We have been overjoyed to hear the news that Qing yin Garden designed by him has won the big prize at the Exposition.

Mr. Wu is friendly, diligent, efficient and intelligent. While working in Yangzhou, he has developed deep friendship with his colleagues and made great contribution to Yangzhou landscape architecture. Today, we've gathered here to celebrate the friendship and the publication of *Yuan Ye Illustrated*. In the meantime, we wish Mr. Wu would achieve greater success in his artistic career.

Friends invited to the banquet are as follows: Mr. Zhao Ming, from Yangzhou Municipal Construction Committee; Mr. Chen Jinggui and Mr. Yang Wenxiang, Yangzhou Gardens bureau; Ms. Zhou Kang, CMC Planning Bureau and Han Tao, deputy director of Planning Bureau; Mr. Ji Wenbin, vice president of Yangzhou architectural design Institute limited company and his Favorite disciple, president Wang Feng, from Yangzhou Tianyi Garden Research Design Institute.

Zhao Ming
Nov 18th,2011
Translator: Benny Xu Qingyu from Yangzhou normal college

跋二

吴肇钊先生是中外园林建设有限公司的总工，在园林界也是著名的园林设计大师级人物。多年来吴君从事园林设计工作从未放下过手中之笔，他的园林创作遍及大江南北。吴肇钊先生同我亦师亦友，经常一起切磋讨论园林，我从中受益匪浅。通读吴君《园冶图释》，深感这是他在设计实践中对《园冶》深刻理解的潜心力作。多年来《园冶》注释有多个版本，描述为多，因每个人理解不同，难以一概全。在当今高节奏发展时代，园林书籍大部分是以照片为主，手绘图纸已经是凤毛麟角，这种快餐式文化的好处是较为直观地看到实物为何，而坏处是一种浮躁心态，难以静下心来钻研历史精髓，而实物往往也难以全面体现原作者的全意。吴肇钊先生的《园冶图释》，将自己的心得体会融汇于图画之中，既满足了读图时代的社会需求，又旁引佐证将多层次对《园冶》的理解用绘图的形式表现出来。《园冶图释》编撰成书此举乃是园林界一件幸事。吴肇钊先生在繁杂的设计工作的同时，还能深入理论研究，实是难能可贵。除学术研究外，仅就此本绘画来说已是一本值得收藏的范本。此书已成为我公司科研成果之一，无论对业界的影响，还是对公司的发展，都将是很重要的贡献。

沈惠身
（中外园林建设有限公司　副总经理）

Postscript 2

Mr. Wu Zhaozhao is the chief engineer of Landscape Architecture Corporation of China as well as a well-known landscape architecture design master in China. Mr. Wu has been involved in landscaping design works for many years and never dropped his drawing works. His creations in landscape architecture can be found across the country.

Mr. Wu Zhaozhao is both my teacher and friend. We often talked about the landscape architecture and I had learned a lot from those discussions. I am deeply impressed by Mr. Wu's *Yuan Ye Illustrated* which is the masterpiece based on his deep understanding of *Yuan Ye* in his daily design practice. There are many versions of notes to *Yuan Ye*, most of them use description to interpret the original works. Since one's understanding is different from another, it is not an easy job to cover all the details. In this time of rapid development most of the landscape architecture books are mainly composed of photos, one can hardly find drawings by hand. On one hand the advantage of this fast-food culture is that the readers can get the contents of the books directly through photos. On the other hand the disadvantage is that the readers read the books in haste and it is difficult for the readers to focus on studying the essence of history in a calm mood. Furthermore, a material thing like photo can hardly reflect the comprehensive meaning of the writer completely. In *Yuan Ye Illustrated* Mr. Wu Zhaozhao expresses his understanding of this famous book through paintings which not only satisfy the current needs of reading books with illustration but also help transform his comprehension of *Yuan Ye* into drawings.

It is truly lucky for the landscape architecture industry to have "Yuan Ye Illustrated" compiled and published. It is really praiseworthy that Mr. Wu Zhaozhao has kept making theoretical researches while undertaking painstaking design works. Apart from the academic research, the paintings in this book is worthy for collection. This book conferred the scientific achievement of our company will certainly make due contribution to the development of Landscape Architecture Corporation of China and produce positive impact on the landscape architecture industry.

Mr. Shen Huishen
Landscape Architecture Corporation of China

译后记

中国古典园林是一种由超现实意境所引发的、精妙的艺术所指导的、细腻的技术所实现的综合体，体现的是一种博大精深的中华文化。吴肇钊先生及其合著者陈艳、吴迪所著此书的目的即是帮助我们及国际友人更好地理解和欣赏古典中国园林设计的奥妙与建造技巧。作为本书的英文译者，我感到了莫大的荣幸，也多谢《中国园林》杂志社曹娟编辑的引荐和中国建筑工业出版社郑淮兵编辑的合作。

我被园林艺术的吸引始于大学时代。在清华读建筑学时代，有一门选修课是姚同珍先生的《园林植物》。我很喜欢并担任了该课的课代表。建筑学在西方是一种人文学科，其覆盖范围很广，涵盖历史、艺术、哲学、心理、工程等。当我开始接触园林时，才知道其涵盖范围更广，因为它不仅包括人工建筑，而且还有山水、树木、花草、虫鱼。对古典中国园林来讲，更重要的还是更多地体现了一种非物质的文化氛围。纯物质的设计相对容易，如果牵扯到非物质方面的设计，如文化和意境营造方面，难度就要大很多。恕我寡闻，在大学的最后一年，我才得知有一本中国古典园林名著《园冶》，这时我也开始对园林设计产生强烈的兴趣。

大学快毕业时，我报考了北京林业大学园林系的研究生。经过加强园林设计学方面的补习，研究生考试时取得了专业第一名的优秀成绩。在复习准备时，我潜心阅读了孙晓翔、孟兆祯等先生们的著作。使我开窍的是1984年的一个夏夜在孟先生家里的手把手的入门辅导。使我尤其印象深刻的是当孟先生步入客厅时被室内植物枝叶碰到脑袋时享受的表情和之后发出的一声感叹："多么美好的感受啊"！之后在北林读研时亦听到孙先生类似的表达："与动植物在一起的生活是人类最为美好的生活。"当时我才二十几岁，即将从建筑院校毕业，还并无如此深刻的感悟。

入北林后，我师从郦芷若先生，花了相当多的时间研究世界古典造园艺术的另一个极端——法国古典园林，以17世纪法国路易十四时代的凡尔赛皇家花园为代表。中西对比，各有千秋。简言之，西方较注重人工体型的塑造，中国更加重视自然意境的追求。孙先生的"生境、画境、意境"三境界论及其名作杭州花港观鱼公园、孟先生的"先立意、后造园"的主张及园林理法及高超的山水设计和叠石功夫对中国园林规划设计教育的指导意义重大。试想一个物质的景观若没有一种非物质的景名提示，无论是明示或暗示，将会是相当乏味、美中不足的。多少年后当我移居到美国南加州，我仍有为我所居住的城市取景名的习惯[1]。

在北林的三年研究生学习和其后五年留校任教的过程中，我受到了很多先辈、同学、同事的指导和影响，除了前面提到的孙、孟、郦先生等外，使我感受深刻的还有张钧成、曹汛先生扎实的古文功底，白日新先生的建筑设计素养，研究生同学包志毅丰富的园林植物知识、王向荣和罗华的灵气，及师兄弟李雄的成熟和全面等。只是因为时空隔离的关系没能有机会与大学长吴肇钊先生晤面，实感遗憾。此次翻译《园冶图释》亦算是对此有所弥补，起码是在心智交流方面。此书的翻译还满足了我多年的夙愿，即把《园冶》从头到尾通读了一遍。

在北林执教五年后我来到了美国，继续深造，主修偏技术的地理信息系统（GIS）。在堪萨斯大学研究生毕业后开始在位于南加州的GIS软件公司esri工作至今。我所在的研发组专攻三维地理信息系统，还算多少与我的老本行建筑设计与园林规划设计有些关联，至少我们的服务和产品可以被规划设计专业人员使用。

关于本书的英译，我有几点解释和体会，希望对英文读者有所帮助。

第一，因为中文和英文分属两大不同语系（汉藏语系及印欧语系），《园冶》的原文使用的是骈体文，故在翻译上有较大难度。有些词语无法完全对译，故我选择保留原始发音。有些词语可以对译，如"园"字，我直接译成 garden，有时亦保留原始发音，如"yuan"。这是因为即便"园"字虽然有其英文对应词 garden，但用此英文词还会是有些微妙的差异的。若把园林统统翻译成 garden 原文的语义范围多少有点缩减或歧义。实际上应该如何翻译"园林"一词目前业界还有争议，但对它的讨论已超出了本序的范围，在此不谈。书名的译法我采用了半音译、半意译的做法，翻译成"*Yuan Ye Illustrated*"。

第二，正如在中文中我们会发现很多外来语一样，在英文中也有越来越多的中文。本书中的有些特有的中文词汇，如亭、台、楼、阁等，虽然有些也有英文对应词，我还是尽量用中文原词的发音，这也算是中文对英文的贡献吧。英文正是有各种各样来自不同文种的贡献才变得如此的丰富和富于生命力。

第三，与英文不同，中文是一种强文脉语言，这意味着它的语义承载力很高。最典型的即是中文成语，四个字即是一整篇故事和道理，如"孟母择邻"等。还有用区区两个字即可阐述一种价值体系，如"逃名"等。如果将这些"幕后的故事"统统翻译讲解出来，将会增加太多的篇幅，故我在译文中一般使用最简要的概括，没有详细的解释。此做法的结果是英文读者可以掌握内容的大概，但还不能透彻地理解全部。若真要全部理解，只能通过学习中文和了解中国文化[2]，别无他法。相信此书的出版也会为传播和普及中国文化和语言起到作用。

中国古典园林，尤其是明清时代的江南的古典园林是特定历史时代的产物，其对神仙境界的追求、咫尺山林的营造、体宜因借的运用、巧夺天工的手法均无与伦比，堪称中国古典私家花园登峰造极之作。此书的出版亦为对世界造园艺术宝藏的一大贡献。在此我谨向学长吴肇钊先生及其合著者陈艳、吴迪表示衷心的感谢和祝贺！

本人才疏学浅，英文亦非母语，虽然尽了很大的努力，但在语言的使用上仍不时感到捉襟见肘。在本书的翻译过程中不免会出现许多谬误，还望读者海涵并指教。

马劲武

2012.8.12. 晨，于美南加州鹅湖

---

1. "鹅（尔斯诺）湖八景"：晴波垂钓、碧水泛舟、斩浪劈波、鲜伞翱翔、云山雾罩、野花飘香、山环水阔、星光交映。此八景的命名对在此购房的华人可以起到很大的指导意义（对此房地产商们应该感谢我给他们带来的生意）。

2. 业余时间在周末我在本地中文学校（河滨华夏中文学校）教中文，现担任该校中文学校校长。我注意到随着中国影响力的增长，越来越多的非华裔学生（儿童和成人）对中文感兴趣，开始学中文。这是非常可喜的事情。

Postscript by the English Translator

Classical Chinese gardens are driven by a hyper-reality conception, guided by artistic principles, and realized by refined construction technologies of their time, and it embodies an extensive and profound Chinese culture. The purpose of this book, co-authored by Mr. Wu Zhaozhao, Chen Yan, and Wu Di, is for us and international readers to better understand and appreciate the wonders and techniques of classic Chinese garden design and construction. Being the English translator of the book, I'm tremendously honored. I appreciate the referral of Cao Juan, an editor of the Chinese Landscape Architecture magazine, and the cooperation with Zheng Huabing, an editor of Chinese Construction Industry Press.

I was first drawn to the subject of garden art in my college days, when I studied architecture at Tsinghua University and took an elective class "Garden Plants" by Professor Yao Tongzhen. I acted as the course representative, as I enjoyed the class very much. In the west, Architecture is a field in humanities, with a wide coverage involving history, art, philosophy, psychology, and engineer etc. When I started to learn about Garden Art or its more modern incarnation Landscape Architecture, I got to know that its coverage is even wider, since it does not only include architecture, but also mountains and waters, as well as fauna and flora. As for ancient classical Chinese gardens, they are more or less embodied and shrouded by a non-physical culture atmosphere. It is relatively easy to do physical design; however, when it's involved with non-physical design, such as construction of culture or artistic conception, the difficulty levels would be much greater. Pardon my ignorance, but I only got to know about "Yuan Ye" during my last year of study in college, and grew much interest since that time.

I applied to the graduate program in Landscape Architecture of Beijing Forestry University(abbreviated as Beilin) before I graduated from Tsinghua University. After some period of intense studies on garden design, I was fortunate to have scored the top points in my graduate entrance exam on landscape architecture design. During my preparation of the exams, I studied hard on the design work of Professors Sun Xiaoxiang, Meng Zhaozhen and others. What seemed to be a moment of epiphany is when I received a hands-on tutoring by Professor Meng at his home on a summer night in 1984. What impressed me significantly is a remark by him when he was brushed by a leaf of an indoor plant when he stepped into the living room, "What a beautiful feeling!", as he was showing a facial expression of extreme enjoyment. Months later when I became a graduate student at Beilin I heard a similar comment by Professor Sun, "The best quality life for man is to live with plants and animals." In my early twenties back then as a new graduate from an architecture school, I was not capable of comprehending this life feeling completely.

After I entered Beilin's graduate school, I studied with Professor Li Zhiruo, who's my supervisor. I took significant amount of time studying the other extreme of the world garden art, the French classical gardens, epitomized and climaxed by the royal gardens of Versailles in Louis XIV's time in the 17th century. Comparing the east against the west, I learned that they each have their own strengths and unique characteristics. A much generalized view is that the occidental gardens emphasize more on man-made physical forms while their oriental counterpart more on seeking natural and/or metaphysical constructs. Professor Sun's tri-state theory – of life, of art, and of mental realm – and his masterpiece design work Huagang Guanyu (Note: Fish Watching at Flower Lough), Professor Meng's assertion of "conception before construction" and his garden design theory and practices, as well as his masterful rockery building skills, play important roles on the education of Chinese landscape architecture planning and design. Imagine if for a physical garden or landscape, there's no non-physical hint or reminder of the scene, whether explicit or implicit, it would be rather boring and less than perfect. Years later after I immigrated to the United States and relocated to Southern California, I still keep a habit of making scenery names for my home city.

During the three years of graduate study at Beilin and five years of teaching thereafter, I had the privilege of working together with many senior researchers, classmates and colleagues, and was influenced by them greatly. Besides the aforementioned Professors Sun, Meng, and Li, I was also impressed by Professors Zhang Juncheng and Cao Xun's mastery of the ancient Chinese language, Professor Bai Rixin's architectural design skills, graduate classmates Bao Zhiyi's rich plant knowledge, Wang Xianrong and Luo Hua's talent in design, and fellow graduate buddy Li Xiong's well-rounded maturity and so on. Due to spatial and temporal isolation, however, I didn't have a chance to meet vis-à-vis with senior fellow alumnus Mr. Wu Zhaozhao. This translation of his work "Yuan Ye Illustrated" can serve as a partial compensation for that lost opportunity, at least at the intellectual communication level. The translation of this book also satisfied my long-cherished wish, that is, reading "Yuan Ye" from cover to cover.

After teaching at Beilin for five years, I came to the United States for further study. I chose to study technology inclined Geographical Information Systems (GIS) and worked for my next graduate degree at the University of Kansas (KU). I have been working at esri in Southern California since graduating from KU. My work group at esri is responsible for research and development of 3D GIS software products, and it more or less relates to what I did before in architecture and landscape architecture planning and design, at least our services and products can be employed by planning and design professionals.

Regarding this book's translation, I'd like to share a few thoughts and explanations, hoping that they'd be helpful for readers.

1. Since Chinese and English belong to different language families, Sino-Tibetan and Indo-European respectively, and the original language format in "Yuan Ye" is in hundreds years old parallel prose style, the translation of the texts presents significant challenges. Some terms cannot be translated thoroughly and thus I chose to transliterate and keep the Chinese original pronunciations. Some words can be translated, such as "Yuan", and I translated it directly into "garden"; however, I sometimes transliterate, too. This is due to the fact that even if there's a counterpart in the English language, "garden" in this case, there's still a subtle difference between "Yuan" and "garden". If I indiscriminately translate all the "Yuan" into "garden", then there'd be semantic loss in its original meanings. In fact there's still on-going debate on how to translate "Yuan Lin" in the profession but discussion about that goes beyond the scope of this postscript. As for the book name, I chose to apply half transliteration and half paraphrase approach and translated it as "Yuan Ye Illustrated".

2. Just as we can find numerous words of foreign origins in Chinese, there're growing vocabularies in English originated from Chinese. For some specific Chinese terms, such as Ting, Tai, Lou, and Ge and so on, even if some of them have the English counterparts, I still transliterate by using the original Chinese pronunciations. This could count as the Chinese contribution to the English vocabularies. It is because of such contributions from various languages that the English language becomes so much enriched and vibrant.

3. Unlike English, Chinese is a high-context language. This means that the Chinese per-unit semantic load is very dense. The most telling example is the Chinese proverb. Typically, a four-character proverb can be a whole story and/or philosophy, such as "Meng Mu Ze Ling". Another example is to just use two characters to carry a value system, such as "Tao Ming". If I chose to tell all these "behind-the-scene stories" then my translation would bloat significantly. Therefore, in my translations I selected to use the most concise summarizations without elaborations. This way the English readers would get the main idea but not necessarily the complete picture. If one wants to understand the whole picture, then he or she would have to learn the Chinese language and its culture in earnest. Unfortunately there's no other alternative. I believe the publication of this book would help to promote the Chinese language and its culture.

The classical Chinese gardens, especially those built in the Ming and Qing dynasties in Southern China, are products of its time. Its quest for celestial and immortal living, its construction of miniature yet expressive landscape, its nature extending and scene leveraging applications, and its mastery of garden building techniques that excel nature are all unparalleled and represent the apex of classical Chinese garden design and construction. The publication of this book will no doubt contribute to the world treasury of garden art. Here I'd like to express my sincere appreciation and congratulations to Mr. Wu Zhaozhao, Chen Yan and Wu Di!

Due to my limited knowledge and the fact that English being not my native language, although with much effort, I still feel inadequate when it comes to language use. There must be errors or mistakes here and there in my translations. I wish readers could understand and I welcome any comments and suggestions.

Jinwu Ma
August 12, 2012
Lake Elsinore, CA, USA

鸣谢

合著《园冶图释》经一年江南奔波、日夜奋战问世。在此首先要感谢关怀此书的前辈学者、同仁们。同时十分感谢校友《中国园林》主编王绍增教授对本书学术研究性的校核。十分感激赏石大师《中国赏石文化发展史》巨著主编，学兄贾祥云教授提供"选石"篇各类石种全部实景内容。

参加本书工作电脑上色：高薇、黄敏；描图：倪定、马强、黄帮辉、罗奕；打字：王春娟、王英芝；扫描打图：肖泽民。

谨此一并致谢。

此书的出版，感谢初玉霞女士鼎力相助。

Acknowledgment

The co-authored *Yuan Ye Illustrated* has been completed after a year of hard work, including traveling throughout southern China and working late at nights. Here I must first thank senior scholars and peers. I also appreciate for the proofreading help from Mr. Wang Shaozeng, editor-in-chief of *Chinese Landscape Architecture* magazine. I appreciate for Professor Jia Xiangyun, the author of *An Evolutionary History of Rock Appreciation in China*, has provided various types of rocks photos.

Gao Wei and Huang Min helped with the computer coloration for the book. Ni Ding, Ma Qing, Huang Banghui and Luo Yi help with the drawing and tracing work. Wang Chunjuan and Wang Yingzhi helped with the typing and Xiao Zemin on scanning and printing.

Here I thank them all for their help.

I'm also indebted to the full help of Ms. Chu Yuxia.

August, 2011